Transcritical CO_2 Heat Pump

Transcritical CO_2 Heat Pump

Fundamentals and Applications

Xin-Rong Zhang
Department of Energy & Resources Engineering
College of Engineering
Peking University
Beijing, China

Hiroshi Yamaguchi
Energy Conversion Research Center
Doshisha University
Kyoto, Japan

WILEY

This edition first published 2021 year

Registered Offices
John Wiley & Sons, Inc., 111 River Street, Hoboken, NJ 07030, USA
John Wiley & Sons Singapore Pte. Ltd, 1 Fusionopolis Walk, #07-01 Solaris South Tower, Singapore 138628

Editorial Office
The Atrium, Southern Gate, Chichester, West Sussex, PO19 8SQ, UK

For details of our global editorial offices, customer services, and more information about Wiley products visit us at www.wiley.com.

Wiley also publishes its books in a variety of electronic formats and by print-on-demand. Some content that appears in standard print versions of this book may not be available in other formats.

Library of Congress Cataloging-in-Publication Data applied for

9781118380048 (Hardback)

Cover Design by Wiley
Cover Image: © Nico Garstman Photography/Shutterstock

Set in 9.5/12.5pt STIXTwoText by SPi Global, Chennai, India
Printed and bound by CPI Group (UK) Ltd, Croydon, CR0 4YY

10 9 8 7 6 5 4 3 2 1

Contents

List of Contributors

Riley B. Barta
Ray W. Herrick Laboratories
Purdue University
West Lafayette
USA

Ammar M. Bahman
Mechanical Engineering Department
College of Engineering and Petroleum
Kuwait University
Kuwaitw

Dipankar N. Basu
Department of Mechanical Engineering
Indian Institute of Technology Guwahati
Assam
India

Lin Chen
Institute of Engineering Thermophysics
Chinese Academy of Sciences
Beijing
China

Feng Cao
School of Energy and Power Engineering
Xi'an Jiaotong University
Xi'an
China

Ramón Cabello
Mechanical Engineering and Construction Department
Jaume I University
Spain

Jesús Catalán-Gil
Mechanical Engineering and Construction Department
Jaume I University
Spain

Brian M. Fronk
George W. Woodruff School of Mechanical Engineering
Georgia Institute of Technology
Atlanta
USA

Eckhard A. Groll
Ray W. Herrick Laboratories
Purdue University
West Lafayette
USA

Rodrigo Llopis
Mechanical Engineering and Construction Department
Jaume I University
Spain

Laura Nebot-Andrés
Mechanical Engineering and Construction Department
Jaume I University
Spain

Daniel Sánchez
Mechanical Engineering and Construction Department
Jaume I University
Spain

Yulong Song
School of Energy and Power Engineering
Xi'an Jiaotong University
Xi'an
China

Hiroshi Yamaguchi
Energy Conversion Research Center
Doshisha University
Kyoto
Japan

Haruhiko Yamasaki
Department of mechanical Engineering
Osaka Prefecture University
Osaka
Japan

Ryohei Yokoyama
Department of Mechanical Engineering
Osaka Prefecture University
Osaka
Japan

Davide Ziviani
Ray W. Herrick Laboratories
Purdue University
West Lafayette
USA

Xin-Rong Zhang
Department of energy and resources engineering
College of engineering
Peking University
Beijing
China

Preface

With the social and economic developments, heat pump becomes more and more important in various energy conversion fields due to its high energy efficiency. However, its working fluids face serious challenges due to global warming and ozone layer depletion since the end of twentieth century. Carbon dioxide (CO_2) is a natural fluid and can provide an excellent energy and environment solutions to replace freon substances, such as chlorofluorocarbons (CFCs) and hydrochlorofluorocarbons (HCFCs). Furthermore, for the past decade, researchers have studied and obtained more and more knowledge about supercritical fluid flow dynamic and heat transfer and also phase change flow and heat transfer. Thus, transcritical thermodynamic cycles using carbon dioxide as working fluid are gaining more and more attention. Transcritical CO_2 heat pump cycles own many unique advantages, such as a higher COP performance and higher temperature for thermal energy provided, etc. These natural advantages of the transcritical CO_2 cycle clearly extend their applications to wider fields and provide better energy and environment solutions to our society.

This book does not aim to provide every aspect of transcritical CO_2 heat pump. But all the key points related to the transcritical CO_2 heat pump cycles and systems are included. Both fundamental knowledge and application are considered and discussed in the main text to help the reader to better understand the transcritical CO_2 heat pump cycles and systems. The fundamental aspects mainly include flow dynamic and heat transfer of supercritical CO_2, evaporative flow and heat transfer of CO_2 liquid-gas fluid, theoretical analysis of transcritical CO_2 compression and expansion processes, and subcooling methods. The book also provides several important applications of the transcritical CO_2 heat pump cycles, such as CO_2 heat pump water heater and CO_2 space heating. The book is expected to be helpful for researchers, postgraduates, engineers and also policy makers, etc.

We thank the many helpers, including the following, for their collecting of documents, reviewing of many publications and checking of texts: Dr. Qiuyun Zheng, Mr. Guanbang Wang, Mr. Zhaorui Peng, Ms. Yisai Gao, Mr. Xuegang Lu, Mr. Junmin Yin, Mr. Bing Fang, Mr. Yudong Zhu. Others who have mainly contributed to some chapters are listed separately. We greatly appreciate these contributions. We also recognize the effort of Mr. Xingyu Shang for the text editing. In addition, the support of the National Key Research and Development Program (2016YFD0400106) and the support from Beijing Engineering Research Center of City Heat are gratefully acknowledged. Finally, as always, we welcome your comments, criticisms and suggestions.

Xin-Rong (Ron.) Zhang
xrzhang@pku.edu.cn
Peking University

Hiroshi Yamaguchi
hyamaguc@mail.doshisha.ac.jp
Doshisha University

1

Introduction

Xin-Rong Zhang

Department of Energy and Resources Engineering, College of Engineering, Peking University, Beijing, China

1.1 Background

1.1.1 Energy Shortage and Energy-Saving Technology – Heat Pump

With the fast development of human society and the rapid expansion of the human population, worldwide energy consumption has grown quickly during the last several decades. As for the year of 2017, primary energy, almost 13 511.2 million tons oil equivalent, was consumed around the world, an average growth rate of 1.7% per year in the period 2006–2016. Among the primary energy sources, up to 85% was non-renewable, such as oil, natural gas, and coal.[1] Huge energy consumption leads to energy shortages as well as increasing carbon emissions, thus causing climate change. Finding, researching and using renewable energy and energy-saving technology become essential and urgent.

Heating is an energy-intensive process in residences, industries and commercial areas, which usually utilizes fossil fuel or electricity as energy sources. Thanks to the development of renewable and energy-saving technologies, the process can be more efficient based on the new ways, and one of the most mature technologies is heat pump. Heat pump is an energy-efficient system which moves heat from low temperature side (heat source) to high temperature side (heat sink) with compressor work; it can supply more heat compared to traditional heaters with equal energy input, providing the potential of heat recovery to reduce primary energy consumption. Nowadays, heat pumps are applied not only as single heating systems, such as for direct heating, space heating, and water heating, but also as multi-function energy conversation systems like simultaneous heating and cooling, product drying and waste heat recovering, etc. As for the heat pump market, it shows considerable development in recent decades; around the world, for instance, the world heat pump market increased by 7.2% and almost two million units were sold in 2013.[2] Therefore, heat pump becomes more and more indispensable in human society.

1 The data in this paragraph come from *BP Statistical Review of World Energy 2018*.

2 Referenced from *Growth in the world heat pump market* by BSRIA Worldwide Marketing Intelligence. www.bsria.co.uk/news/article/growth-in-the-world-heat-pump-market.

Transcritical CO_2 Heat Pump: Fundamentals and Applications,
First Edition. Xin-Rong Zhang and Hiroshi Yamaguchi.

1.1.2 Heat Pump Challenges and Natural Refrigerants

However, with the growing concerns for the environment, heat pump technologies are facing new challenges, especially the choice and use of refrigerant. Refrigerant is the blood of a heat pump system which transfers heat from sources to sinks with phase changing. Since synthetic refrigerants (i.e. chlorofluorocarbons [CFCs]) were invented and used from the 1930s, this kind of new refrigerant was widely used and dominated the market for several decades later due to its non-toxic, non-flammable and high efficiency for the corresponding cycles, while the situation changed in the 1970s when people started to notice that chlorine and bromine related compounds (CFCs, hydrochlorofluorocarbons [HCFCs]) caused ozone depletion, which was harmful to earth life [1]. One of the most remarkable phenomena is the existence of ozone holes (Figure 1.1 shows the area of the ozone hole in the southern hemisphere[3]). Hence, based on ozone layer protection, the Montreal Protocol (1987) [2] stated the schedule of phasing out of CFCs and HCFCs, thus the production and the use of such kind of refrigerants was restricted. The replacement refrigerants (hydrofluorocarbons [HFCs]) that were proposed and popularized later, were regulated as well (Kyoto Protocol, 1997 [3]), because they emit strong greenhouse gases which can enhance global warming and cause related environmental issues (Figure 1.2 shows

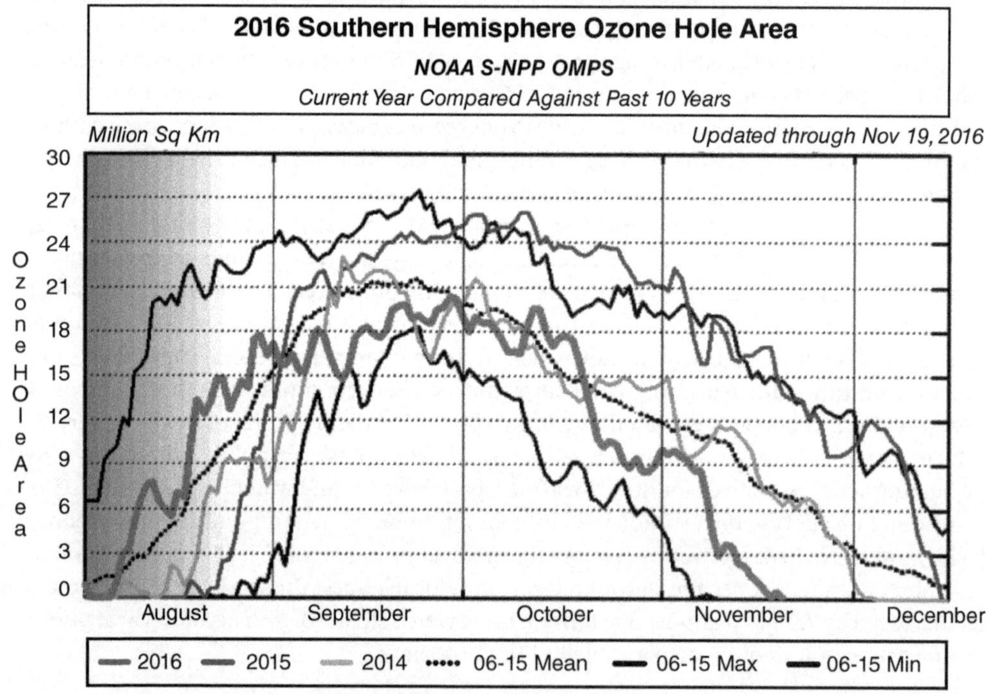

Figure 1.1 The current trend of Southern Hemisphere Ozone Hole Area. Source: Referenced from NOAA National Weather Service Climate Prediction Center.[3]

3 Referenced from NOAA National Weather Service Climate Prediction Center, USA. https://www.cpc.ncep.noaa.gov/products/stratosphere/sbuv2to/ozone_hole.shtml.

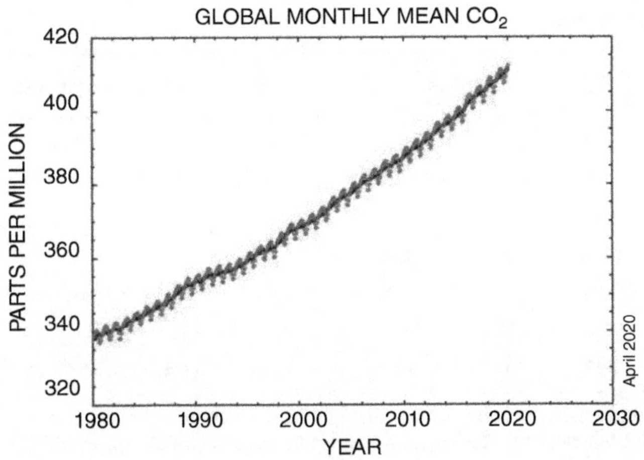

Figure 1.2 The content variation of carbon dioxide in atmosphere (1980–2020). Source: Referenced from Earth System Research Laboratory (ESRL) Global Monitoring Division.[4]

the content variation of CO_2 in the atmosphere[4]). With the pressing need to satisfy the existing environmental laws and protocols, researching and using new refrigerants which cause no harm to the ozone layer and prevent global warming are current trends, especially for heat pump/refrigeration/air-conditioner manufacturers, refrigerant suppliers, environmental experts, thermodynamic researchers and governments.

To address the environmental concerns, the use of natural refrigerants has grown in recent years. Natural refrigerants, rather than man-made, were widely used from the 1800s to the 1930s until the invention of synthetic refrigerants. With the development of design and manufacturing technology, the weaknesses of low-efficiency and lack of safety which troubled people in the early years were overcome gradually, so that together with the advantages of environmental friendliness, natural refrigerants are thus in renaissance now.

1.1.3 One of the Most Potential Natural Refrigerants – Carbon Dioxide (CO_2)

As one of the main natural refrigerants, carbon dioxide (CO_2, R744) gains more attention nowadays due to its environmental friendliness. As for CO_2 itself, it has a relatively low impact on climate change and no impact on ozone. Table 1.1 shows the properties of some common refrigerants, in which ODP (ozone depletion potential) is an index to reflect the impact of refrigerant on the ozone layer, and GWP (global warming potential) is to measure the potency of a greenhouse gas compared to CO_2 over a certain period. It can be found that the ODP of CO_2 is 0 which is the same as that of HFCs and the GWP of CO_2 is about 0.01–0.7% to that of CFCs, HCFCs, and HFCs, showing less impact on global warming with the same mass. Although CO_2 is the major greenhouse gas leading to climate change due to its higher content in the atmosphere compared to other refrigerants, the substitution for

4 Referenced from Earth System Research Laboratory (ESRL) Global Monitoring Division, USA. https://www.esrl.noaa.gov/gmd/ccgg/trends/gl_full.html#.

Table 1.1 Properties of various refrigerants.

Category	CFCs	HCFCs	HFCs		HCs	Natural refrigerants	
Refrigerant	R12	R22	R134a	R152a	R290	R717	R744
Chemical formula	CCl_2F_2	$CHClF_2$	CH_2FCH_3	CH_3CHF_2	$CH_3CH_2CH_3$	NH_3	CO_2
ODP	1	0.05	0	0	0	0	0
GWP (100-yr)[a]	10 200	1760	1300	138	3	0	1
Flammability/toxicity	N/N	N/N	N/N	Y/N	Y/N	Y/Y	N/N
Critical temperature (°C)	112.0	96.0	101.1	113.3	96.7	133.0	31.1
Critical pressure (MPa)	4.11	4.97	4.07	4.52	4.25	11.42	7.38

a) GWP references from IPCC (International Panel on Climate Change) Fifth Assessment Report. Working Group I Report "Climate Change 2013: The Physical Science Basis." Chapter 8. Appendix 8A.

high GWP refrigerants is still effective for greenhouse gas reduction since CO_2 as a refrigerant comes from other industrial processes, thus the net global warming impact is zero [4].

Besides the environmental advantages, CO_2 also has satisfactory safe, thermodynamic and economic properties.

- **Safety.** CO_2 is non-toxic and non-flammable, which is different from other natural refrigerants such as ammonia (NH_3) and sulfur dioxide (SO_2).
- **Thermodynamic benefits.** CO_2 has a relatively low critical temperature (31.1 °C) and high operating pressure (e.g. its critical pressure is 7.38 MPa, which is one to two times higher than the common synthetic refrigerants). The former characteristic allows the CO_2 heat pump to operate as a transcritical cycle easily; that is to say the CO_2 would transfer heat partly above the critical temperature and be in supercritical state in the heat rejection process. Compared to a system with refrigerant operating as a subcritical cycle, a transcritical cycle has some unique features, like the single-phase gas cooling process that occurs when it rejects heat, which is different from the condensation process in subcritical cycles, and gas cooling has a larger temperature glide which fits the water/air heating process better. Another characteristic, high operating pressure, brings challenges to component design and manufacture because it requires robust components and a stable compressor. This caused problems in the period before the 1930s, and thus restricted the use of the CO_2 cycle in higher ambient conditions due to extremely high pressure. However, the problem can be solved nowadays with the development of material research, component design and manufacture.

  Besides the thermodynamic properties mentioned above, CO_2 also has some different properties compared with conventional refrigerants. Its volumetric capacity is quite large which contributes to use of smaller components to achieve a more compact system [5], and the larger slope of vapor pressure versus saturation temperature results in smaller temperature change with a given pressure drop which allows higher mass flux in evaporation. Furthermore, the thermal conductivity (k) of CO_2 is relatively high and the k of CO_2 liquid and vapor is respectively 20% and 60% higher than that of R134a liquid and vapor [6], indicating better heat transfer.

- **Economy.** Carbon dioxide is economical as it is available in the air and a lot of industrial processes, leading to low cost.

As for vapor compression heat pumps, **cycle performance** is worthy of attention due to energy efficiency, compression ratio, etc. In the early period (before 1930s), CO_2 was mainly used as a refrigerant in a subcritical cycle, while the cycle had low efficiency when condensation temperature increased due to the low critical temperature. Since transcritical CO_2 heat pump cycles were proposed in 1980s, energy efficiency has been demonstrated as being as competitive as conventional synthetic refrigerant cycles, especially when compared over seasons and adding supplementary heat [6]. As for the compression ratio of transcritical CO_2 heat pumps, even though it has a relatively high operating pressure, the pressure ratio is smaller than with common refrigerants. For instance, the compression ratio of a transcritical CO_2 cycle is almost half of that of a R134a cycle, thus improving the compressor efficiency directly.

Transcritical CO_2 heat pumps can be widely applied in residential, commercial and industrial areas. In Japan, a transcritical CO_2 heat pump water heater has already been commercialized named "Eco Cute". More than three million have been installed since 2011 [7]. Besides, the heat pump also has potential to be used in space heating, automotive air conditioning systems and drying processes, indicating its various multifunctional uses.

1.1.4 Motivation for This Book

Based on the foregoing, it shows that the CO_2 heat pump is of great significance to be researched and applied due to its natural, stable, environmentally-friendly refrigerant properties and also its competitive cycle performance. Thus, it is quite important to study and understand its thermodynamic mechanism, operating characteristics and suitable applications through a systematic methodology, especially for people who have great interest in this topic, such as college students, technical staff and related researchers. After decades of development in this field, it is mature and essential to summarize related basic fundamentals and research results in one professional book, such as this one.

This book has mainly been divided into two parts to illustrate CO_2 heat pumps: fundamentals and applications. In this chapter, simplified introductions to these two parts (Sections 1.2 and 1.3) and also a clear guide to each chapter (Section 1.4) will be shown next to provide a basic understanding for readers.

1.2 Fundamentals

Common transcritical CO_2 heat pumps operate on the principle of vapor compression cycles, which have various configurations after years of development. The original and simplest cycle is shown in Figure 1.3a which comprises one gas cooler, one expansion valve, one compressor and one evaporator, and which can be modified as an internal heat exchange cycle, two-stage cycle or cycle with expander, etc. Thus, this basic cycle is concentrated to illustrate the science mechanisms in transcritical CO_2 heat pumps. In this section, the first part mainly describes the thermodynamic processes in the basic

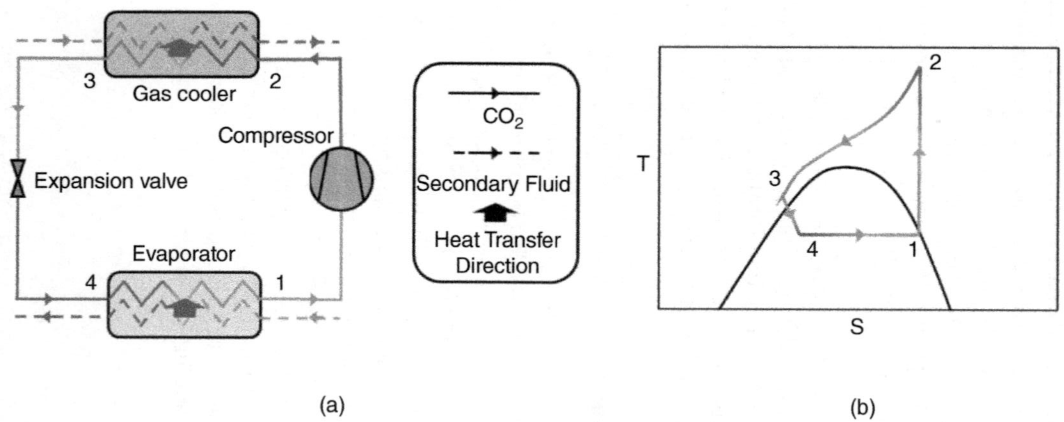

Figure 1.3 (a) A schematic of the basic transcritical CO_2 heat pump cycle. (b) T-s diagram of the corresponding cycle.

cycle and the characteristics of typical processes. The second part introduces performance characteristics for the whole cycle, and the last part gives a brief overview of cycle modification.

1.2.1 Operating Processes of the Basic Transcritical CO_2 Cycle

The basic cycle and its corresponding T-s diagram are respectively shown in Figure 1.3a and b. The cycle is composed of four processes as follows:

1–2: Isentropic compression process. CO_2 from the evaporator is compressed from saturated vapor to super-heated gas in supercritical state.
2–3: Isobaric cooling process. The supercritical CO_2 enters the gas cooler and rejects heat by gas cooling without phase change.
3–4: Adiabatic expansion process. The high-pressure CO_2 is expanded adiabatically to low pressure as vapor-liquid mixture through expansion valve.
4–1: Isobaric evaporation process. Two-phase CO_2 flows into the evaporator and absorbs heat from outside heat source under a subcritical pressure. It is then evaporated to a saturated vapor state at the outlet of the evaporator and enters the compressor again to repeat the process 1–2.

Each process in the cycle shows unique heat transfer and flow characteristics, which attracted researchers so much because of their help with component design and cycle optimization. A general view of the two-heat exchange process is summarized below:

- **Heat rejection process (2–3)** is the most particular process in transcritical cycles because heat is rejected by single phase cooling instead of condensation as in conventional subcritical cycles (therefore, the heat exchanger is called a gas cooler instead of a condenser).
  For this process, special heat transfer and flow characteristics can be observed due to the sharp changes of thermodynamic and transport properties[5] of CO_2, especially when it works in the region near pseudo-critical temperature (T_{pc}).[6] In the near-critical region, the fluid behaves with high expansion and low thermal diffusion, bringing new features in fluid flow and convective structures, such as a piston effect (a kind of thermal relaxation process) in micro-channels. These unique mechanisms of fluid flow and heat transfer play an important role in the heat rejection process in heat pumps.
  Research has focused on the unique mechanisms since the 1960s, usually in the in-tube heating/cooling process, like Shitsman [8], Krasnoshchekov *et al* [9], Shiralkar and Griffith [10], Kurganov *et al* [11], Jiang *et al* [12], etc. These in-tube researches contributed to the development of correlations, and the design/optimization/control/operation of conventional channel/mini-channel/micro-channel[7] gas coolers.

---

5 The thermodynamic properties include density (ρ) and specific heat (C_p), the transport properties include viscosity (μ) and thermal conductivity (k).
6 The pseudo-critical temperature (T_{pc}) is defined as the temperature at which the specific heat (C_p) reaches a peak under isobar condition.
7 As the classification of channels given by Kandlikar [13] and Kandlikar and Grande [14], the hydraulic diameters (D_h) of conventional channels, mini-channels, and micro-channels are respectively in the range of >3, 0.2–3 and 0.01–0.2 mm.

Because of the non-uniform distribution of properties in tubes, heat transfer deterioration or enhancement will be observed in specific conditions, which can be affected by heat flux, mass flow rate, inlet temperature, wall roughness, flow direction and tube diameter. The feature of non-uniformity leads to buoyancy force and flow acceleration bringing unique characteristics to the mechanisms but also difficulties with research. However, with the presence of non-dimensional parameters, buoyancy parameter Bo^* and flow acceleration parameter Kv_T, it becomes helpful and convenient to estimate their effects on heat transfer in theoretical and practical analysis. The development of heat transfer and pressure drop correlations is quite essential for gas cooler design, while the traditional single-phase correlations are not accurate here due to the varying properties (This conclusion has been verified by lots of research). It should be noted that the heat transfer correlation is not the same for heating and cooling, as for the gas cooling process. Krasnoshchekov *et al* [9] first carried out the heat transfer correlation through experimental study. The correlation considered the difference between wall temperature properties and bulk temperature properties which reflects the property variation in radial direction, thus improving the accuracy. Baskov *et al* [15], Petrov and Popov [16], Yoon *et al* [17], etc. also proposed different correlations adapting to different conditions showing relative accuracy. For pressure drop of the cooling process, the correlations proposed by Petrov and Popov [16], etc. were widely used.

- In **evaporator (4–1)**, CO_2 vaporizes during the flow boiling process. Compared to conventional HFCs, CO_2 shows a higher heat transfer coefficient and a lower pressure drop, indicating better overall performance.
  During the boiling of CO_2, the heat transfer coefficient decreases with increasing quality, and a sharp drop occurs in critical quality at which point dryout phenomenon occurs (i.e. the liquid film breaks down). Boiling heat transfer can be affected by heat flux, mass flux and evaporation temperature. In the region with low qualities, nucleate boiling dominates the heat transfer process, showing more sensitivity to heat flux rather than mass flux. And after dryout at high qualities, convective boiling is the major heat transfer process, with the heat transfer coefficient being influenced more by mass flux. Further, the dryout quality is mainly determined by mass flux [18, 19]. As for the pressure drop characteristic, it increases with increasing mass flux while decreasing with increasing evaporation temperature [17].
  Flow pattern is another interesting topic for two-phase flow. During the boiling process, slug flow can be observed at low qualities and annular flow with entrainment of droplets can be observed at high qualities [20].
  Several researchers have focused on the effect of lubricant on heat transfer, which is essential to be discussed in vapor compression cycles. Rieberer [21] concluded that heat transfer can be influenced by compressor lubricant and the nucleate heat transfer may be weakened. Zhao *et al* [22] concluded that the decrease rate of heat transfer during boiling can be higher with increasing concentration of lubricant.
  Like the development of correlations in the gas cooling process, evaporation correlation is of the same importance to evaporator design and operation. Boiling flow correlations have been developed a lot but results have shown that generalized correlations were not accurate for CO_2, therefore correlations for CO_2 should be developed specifically [5]. The

correlations presented by Jung *et al* [23], Cheng *et al* [24], Fang [25], etc. showed better predictability after comprehensive comparison [26, 27].

1.2.2 Characteristics of Transcritical CO_2 Cycles

Based on the unique properties for the near-critical CO_2 resulting in different heat exchange processes, several peculiarities of transcritical CO_2 heat pump cycles can be revealed.

- **The first characteristic is the large temperature glide of heat rejection**, which is the temperature difference between the gas cooler inlet and outlet. As Figure 1.4 shows, the temperature profile of gas cooling can match the heated curve of secondary fluid (Figure 1.4b) better than condensation condition (Figure 1.4a) due to the larger temperature glide, implying less irreversible losses in gas cooler. Thus, the transcritical heat pump can operate more effectively. In other words, transcritical CO_2 heat pumps are more suitable than subcritical heat pumps in heating applications. Further, better performance can be obtained with larger temperature glide which can be operated by reducing the inlet temperature of the secondary fluid [5].
- **Another characteristic is the existence of the optimum high-side pressure (heat rejection pressure).** In the supercritical region, temperature and pressure are two independent parameters, implying that the high-side pressure can be adjusted with the fixed cooling outlet temperature; thus, it is important to control high-side pressure to give a better performance. Due to the special thermodynamic properties in the supercritical region, the optimum high-side pressure can be obtained. As Figure 1.5 shows, with the increase of high-side pressure ($\Delta P_{rejection}$), the heating capacity increases for a given gas cooler outlet temperature. The increased capacity ($\Delta q_{heating}$) can compensate the increased compressor work (Δw) with the high-side pressure near the critical pressure, therefore the coefficient of performance (COP) increases. With thc much higher pressure, the isotherm becomes steeper, leading to a smaller increase in heating capacity with the same pressure increment. The increased capacity cannot compensate for the increased compressor work anymore, hence the COP decreases and the optimal high-side pressure exists.

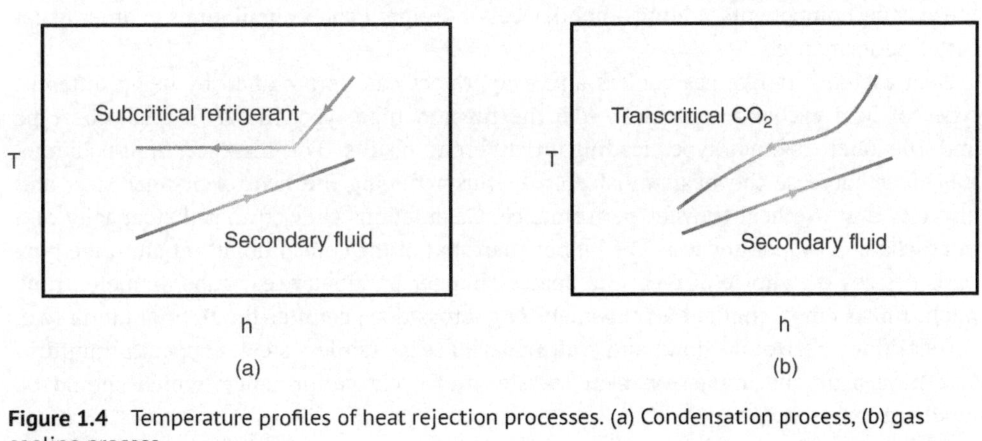

Figure 1.4 Temperature profiles of heat rejection processes. (a) Condensation process, (b) gas cooling process.

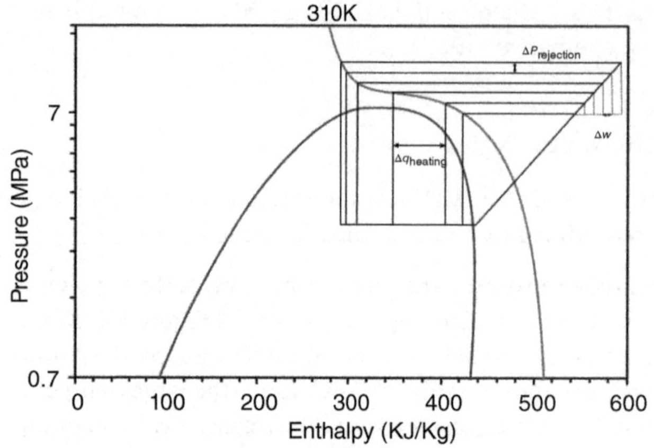

Figure 1.5 The variation of heating capacity and compressor work with the identical high-side pressure increment illustrated in the P-h diagram of transcritical carbon dioxide heat pump cycle.

- **In transcritical CO_2 cycle, the effect of evaporation temperature on heating capacity and COP is smaller than that of other conventional subcritical cycles** which indicates the transcritical CO_2 heat pump can supply heat more stably in conditions with varying heat source temperatures (for instance, air source transcritical CO_2 heat pumps can perform better with varying ambient temperatures over a year). Therefore, when compared with conventional heat pumps by considering supplementary heat, the transcritical CO_2 heat pump may have higher overall heating COP (i.e. the ratio of heat pump heating load plus supplementary heating load to the power input) due to the smaller requirement on supplementary heating devices [6].

1.2.3 Modifications of Transcritical CO_2 Cycles

Based on the basic CO_2 cycle, several modifications to cycles can be obtained by modifying the existing components, adding other devices or designing new circulations in order to get better performance.

Heat exchangers like gas coolers and evaporators can be modified by using different types of heat exchange channels, like the fin and tube type, the shell and tube type and the microchannel type, leading to different results. For instance, microchannel tubes can increase the heat transfer area, thus reducing the heat exchanger size, and also can improve heat transfer performance. Calculations showed that the capacity of a microchannel evaporator was 33% higher than that of the conventional fin-and-tube type [28]. Besides the choice of type, the heat exchanger parameter (e.g. tube diameter, tube pitch, fin spacing), channel arrangement (e.g. cross-flow, counter-flow), orientation (e.g. vertical flow, horizontal flow) and wall material (e.g. stainless steel, copper, aluminum) also have a unique impact on heat transfer and cycle performance, which should be modified comprehensively.

Adding a suction line heat exchanger (SLHX) is a common method of preheating the vapor before compression and subcools the cooled fluid at the outlet of the gas cooler. This operation can improve the COP by 7% through simulation analysis [5]. Besides, adding compressors is another basic modification which makes the cycle run as a multi-stage cycle to improve compressor efficiency by reducing the pressure ratio of each stage. Further, the arrangement of other devices like flash tanks and intercoolers can cool the exhausted gas at outlet of the low-pressure compressor, therefore enhancing energy efficiency.

1.3 Applications

As mentioned in Section 1.2, transcritical CO_2 heat pumps have more advantages in heating compared to subcritical heat pumps, especially for applications which need a large temperature increase. Transcritical CO_2 heat pumps have the potential to be applied in various areas such as residential heating and commercial-size heating to satisfy the demand of hot water supply or hot air supply, and some types have already been popularized on the market such as residential CO_2 heat pump water heaters. Next, several applications will be listed and briefly introduced.

- **Transcritical CO_2 heat pump water heater** (as shown in Figure 1.7) has attracted people since the late 1980s due to its good performance in the supply of high temperature water with good COP, which benefited from well matching in the heat rejection process. This kind of water heater was commercialized in Japan and named "Eco Cute" in 2001. As Figure 1.6 shows, sales of Eco Cute in Japan grew rapidly and the cumulative number of installed units exceeded three million by 2011 with more than 400 000 sold every year by the leading companies like Denso, Sanyo, Sanden, and Mitsubishi Electric [7]. Based on its rapid growth, Japan established standard specifications for heat pump water heaters and it is the only country for the refrigerant CO_2. The leading manufacturers continue to innovate products to meet different demands such as making heat pumps that operate in cold regions, [29] or enhance the performance of key components like compressors and gas coolers [7]. Compared to the mature market in Japan, the European market seems new and underexploited. Besides the domestic water heater market, some commercial

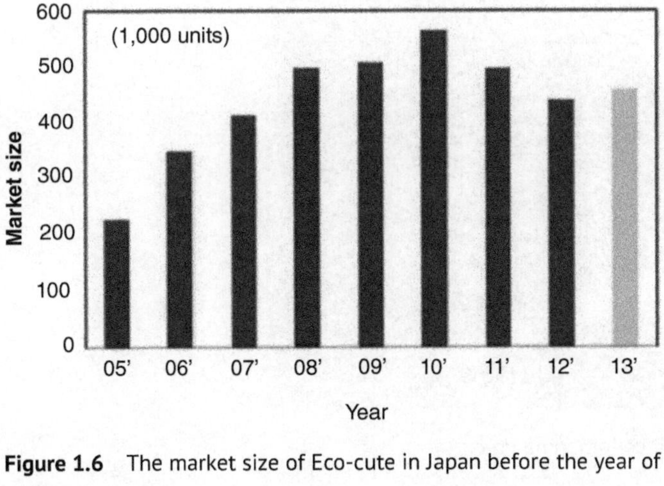

Figure 1.6 The market size of Eco-cute in Japan before the year of 2013 [7].

Figure 1.7 Schematic of Eco-cute products [7].

applications of CO_2 heat pumps have been developed in Europe to supply hot water for hotels, restaurants, hospitals, and schools, and some examples have been constructed in Ireland, France, Denmark, and Switzerland since 2012 [30].

As for the transcritical CO_2 heat pump water heater, though it has been commercialized successfully, it still needs to be researched, optimized, and developed. It has been found that the inlet and outlet water temperature and ambient temperature can affect the optimal discharge pressure and heating COP. For example, a higher ambient temperature and a lower inlet water temperature lead to a higher overall COP [31]. Besides the optimization of operating parameters, the modification of the operating cycle and the management of water storage and heating also need to be carefully considered.

- **Heating for vehicles,** especially in low ambient temperatures, is a current trend due to the compact structure of CO_2 heat pumps, environmentally friendly benefits and suitable thermodynamic properties of carbon dioxide. Because of the finite spaces in vehicles, compact heating devices with large heating capacity are required (Figure 1.8).

Figure 1.8 Air outlet of a car air conditioning unit.

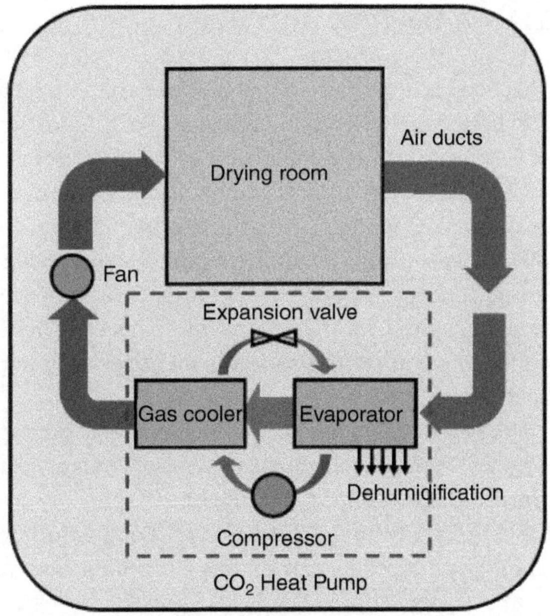

Figure 1.9 The schematic of a CO_2 heat pump dryer.

Meanwhile, regulations in the EU, Japan, and the US aim to phase out HFCs in mobile air conditioning systems [32], thus CO_2 is considered as one of the most suitable refrigerants for substitution.

When using CO_2 heat pumps in vehicles, it has been proven that the temperature can increase rapidly and the heating-up time can be reduced by 50% compared to that of the conventional heat cores [33]. Modifying the heat pump system by adding exhaust heat recovery and utilizing cabin return air were effective ways to adjust heating capacity; the former improving capacity by 100% compared to conventional heating [34] and the latter reducing heating load thus saving electricity power [35]. Further, heating performance can be influenced by outdoor temperature, outdoor air velocity, indoor temperature, indoor air flow rate, CO_2 charge volume and compressor speed [36], indicating that further optimization can be achieved for better energy efficiency.

- **Transcritical CO_2 heat pump applied in drying** (as shown in Figure 1.9) has great potential owing to the good match of gas cooling to the air heating process, therefore leading to a higher air temperature and a larger moisture extraction rate. Compared with conventional refrigerants and conventional heaters, CO_2 heat pumps have been tested and have proved competitive. Schmidt *et al* [37] concluded that the higher compressor efficiency can compensate for the large throttling losses for the CO_2 cycle, resulting in an equivalent or even better performance compared with the R134a cycle. Klöcker *et al* [38] compared two CO_2 heat pump prototypes with traditional electrical heaters for drying experimentally, showing that even 53% of energy can be saved. Moreover, in CO_2 heat pump drying, the optimization of operating conditions is quite essential, such as air temperature, air flow rate, flow bypass ratio, etc., which can influence both the heat pump efficiency and drying performance.

1.4 A Guide to This Book

This book is not expected to include everything related to transcritical CO_2 heat pump, but is expected to focus on the main parts of transcritical CO_2 heat pump from the two aspects of fundamentals and applications. Chapter 1 provides background and motivation for the book. Chapter 2 presents fundamental concept and working principle for transcritical CO_2 heat pump. Chapters 3–6 cover the fundamentals of the four basic processes of transcritical CO_2 heat pump, cooling, evaporation, expansion, and compression, respectively. Some design concepts are also involved. Chapter 7 is specially included to cover the most relevant study on subcooled CO_2 cycles and aims to solve the main drawback of transcritical CO_2 cycles. Chapters 8–10 focus on some important application examples of transcritical CO_2 heat pump cycles.

Here, it should be mentioned that the basic principle of transcritical CO_2 heat pump is not only seen in Chapter 2, but also in other chapters. Similarly, CO_2 thermophysical properties are also presented in some different chapters. However, the CO_2 properties and principles are explained from the different views and requirements for the different chapters, such as from the phase change, from the subcooling, and from the cycle.

References

1 Molina, M.J. and Rowland, F.S. (1974). Stratospheric sink for chlorofluoromethanes: chlorine atom-catalysed destruction of ozone. *Nature* 249 (5460): 810–812.

2 Protocol, M. (1987). Montreal protocol on substances that deplete the ozone layer. Washington, DC: US Government Printing Office, 26: 128–136.

3 Protocol, K. (1997). United Nations framework convention on climate change. Kyoto Protocol, Kyoto, 19: 497.

4 Petter, N. (2002). CO_2 heat pump systems. *International Journal of Refrigeration* 25 (4): 421–427.

5 Austin, B.T. and Sumathy, K. (2011). Transcritical carbon dioxide heat pump systems: a review. *Renewable and Sustainable Energy Reviews* 15 (8): 4013–4029.

6 Kim, M. (2004). Fundamental process and system design issues in CO_2 vapor compression systems. *Progress in Energy and Combustion Science* 30 (2): 119–174.

7 Zhang, J.-F., Qin, Y., and Wang, C.-C. (2015). Review on CO_2 heat pump water heater for residential use in Japan. *Renewable and Sustainable Energy Reviews* 50: 1383–1391.

8 Shitsman, M.E. (1963). Impairment of the heat transmission at supercritical pressures. *Teplofizika Vysokikh Temperatur* 1 (2): 267–275.

9 Krasnoshchekov, E.A., Kuraeva, I.V., and Protopopov, V.S. (1970). Local heat transfer of carbon dioxide at supercritical pressure under cooling conditions. *Teplofizika Vysokikh Temperatur* 7 (5): 922–930.

10 Shiralkar, B.S. and Griffith, P. (1969). Deterioration in heat transfer to fluids at supercritical pressure and high heat fluxes. *ASME Journal of Heat Transfer 91(1)*: 27–36.

11 Kurganov, V.A. and Kaptilnyi, A.G. (1993). Flow structure and turbulent transport if a supercritical pressure fluid in a vertical heated tube under the conditions of mixed

convection. Experimental data. *International Journal of Heat and Mass Transfer* 36 (13): 3383–3392.

12 Jiang, P.X., Ren, Z.P., and Wang, B.X. (1995). Convective heat and mass transfer of water at supercritical pressures under heating or cooling conditions in vertical tubes. *International Journal of Thermal Sciences* 4: 15–25.

13 Kandlikar, S.G. (2002). Fundamental issues related to flow boiling in minichannels and microchannels. *Experimental Thermal and Fluid Science* 26 (2–4): 389–407.

14 Kandlikar, S.G. and Grande, W.J. (2003). Evolution of microchannel flow passages – thermohydraulic performance and fabrication technology. *Heat Transfer Engineering* 24 (1): 3–17.

15 Baskov, V.L., Kuraeva, I.V., and Protopopov, V.S. (1977). Heat transfer with the turbulent flow of a liquid at supercritical pressure in tubes under cooling conditions. *Teplofizika Vysokikh Temperatur* 15 (1): 96–102.

16 Petrov, N.E. and Popov, V.N. (1985). Heat transfer and resistance of carbon dioxide being cooled in the supercritical region. *Thermal Engineering* 32 (3): 131–134.

17 Yoon, S.H., Cho, E.S., Hwang, Y.W. et al. (2004). Characteristics of evaporative heat transfer and pressure drop of carbon dioxide and correlation development. *International Journal of Refrigeration* 27: 111–119.

18 Sun, Z. and Groll, E. (2000). CO_2 flow boiling in horizontal tubes. Internal report No. HL-2001-8, Ray W. Herrick Laboratories, Purdue University.

19 Yun, R., Kim, Y.C., Kim, M.S., and Choi, Y.D. (2003). Boiling heat transfer and dryout phenomenon of CO_2 in a horizontal smooth tube. *International Journal of Heat and Mass Transfer* 46: 2353–2361.

20 Pettersen, J. (2002). Flow vaporization of CO_2 in microchannel tubes. PhD Thesis. Norwegian University of Science and Technology, Norway.

21 Rieberer, R. (1998). CO_2 as working fluid for heat pump. PhD Thesis. Institute of Thermal Engineering, Graz University, Austria.

22 Zhao, Y., Molki, M., and Ohadi, M.M. (2002). Flow boiling of CO_2 with miscible oil in microchannels. Technical and Symposium Papers 4505, ASHREA Winter Meeting, Atlantic city, USA.

23 Jung, D.S., Radermacher, R., McLinden, M., and Didon, D. (1989). A study of flow boiling heat transfer with refrigerant mixtures. *International Journal of Heat and Mass Transfer* 32: 1751–1764.

24 Cheng, L., Ribatski, G., Moreno Quiben, Q., and Thome, J.R. (2008). New prediction methods for CO_2 evaporation inside tubes: part I – a two-phase flow pattern map and a flow pattern based phenomenological model for two-phase flow frictional pressure drops. *International Journal of Heat and Mass Transfer* 51: 111–124.

25 Fang, X.D. (2013). A new correlation of flow boiling heat transfer coefficients for carbon dioxide. *International Journal of Heat and Mass Transfer* 64: 802–807.

26 Fang, X.D., Zhou, Z.R., and Li, D.K. (2013). Review of correlations of flow boiling heat transfer coefficients for carbon dioxide. *International Journal of Refrigeration* 36: 2017–2039.

27 Mastrullo, R., Mauro, A.W., Rosato, A., and Vanoli, G.P. (2010). Carbon dioxide heat transfer coefficients and pressure drops during flow boiling: assessment of predictive methods. *International Journal of Refrigeration* 33: 1068–1085.

28 Yun, R., Kim, Y., and Park, C. (2007). Numerical analysis on a microchannel evaporator designed for CO_2 air-conditioning systems. *Applied Thermal Engineering* 27: 1320–1326.

29 Vanaga, G. (2014). Very latest "Eco Cute" innovations from HVAC&R Japan.

30 Maratou, A., Lira, J.T., Jia, H., and Masson, N. (2012). CO_2 heat pumps in Europe market dynamics & legislative opportunities.

31 Fernandez, N., Hwang, Y., and Radermacher, R. (2010). Comparison of CO_2 heat pump water heater performance with baseline cycle and two high COP cycles. *International Journal of Refrigeration* 33: 635–644.

32 Andersen, S. Chowdhury, S., Craig, T. et al. (2017). Comparative manufacturing and ownership cost estimates for secondary loop mobile Air conditioning systems (SL-MACs). SAE International.

33 Hammer, H. (2000). Carbon dioxide (R744) as supplementary heating device SAE Phoenix Forum 2000.

34 Kim, S.C., Kim, M.S., Hwang, I.C., and Lim, T.W. (2007). Heating performance enhancement of a CO_2 heat pump system recovering stack exhaust thermal energy in fuel cell vehicles. *International Journal of Refrigeration* 30 (7): 1215–1226.

35 Wang, D., Yu, B., Li, W. et al. (2018). Heating performance evaluation of a CO_2 heat pump system for an electrical vehicle at cold ambient temperatures. *Applied Thermal Engineering* 142: 656–664.

36 Wang, D., Yu, B., Hu, J. et al. (2018). Heating performance characteristics of CO2 heat pump system for electrical vehicle in a cold climate. *International Journal of Refrigeration* 85: 27–41.

37 Schmidt, S.L., Klöcker, K., Flacke, N., and Steimle, F. (1998). Applying the CO_2 transcritical process to a drying heat pump. *International Journal of Refrigeration* 21: 202–211.

38 Klöcker, K., Schmidt, E.L., and Steimle, F. (2001). Carbon dioxide as a working fluid in drying heat pumps. *International Journal of Refrigeration* 24: 100–107.

2

Current Development of CO_2 Heat Pump

Hiroshi Yamaguchi[1] and Xin-Rong Zhang[2]

[1] *Energy Conversion Research Center, Department of Mechanical Engineering, Doshisha University, Kyoto, Japan*
[2] *Department of Energy and Resources Engineering, College of Engineering, Peking University, Beijing, China*

2.1 Introduction

The usage of the CO_2 heat pump is expanding rapidly with state-of-the-art technology in many fields of industry. This is solely due to awareness of the immediate demand for the global environment. Nowadays the new technology is shifting more and more to CO_2-based appliances. A CO_2 heat pump is certainly the one that is the most promising and sustainable technology which is introduced to a greater extent in this chapter.

For more than 200 years since the industrial revolution period, the thermodynamic heat pump cycle, or in other words refrigeration cycle, has been used in many industrial fields for various purposes [1, 2]. Until more recent times there have been no concerns about the effects of working fluid used in the cycle. However, the popular working fluids (chemically synthesized) in the thermodynamic cycle, such as Chlorofluorocarbons (CFCs) and Hydrochlorofluorocarbons (HCFCs), have had much effect on depletion of the ozone layer and are the cause of the ozone hole in the Antarctica area found in the late twentieth century [3, 4].

The 1987 Montréal Protocol on Substances that Deplete the Ozone Layer was first established with the agreement of 47 countries for the protection of the ozone layer [5]. The primary purpose of the protocol deals with phasing out of some of the refrigerants and their production which are the main reason for ozone depletion. Since the phasing out of CFCs, the ozone hole in the Antarctica area has been found to be recovering slowly [6].

The international agreement of industrial countries to phase out the use of HCFCs under the Montréal Protocol started in 2013 by aiming to eliminate the use of HCFCs by 2030 and 2040 in developed and developing countries, respectively [7]. The timeline of the Montréal Protocol is drawn in Figure 2.1 [5, 8]. The 1997 Kyoto Protocol, the world's first global warming and climate change treaty, is a binding agreement to reduce greenhouse gas (GHG) emission levels with 192 parties' agreement. In 2012, the second commitment period (2013–2020) had undertaken to reduce 20% of the 1990 GHG emission level by 2020. In order to reduce global warming and the greenhouse effect, Hydrofluorocarbon (HFCs) had been recommended to replace CFCs and HCFCs, which were widely used in

Transcritical CO_2 Heat Pump: Fundamentals and Applications,
First Edition. Xin-Rong Zhang and Hiroshi Yamaguchi.

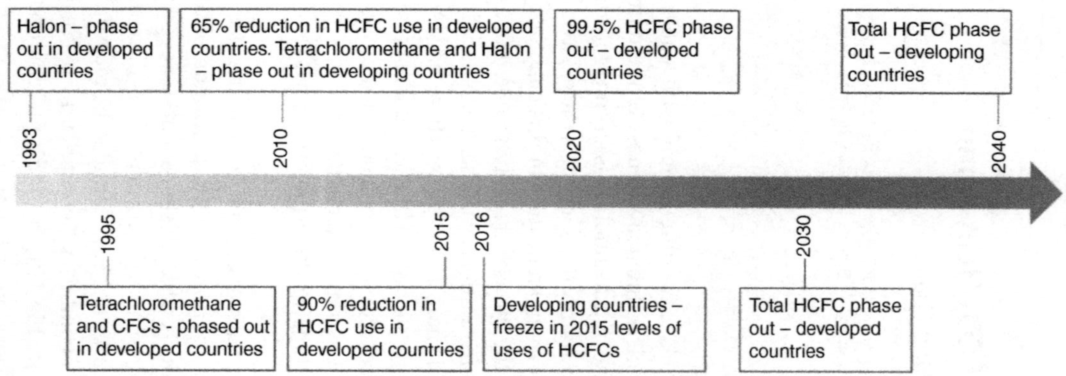

Figure 2.1 Timeline of Montréal Protocol [5, 8].

various industries for many years [9]. HFCs are commonly used in air conditioning and refrigerant systems, which do not affect the Ozone Depletion Potential (ODP). The ODP of the refrigerant is the relative degradation amount to the ozone layer with CFC-11 as a datum reference, where CFC-11 has a constant ODP of 1.0. However, HFCs have a very high Global Warming Potential (GWP), where GWP means the ratio of the effect of the refrigerant that will cause global warming by comparison with a similar mass of carbon dioxide (CO_2). Subsequently, a 2–4% increase of HFCs is forecasted in the overall climate forcing impact by 2050 [10].

In 2015, the 21st session of the Conference of the Parties to the United Nation Framework Convention on Climate Change, the so-called "Paris Agreement," suggested procedures to reduce carbon emissions at the earliest and aimed to respond to the global climate change threat by lowering the global average temperature rising in this century to well below 2°C above the pre-industrial level, and to limit temperature rising further below to 1.5°C [11]. To achieve the goal of the Paris Agreement, the natural working fluids such as ammonia (R-717 or NH_3) and carbon dioxide (R-744 or CO_2) have been recommended to be used as a refrigerant in industrial applications instead of HFCs. Especially, CO_2 was also listed in the required corporate control in the Kyoto Protocol. Furthermore, from "Act on Rational Use and Proper Management of Fluorocarbons" 2015, Ministry of the Environment, Government of Japan, states that CO_2 is one of the recommended alternative refrigerants to use to combat the world's emission crisis [12].

With regard to the prevention of global warming and the greenhouse effect, the use of CO_2 as a natural working fluid has been given much attention for decades by taking into account that CO_2 itself is environmentally friendly when compared with other working fluids, where ODP and GWP of CO_2 are defined by 0 and 1, respectively. The characteristics and properties of various working fluids are listed in Table 2.1. CO_2 is

Table 2.1 Characteristic of some representative working fluids [13].

Properties	R-12	R-22	R-134a	R-407C	R-410A	R-717	R-290	R-744
ODP/GWP	1/8500	0.05/1700	0/1300	0/1600	0/1900	0/0	0/6	0/1
Flammability/toxicity	N/N	N/N	N/N	N/N	N/N	Y/Y	Y/N	N/N
Molecular mass (kg $kmol^{-1}$)	120.9	86.5	102	86.2	72.6	17	44.1	44
Critical pressure (MPa)	4.11	4.97	4.07	4.64	4.79	11.42	4.25	7.38
Critical temperature (°C)	112.0	96.0	101.1	86.1	70.2	133.0	96.7	31.1
Reduced pressure[a)]	0.07	0.1	0.07	0.11	0.16	0.04	0.11	0.47
Reduced temperature[b)]	0.71	0.74	0.73	0.76	0.79	0.67	0.74	0.9
Refrigerant capacity[c)] (kJ m^{-3})	2734	4356	2868	4029	6763	4382	3907	22 454

R-12: dichlorodifluoromethane; R-22: chlorodifluoromethane; R-134a: tetrafluoroethane; R-407C: ternary mixture of difluoromethane/pentafluoroethane/tetrafluoroethane (23/25/52, %); R-410A: binary mixture of difluoromethane/pentafluoroethane (50/50, %); R-717: ammonia; R-290: propane; R-744: carbon dioxide.

a) Ratio of saturation pressure at 0°C to critical pressure.
b) Ratio of 273.15 K (0°C) to critical temperature in Kelvin.
c) Volumetric refrigeration capacity at 0°C.

also classified as non-flammable, non-toxic, chemically inactive, and inexpensive as well. Also, the vapor pressure and volumetric refrigeration capacity of CO_2, which are 3485 kPa and 22 545 kJ m^{-3} at 0°C, are much higher when compared with other common working fluids available on the market. It means that by using CO_2 in the thermodynamic cycle, the system performance and energy transport with the working fluid would be much improved. The utilization of CO_2 in green technology applications has been researched and developed by various research groups and applied to various fields such as chemical engineering, industrial engineering, mechanical engineering, environmental engineering, electrical engineering, nuclear engineering, and so on.

2.2 CO_2 Properties

CO_2 has been regarded as a promising next-generation working fluid for various purposes due to its ecologically and environmentally safe properties [14], as has been explained in the previous section. Although dry ice solid-gas state is a unique property of CO_2, which is used in the heat pump cycle as a cooling/refrigeration mechanism (in this chapter), it is mentioned here that the supercritical state of CO_2 is also an exciting phase for use in the power generation cycles, such as the Rankine cycle [15], as an application of renewable energy to electric power and thermal energy generation. The challenging heat pump system, as described in this chapter, can achieve the temperature below the triple point temperature of CO_2 as −56.6°C [16]. As displayed in Figure 2.2a and b for respectively the Mollier diagram and P-T Phase diagram, the critical point of CO_2 is marked at temperature of 31.1°C and pressure of 7.38 MPa, while the sublimation point and the triple point are also marked respectively at −78.5°C and 0.101 MPa, and −56.6°C and 0.518 MPa of temperature and pressure.

In an ordinary power cycle, the ideal gas is usually agreed to explain the change of the thermodynamic state in a power cycle. However, at the point of high temperature and high pressure such as the supercritical state, the behavior of real gas deviates significantly from the ideal gas [17].

For one mole of real gas;

$$\frac{\boldsymbol{PV}}{\boldsymbol{RT}} \rightarrow \mathbf{1}_{\boldsymbol{limitP\rightarrow 0}} \tag{2.1}$$

where P is pressure, V is volume, R is gas constant and T is temperature.

It can be seen from the above equation that when the pressure of the gas is close to zero, the behavior of real gas will be similar to the ideal gas, with which the ideal gas law can be defined as

$$\mathrm{Z} = \frac{\textit{molar volume of real gas at same T and P}}{\textit{molar volume of ideal gas at same T and P}} = \frac{PV}{RT} \tag{2.2}$$

For real gases, Z may be higher or lower than one. If the value of Z is close to one, it means the real gas behaves like an ideal gas. If the value of Z is higher than one, it means the gas is less compressible. On the other hand, gas is more compressible when Z is less than one.

In this section, the compressibility factor for carbon dioxide (CO_2) is calculated using the equation of state (EOS). Two EOSs are typically provided, which are Peng-Robinson EOS

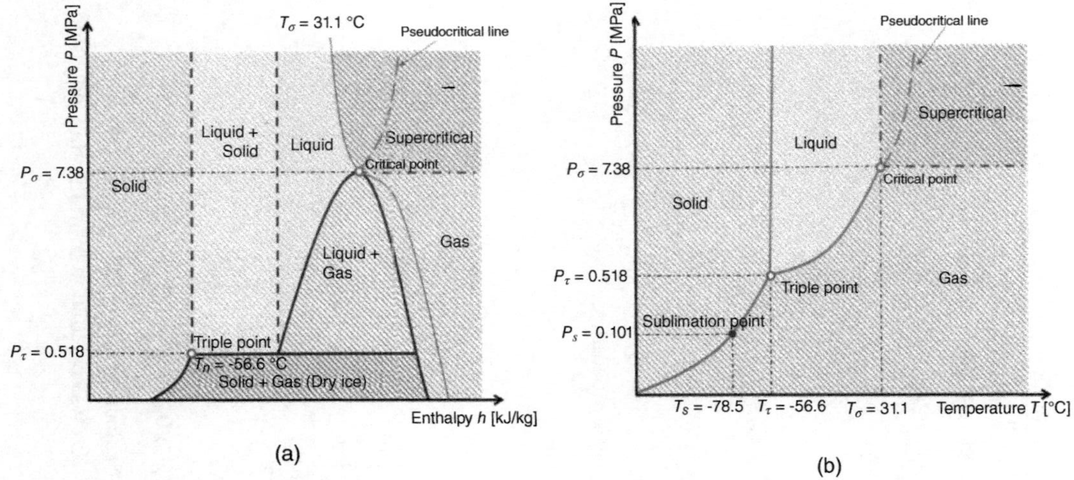

Figure 2.2 (a) Mollier (*P-h*) diagram and (b) Phase (*P-T*) diagram of CO_2.

[18] and Angus EOS [19]. The Peng-Robinson EOS (i) is derived theoretically, while the Angus EOS (ii) is obtained from the curve fitting of the experimental data. Also, the Angus EOS is the base equation for the database PROPATH [20], which is used to find the value of CO_2 in a review of the present work.

(i) CO_2 compressible factor calculated from the Peng-Robinson EOS
The Peng-Robinson EOS [18];

$$P = \frac{RT}{V-b} - \frac{a(T)}{V(V+b)+b(V-b)} \tag{2.3}$$

which can be written in term of Z as

$$Z^3 - (1-B)Z^2 + (A-3B^2-2B)Z - (AB-B^2-B^3) = 0 \tag{2.4}$$

where

$$A = \frac{aP}{R^2T^2} \tag{2.5}$$

$$B = \frac{bP}{RT} \tag{2.6}$$

$$C = \frac{PV}{RT} \tag{2.7}$$

The term of A and B at any temperature;

$$\mathrm{a}(T) = a(T_c)\alpha(T_r, \omega) \tag{2.8}$$

$$\mathrm{b}(T) = b(T_c) \tag{2.9}$$

where,

$$\mathrm{a}(T_c) = 0.45724\frac{R^2T_c^2}{P_c} \tag{2.10}$$

$$b(T_c) = 0.07780\frac{RT_c}{P_c} \tag{2.11}$$

and

$$\alpha(T_r, \omega) = 1 + (T_r^{1/2})(0.37464 + 1.54226\omega - 0.26992\omega^2) \tag{2.12}$$

(ii) CO_2 compressible factor calculated from the Angus EOS
The compressibility for CO_2 was calculated from the state equation;

$$P = \rho RTZ$$

In which Z (Angus EOS) [19] is given as

$$Z = 1 + \frac{\rho}{\rho_c}\sum_{i=1}^{6}\sum_{j=1}^{9} a_{ij}\left(\frac{\rho}{\rho_c} - 1\right)^i\left(\frac{T_c}{T} - 1\right)^j \tag{2.13}$$

where $\rho_c = 468\ \mathrm{kg\ m^{-3}}$ and $T_c = 304.2\ \mathrm{K}$, which are CO_2 critical density and critical pressure, respectively, the coefficients a_{ij} of Angus EOS for CO_2 is tabulated in Table 2.2.

Table 2.2 Coefficients for a_{ij} [19].

	0	1	2	3	4	5	6
0	−0.725854437	−1.68332974	0.259587221	0.376945574	−0.67075537	−0.871456126	−0.149456928
1	0.447869183	1.26050961	5.96957049	15.4645885	19.4449475	8.64880497	0
2	−0.172011999	−1.83458178	−4.61487677	−3.81121926	3.6171349	4.92265552	0
3	0.004463049	−1.76300541	−11.1436705	−27.8215446	−27.168572	−6.42177872	0
4	0.255491571	2.37414246	7.50925141	6.61133318	−2.4266321	−2.57944032	0
5	0.05946673	1.16974683	7.4370641	15.0646731	9.57496845	0	0
6	−0.14796001	−1.69233071	−4.68219937	−3.13517448	0	0	0
7	0.013671044	−0.10049233	−1.63653806	−1.87082988	0	0	0
8	0.039228458	0.441503812	0.88674197	0	0	0	0
9	−0.01198721	−0.084605195	0.046456437	0	0	0	0

In comparison, it has often been said that the Angus EOS has more advantages compared with Peng-Robinson EOS in accuracy of representing measured CO_2 data. In this section, all properties of CO_2 are obtained and calculated based on the PROPATH [20], which have been cited in the IUPC Table [19].

It is useful to consider the unique property of CO_2 when used in a thermodynamic (power or heat pump) cycle. The supercritical state of CO_2 at 9 MPa, which is a representative operation pressure in a transcritical power cycle system [15], has the remarkable energy transfer (heat transfer) characteristic in which its thermo-physical properties exhibit rapid variations with a change in temperature, especially near the pseudo-critical point (around 312 K) as indicated in Figure 2.3. It varies in its properties, such as flow viscosity, thermal conductivity, and density, as the temperature is increased across the pseudocritical point at the given pressure, and all properties (except for c_p) decrease significantly. On the other hand, the specific heat (c_p) reaches the peak at the pseudocritical point. This characteristic brings many different features of heat transfer from the constant properties of ordinary fluids. It should be noted that, when the temperature increases to higher than the pseudocritical region, the properties of the CO_2 slowly change, i.e. specific heat and density become smaller, while viscosity and thermal conductivity increase along with the temperature variation due to the reason that supercritical CO_2 is highly compressible. The change of characteristic properties of supercritical CO_2 at 9 MPa is shown in Figure 2.3, calculated from PROPATH [20].

The forced convection heat transfer is a higher mode of transferring the thermal energy of working fluid in a thermodynamic cycle since the heat transfer efficiency in the forced convection can be largely increased by operating the working fluid at a supercritical region. In the supercritical region, however, the working fluid behaves differently from an ordinary power cycle operation condition. The significant change in the fluid properties enhances the convection heat transfer coefficient of the working fluid, particularly owing to the change in specific heat (c_p) as demonstrated in Figure 2.3 [21].

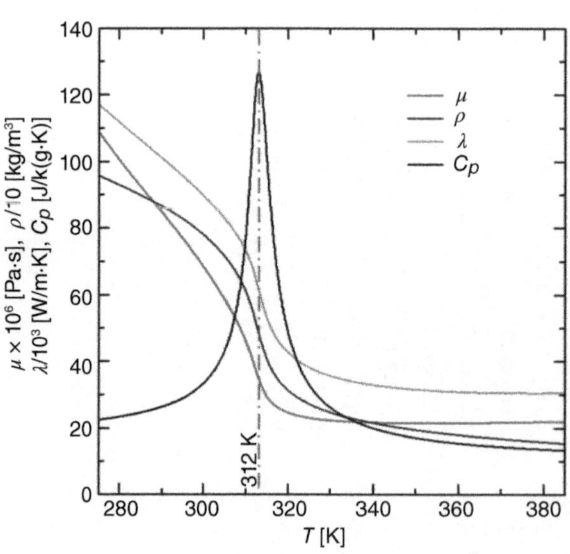

Figure 2.3 Variations of thermo-physical properties with the temperature at 9 MPa, density $\rho \times 10^{-1}$ (kg m^{-3}), specific heat c_p (J kg^{-1} K^{-1}), thermal conductivity $\lambda \times 10^{-3}$ (W m^{-1} K^{-1}) and dynamic viscosity $\mu \times 10^6$ (Pa s).

2.3 Working Principle of Transcritical CO_2 Heat Pump

In a thermodynamic cycle, heat is rejected by the working fluid at a high temperature and received at a low temperature, while a required amount of work is given from outside of the cycle. It is called *heat pump* or *refrigeration* cycle. The term "**heat pump**" is usually applied to a machine, whose principal purpose is to supply heat at an elevated temperature, and the term "**refrigerator**" to one whose purpose is extraction of heat from a cold space. This distinction in terminology is arbitrary because a heat pump and refrigerator are identical in principle and it is possible to use one machine to fulfill the function of a heat pump and refrigerator simultaneously. If the series of processes, which makes up a reversible power cycle, are plotted on a P-h and T-s diagram, the enclosed area in the T-s diagram (Figure 2.4a) is traced out in a clockwise sense, indicating that the net work done is positive. The negative net work of a reversible refrigeration cycle is proportional to an area traced out by processes in an anticlockwise sense. A reversed Carnot cycle, using a wet vapor as a working fluid, is typically shown in the T-s diagram of Figure 2.4a. Vapor is compressed isentropically from low pressure and temperature (state 1) to a higher pressure and temperature (state 2) and is passed through a condenser, in which it is condensed at constant pressure to state 3. The fluid is then expanded isentropically to its original pressure (state 4) and is finally evaporated at constant pressure to state 1. Note that dashed points 2′ and 3′ are the points when the cycle (heat pump) is used in a transcritical cycle.

The criterion of performance of the cycle, expressed as the ratio output/input, depends on what is regarded as the output. In a refrigerator, the objective is to extract the maximum amount of heat Q_{41} in the evaporator for a net expenditure of work W. Therefore the

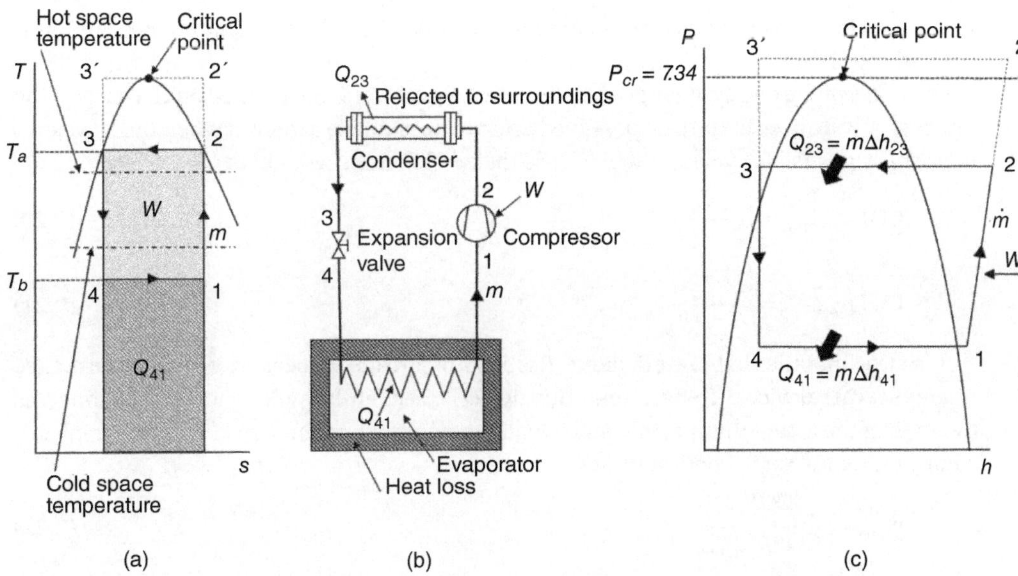

Figure 2.4 (a) T-s diagram of the reversed Carnot cycle. (b) Schematic representation of the refrigeration or heat pump cycle. (c) P-h diagram of the refrigeration or heat pump cycle.

coefficient of performance (COP) of a refrigerator is defined as:

$$\mathrm{COP}_{ref} = -\frac{Q_{41}}{W} \tag{2.14}$$

Since W is negative and Q_{41} positive with our sign convention, the negative sign is introduced to make COP_{ref} a positive number. In a heat pump, it is necessary to obtain the maximum amount of heat Q_{23} from the condenser for the net expenditure of work, W and defined by:

$$\mathrm{COP}_{hp} = \frac{Q_{23}}{W} \tag{2.15}$$

The relation between these two coefficients of performance can be established by applying the first law of thermodynamics. Thus:

$$(Q_{41} + Q_{23}) - W = 0 \tag{2.16}$$

and hence:

$$\mathrm{COP}_{hp} = \mathrm{COP}_{ref} + 1 \tag{2.17}$$

Inspection of the areas representing Q_{41}, Q_{23} and W in T-s diagram (Figure 2.4a) shows that COP_{ref} maybe greater than unity, and that COP_{hp} must always be greater than unity.

In Figure 2.4, Q_{23} and Q_{41} are obtained as follows from (a) T-s and (c) P-h diagram respectively:

$$Q_{23} = \dot{m}\Delta s_{23} T_a, Q_{41} = \dot{m}\Delta s_{23} T_b, W = \dot{m}\Delta s_{23}(T_a - T_b) \tag{2.18}$$

and

$$Q_{23} = \dot{m}\Delta h_{23}, Q_{41} = \dot{m}\Delta h_{41}, W = \dot{m}\Delta h_{12} \tag{2.19}$$

where $\dot{m}$ is the mass flow rate, circulating in the cycle, Δh is the enthalpy change, and Δs is the entropy change.

The COPs of the two systems given in Eqs. (2.14) and (2.15) are defined on the basis of the first law only. However, the best possible performances can be achieved when these systems perform as reversible cycles, in which case their efficiencies would be:

$$\mathrm{COP}_{Carnot,ref} = \frac{T_b}{T_a - T_b} \tag{2.20}$$

$$\mathrm{COP}_{Carnot,hp} = \frac{T_a}{T_a - T_b} \tag{2.21}$$

It is clear that the first-law efficiency itself is not a realistic measure of the performance of engineering devices. To overcome this defect, a second-law efficiency η_{II} is defined as the ratio of the actual thermal efficiency to the maximum possible (reversible) thermal efficiency under the same conditions [22]:

$$\eta_{II,ref} = \frac{\mathrm{COP}_{ref}}{\mathrm{COP}_{Carnot,ref}} \tag{2.22}$$

$$\eta_{II,hp} = \frac{\mathrm{COP}_{hp}}{\mathrm{COP}_{Carnot,hp}} \tag{2.23}$$

The second law efficiency which is also known as exergy efficiency shows us the deflection ratio of a cycle from a reversible one that has the possible maximum cycle efficiency. In other words, the exergy éfficiency is a measure of approximation to reversible operation. From this view, it can be also defined as;

$$\eta_{II}\text{or } \varepsilon = \frac{\textit{Exergy recovered}}{\textit{Exergy consumed}} = 1 - \frac{\textit{Exergy destroyed}}{\textit{Exergy consumed}} \tag{2.24}$$

Equation (2.24) shows that when calculating the exergy efficiency, it is very important to determine how much exergy or work potential is recovered or consumed during a process [22]. Exergy is defined as the maximum amount of work that can be produced by a system when it comes to equilibrium with a reference environment. Exergy analysis is a method that uses the conservation of mass and conservation of energy together with the second law of thermodynamics for the design and analysis of energy systems [23]. Exergy analysis applied to a system describes all losses both in the components of the system and in the whole system. With the help of exergy analysis, the magnitude of these losses or irreversibilities and their order of importance can be understood. With the use of irreversibility, which is a measure of process imperfection, the optimum operating conditions can easily be determined. In addition, exergy analysis can indicate the possibilities of thermodynamic improvement potentials of the process under consideration [24].

Back to Figure 2.4a,c, it has been mentioned that points $2'$ and $3'$ behave with higher pressure than the critical value ($P > P_c$), meaning that the heat pump cycle operates as a transcritical cycle. The word "transcritical" implies a processcrossing critical point. In detail, the evaporation process still operates at a subcritical state, while the compressor compresses the working fluid into supercritical state to reject heat. This reveals that the largest difference between the transcritical cycle and previous conventional cycle is that the heat rejection process operates at supercritical levels. Because there is no phase change during the heat rejection process in supercritical state, this process is called the gas cooling process instead of the condensation process. Similarly, the component to reject heat is called the gas cooler instead of the condenser.

The gas cooling process is the most special process in transcritical cycles. In traditional subcritical cycles, there is a small temperature difference of refrigerant between the condenser inlet and outlet during the condensation process, due to the law of phase-change process. Unlike the condensation process, there is a larger temperature difference between the gas cooler inlet and outlet during cooling in the supercritical region. The temperature difference is also called temperature glide. Thanks to the large glide of the gas cooling process, the heat transfer between the refrigerant and the secondary fluid matches better (compared in Figure 2.5a,b), giving higher heat transfer efficiency and also higher COP.

Exergy is the part of energy which can be utilized to the maximum under given conditions. Exergy destruction can be used to measure the irreversibility of an energy transfer process. In a gas cooling process, the exergy destruction of the heat transfer process between the refrigerant and the secondary fluid can be calculated as following.

For the heat transfer process between a hot flow and a cold flow, an infinitesimal segment is concentrated in which the two-flow can be considered as the constant temperature flow. Assuming that hot flow has a temperature of T_H, cold flow has a temperature of T_L, the heat transfer rate is δQ, and there is no heat loss during heat transfer. The exergy destruction can be expressed as below.

$$\delta E = \delta E_H - \delta E_L = \left(1 - \frac{T_0}{T_H}\right)\delta Q - \left(1 - \frac{T_0}{T_L}\right)\delta Q = T_0\left(\frac{T_H - T_L}{T_L T_H}\right)\delta Q \qquad (2.25)$$

where E represents exergy destruction, E_H represents the exergy of high temperature flow, E_L represents the exergy of low temperature flow. It can be noticed from Eq. (2.25) that a larger temperature difference between hot and cold flow can lead to a larger exergy destruction, which represents lower heat transfer efficiency. Thus, as shown in Figure 2.5a,b, the temperature difference of transcritical flow always maintains at a low level compared to that of subcritical flow during the heat rejection process, implying higher heat transfer efficiency of transcritical flow. The thermodynamic advantage of transcritical heat pump emerges from this view.

Large temperature glide of the transcritical cycle also subserves cycle performance. Increasing temperature glide can increase the COP of heat pumps, as confirmed by several simulations, as well as theoretical and experimental studies. This characteristic indicates the way to improve cycle performance, that is to reduce the inlet temperature of secondary fluid.

Gas cooling pressure is also an interesting task for transcritical heat pump cycles. It can be seen that the temperature and pressure are two independent parameters for supercritical fluids, and the pressure can be adjusted for better cycle performance. In a transcritical cycle, the simulation results show the existence of optimum gas cooling pressure under given gas cooler outlet temperature, which gives directions for cycle control and optimization.

Due to the low critical temperature of CO_2 (31.1°C), CO_2 heat pump cycle can be easily operated as the transcritical cycle within common operating temperature ranges. In addition, the special process of transcritical cycle, gas cooling process, gives special advantages for heating demand. Therefore, transcritical CO_2 heat pump cycle has been recognized as the most suitable high-energy efficiency heating device for systems such as water heaters, space heaters and hydronic floor heaters.

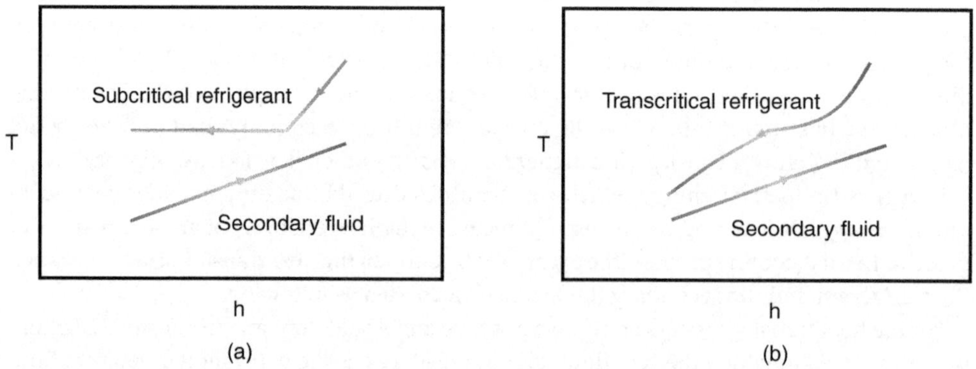

Figure 2.5 Temperature profiles of heat rejection processes. (a) Condensation process, (b) gas cooling process.

2.4 A Brief History of CO_2 Heat Pump

The use of CO_2 as a cooling agent was first reported in 1835 by a French chemist named Thilorier [25]. He discovered dry ice (solidified CO_2) by merely observing the large amount of liquid CO_2 in the cylinder. During his experiments, the liquid CO_2 evaporated and left the dry ice (without liquid) at the bottom of the container. Thilorier's original experimental setup is shown in Figure 2.6a [26]. This observed phenomenon was called sublimation by chemists. However, dry ice was only employed and observed in the laboratory without being used in any application. Later, in 1897, the first patent for dry ice was granted to an English medical doctor named Herbert Samuel Elworthy [28].

The use of CO_2 in refrigeration systems was developed in the nineteenth century. The first instance of use of refrigerant CO_2 in a vapor compression system was reported in 1850 by an American engineer named Alexander Catlin Twining. In 1867, the actual refrigeration system using CO_2 was built by an American inverter, Thaddeus Lowe, for the purpose of ice making [29]. Refrigeration systems using CO_2 as a working fluid have been extended and

Figure 2.6 (a) Original apparatus for liquefying CO_2 by Thilorier [26]. (b) Patent on CO_2 system by Lorentzen [27].

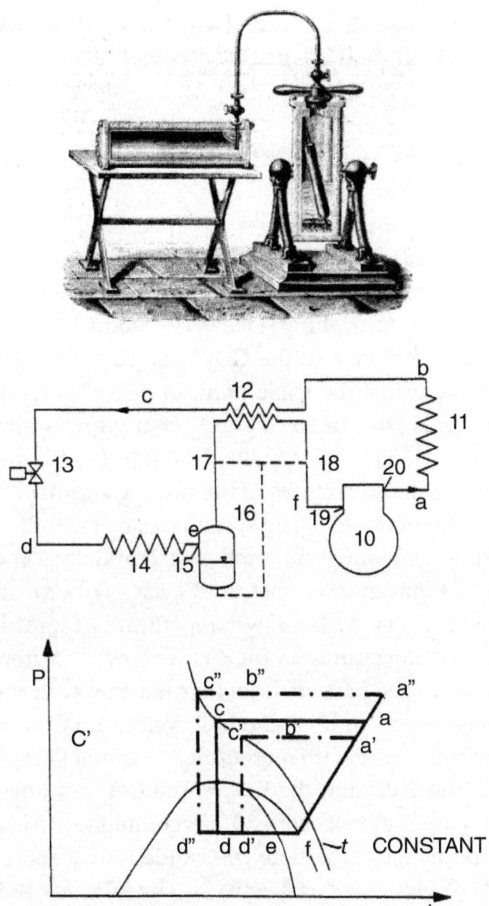

developed by many researchers from the 1880s to the beginning of the 1900s, with the aim of developing extensive usage in marine (refrigerant aboard ship) and general applications [30, 31].

However, in the late nineteenth century, utilization of CO_2 in refrigeration systems as the working fluid ceased and was replaced by the Fluorocarbon refrigerants [32]. Fluorocarbon had gained much attention due to its higher COP when used in refrigeration systems compared with working fluid CO_2, due to the low operating pressure of the Fluorocarbon refrigeration system [33, 34]. However, due to concerns about the effect on the environment and the global warming crisis, as mentioned above in the Montréal and Kyoto Protocols, CFC and HCFC compounds are due to be banned as stated earlier.

After almost a century of phasing out of the CO_2 refrigeration system, in the late 1980s the advantages of CO_2 properties in refrigerant systems were re-discovered and highlighted again [35, 36]. In 1993, the "father" of CO_2 refrigeration, a Norwegian professor, Lorentzen, proposed the feasibilities of high efficiency design of the CO_2 refrigerant system with many proposals and suggestions for improving the system. The patent for the CO_2 system is shown in Figure 2.6b [27, 37]. Later, in 2001, professor Lorentzen extended the use of CO_2 working fluid to a commercialized heat pump and mobile air conditioning unit, known as the "ecological cute" (Eco Cute) [38]. The Eco Cute was introduced in 2001 and rapidly developed, especially in the Japanese market [39].

2.5 CO_2 Cascade Heat Pump System

As described earlier on the state of developing the CO_2 solid-state application, the CO_2 transcritical compression thermodynamic cycle, or so-called CO_2 heat pump, has been given much attention in attempts to combat global warming. For decades, many studies on the CO_2 heat pump have been conducted, both theoretically and experimentally [40, 41]. However, most of the CO_2 heat pump studies would not reach the refrigeration temperature below the triple point of −56.6°C due to restrictions of thermodynamic limits and devices [40]. In 2009, a CO_2 heat pump system which was able to achieve a temperature below −56.6°C for biomedical and food industries was purposed by Yamaguchi *et al.* by using the advantages of the ultra-low temperature CO_2 in the cascade refrigeration system (Figure 2.7) [43]. The system can be extended for use in various refrigeration applications such as biomedical applications and food industries [44]. The system is combined with the high- and low-pressure cycles (LPCs), as the P-h diagram shows in Figure 2.8, for achieving the ultra-low-temperature CO_2 refrigeration cycle. It is to be mentioned that the high temperature in the CO_2 exothermic process of the high-pressure cycle (HPC) can be solely used for industrial thermal energy usage such as for heating water. To combine both advantages in high and low cycles, the CO_2 cascade heat pump can give a relatively good performance with high utility potential [42].

The HPC and the LPC of the CO_2 cascade heat pump system (A′B′C′D′ and ABCD in Figure 2.8, respectively) are combined with the brine fluid with evaporator of the HPC and at the same time the condenser of the LPC as shown in processes A to B and C to D of Figure 2.8, respectively. The HPC is used for cooling of the brine, supplying thermal energy as an output to the user, and an LPC is used to cool down the refrigerated space to an

Figure 2.7 Outlook of CO_2 cascade heat pump system [42].

ultra-low-temperature below −56.6°C by the CO_2 solid-gas two-phase flow. The operating of an HPC can cool the brine fluid, and the brine can cool CO_2 in the condenser of an LPC to a temperature low enough to obtain powdered dry ice through an expansion valve. The operational state of an LPC is expected to be kept stable with the help of an HPC, which can cool the brine fluid and makes the brine temperature stable. In an HPC, CO_2 can be pressurized into high temperature and highpressure supercritical state by compression. The first condenser is cooled by tap water from its ambient state and is heated to (above) 130°C. Moreover, in a HPC, through the ejector and the gas/liquid separator, CO_2 is finally cooled down to −20°C in the third condenser, where CO_2 in the LPC is cooled down through the brine fluid from an HPC.

In the LPC, as shown in Figure 2.9 the cycle is composed of three condensers, an expansion valve, an evaporator/sublimator (test section) and a compressor. As depicted schematically in Figure 2.9, three condensers are arranged in series in order to sufficiently cool CO_2 below −20°C, because the discharge temperature of the compressor reaches approximately 130°C [45]. The first and second condensers are tube-in-tube heat exchangers, and are cooled by tap water. The third condenser is a plate-type heat exchanger made of stainless-steel tube, which is cooled by the brine channel connected with the evaporator in the HPC as stated previously.

The performance evaluation of the CO_2 cascade heat pump system has been investigated and the results show that the combination of high- and low-pressure cycles in the system led

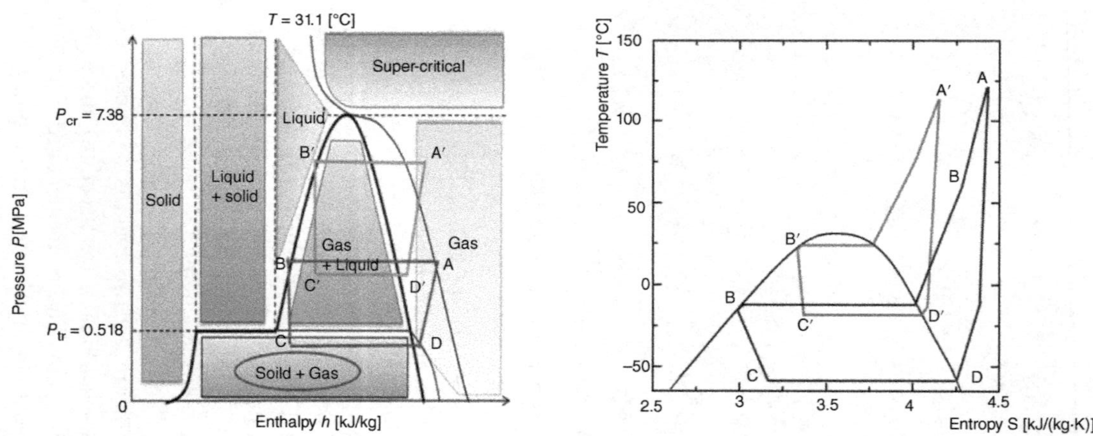

Figure 2.8 Thermodynamic cycle of the CO_2 cascade heat pump system including a high-pressure cycle (HPC) and a low-pressure cycle (LPC).

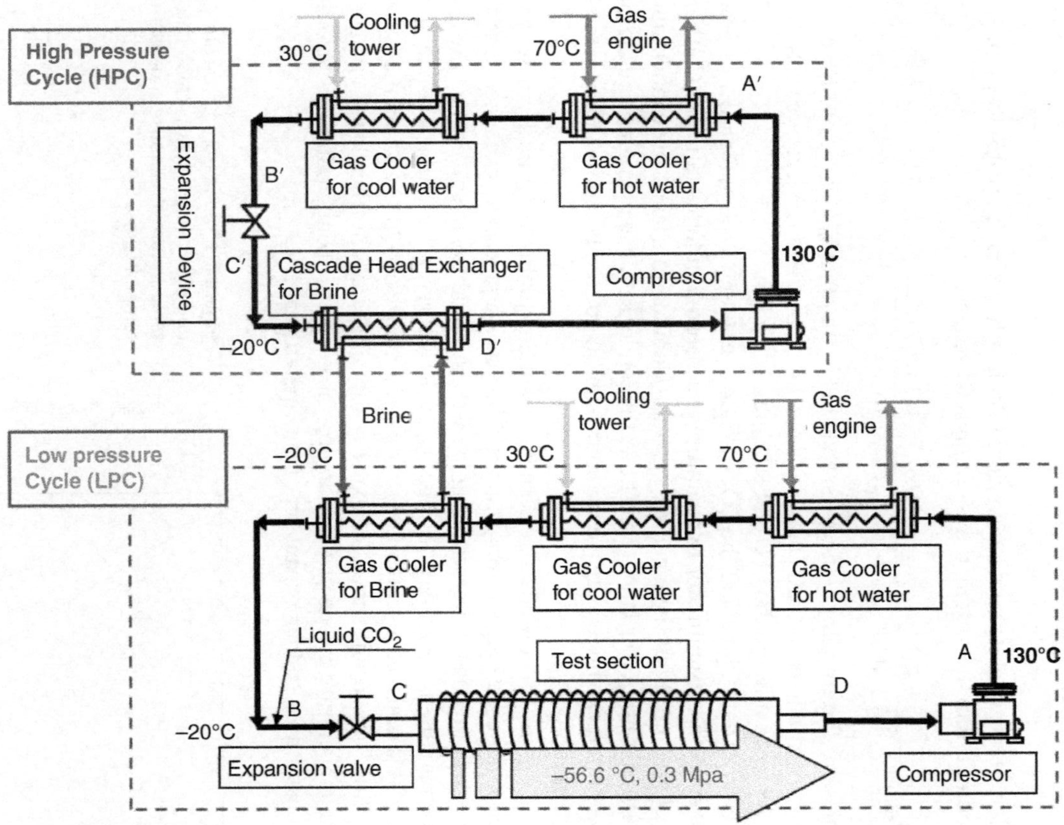

Figure 2.9 Schematic diagram of the CO_2 cascade system.

to higher efficiency altogether. The system also gives better efficiency compared with other methods using conventional working fluid. Deriving the performance from the components stated previously, the overall performance of the system will be given by estimated COP values. The COP can be calculated as the ratio of the cooling capacity in the system to the work input or the compressor work. The COP of the LPC can be written as:

$$\mathrm{COP}_{\mathrm{LPC}} = \frac{\Delta h_{lcool}}{W_l} = \frac{h_D - h_C}{h_A - h_D} \tag{2.26}$$

When the COP of the whole system is considered, with only the cooling capacity in the LPC as a useful output from the system, the equation is written as;

$$\mathrm{COP}_{\mathrm{system}} = \frac{\Delta h_{lcool}}{W_l + W_h} = \frac{\dot{m}_l(h_D - h_C)}{\dot{m}_l(h_A - h_D) + \dot{m}_h(h_{A'} - h_{D'})} \tag{2.27}$$

where h represents the enthalpy at the given state in Figure 2.9, $\Delta h_{\mathrm{lcool}}$ is the cooling capacity in the system, $\dot{m}_\mathrm{l}$ and $\dot{m}_\mathrm{h}$ are CO_2 mass flow rate in the LPC and HPC, respectively. Here W_l and W_h are work input to the LPC and HPC, respectively.

As to the current development of the CO_2 cascade heat pump system, Figure 2.10 shows the effective evaporator/sublimator temperature (the saturation temperature of CO_2) and the measured pressure P_1 at the tapered evaporator/sublimator against the condensation temperature. In this figure, the lateral axis shows the condensation temperature and the longitudinal axes represent the effective evaporator/sublimator temperature and, separately, the measured pressure P_1.

As shown in Figure 2.10, the effective evaporator/sublimator temperature decreases with the decrease of the condensation temperature. This can be physically explained that the condensation pressure also decreases along the wet saturated steam curve when the condensation temperature is decreased. Resultantly, the lowest cryogenic refrigeration temperature of −66.3°C is achieved at the condensation temperature of −30°C without the system operation failing. This result shows a capability for achieving a further lower cryogenic refrigeration temperature by effectively changing the geometric configuration of the evaporator/sublimator from sudden expansion to the tapered shape.

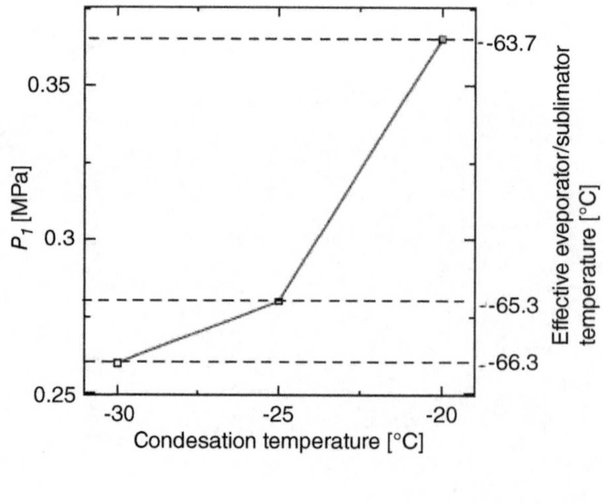

Figure 2.10 Measured pressure P_1 and effective evaporator/sublimator temperature at various condensation temperatures.

After confirming the fact that the lower cryogenic refrigeration temperature is achieved by installing the tapered evaporator/sublimator into the CO_2 ultra-low temperature cascade refrigeration system, the system performance of the LPC with the tapered evaporator/sublimator is estimated as follows. Figure 2.11 shows the COP of the LPC at various condensation temperatures. In this figure, the lateral axis shows the condensation temperature and the longitudinal axis represents the COP. It was found that the COP is indeed enhanced with the decrease of the condensation temperature. In Figure 2.11b, COP_{carnot} was calculated using Eq. (2.20), where T_a is taken 5°C higher than the condensation temperature for representing the high temperature source. As seen from the figure, with the increasing of T_a, COP_{carnot} decreases which is also obvious from Eq. (2.20). This is because the increase of T_a leads to an increase in the heat energy to be removed from the system. Figure 2.11c shows the second law or exergy efficiency of the refrigeration system

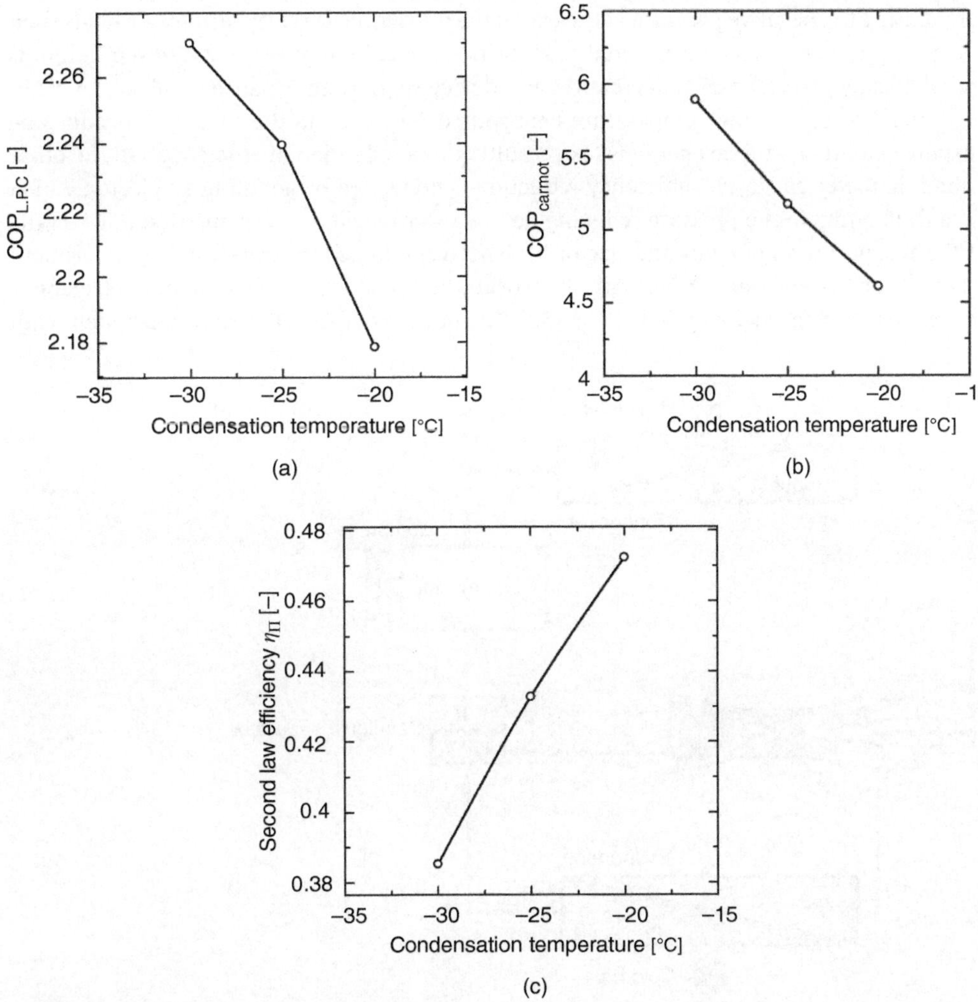

Figure 2.11 (a) Variation of COP_{LPC}, (b) variation of COP_{carnot}, (c) variation of η_{II}.

η_{II} changing with T_a. Referring to Eq. (2.22), second law efficiency is the ratio of actual COP to COP_{carnot}. As both performance indicators decrease with T_a, the decrement ratio of COP_{carnot} is much higher than COP which leads to an increase in second law efficiency.

However, it is reported that the blocking phenomena of dry-ice occurred in the evaporator/sublimator (test section) of the current design, with which the system is much affected, resulting in low efficiency and ultimately failure in operation [46]. The novel swirl promoter is planned to be introduced and installed in the CO_2 cascade heat pump, and is believed to give better efficiency [47].

2.6 Advanced CO_2 Heat Pump System with an Ejector

An ejector is a fluid pump that ejects steam or the like from a nozzle and sucks in another fluid using the negative pressure at the exit of the jet section. It is used in various fields such as gas equipment. In a conventional refrigeration cycle, high pressure liquid refrigerant is adiabatically expanded by a decompression device such as an expansion valve or a capillary tube to obtain a low temperature heat source. However, in the course of its adiabatic expansion, much kinetic energy is lost, resulting in a reduction of efficiency. On the other hand, in the ejector cycle, efficiency reduction is prevented by adopting an ejector, which is a fluid pump in the pressure reducing device. Over recent decades, many studies on the CO_2 heat pump system with the ejector have been conducted theoretically and experimentally [48, 49]. Denso succeeded with the world's first practical application of a refrigerator equipped with an ejector cycle in 2003 [50]. The outline of the CO_2 heat pump system with

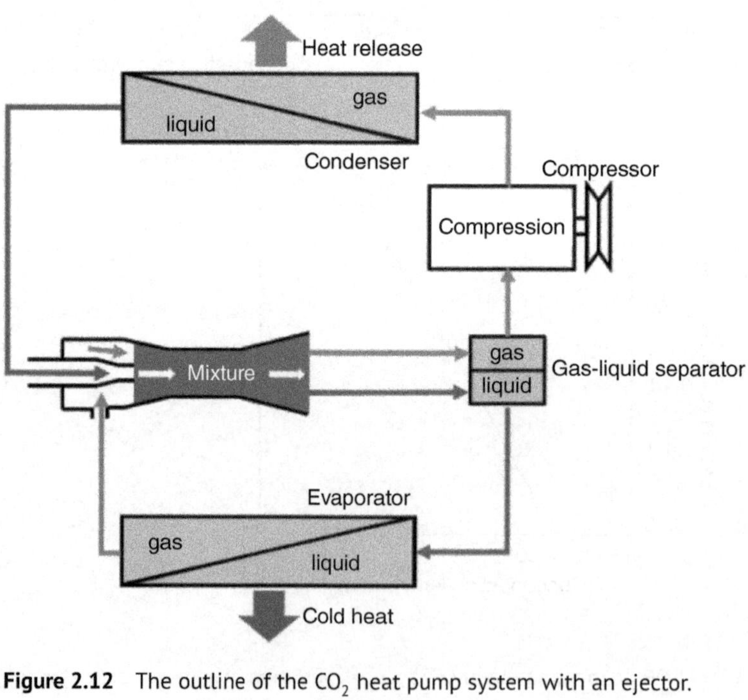

Figure 2.12 The outline of the CO_2 heat pump system with an ejector.

an ejector is shown in Figure 2.12. The high-pressure refrigerant at the condenser outlet is depressurized at the nozzle part and the refrigerant at the outlet of the evaporator is sucked. The high-speed refrigerant decompressed at the mixing part and the low-speed refrigerant sucked from the evaporator outlet are mixed, reducing the flow velocity at the diffuser section with the enlarged flow area. A gas-liquid separator is often installed at the tip of the ejector to separate the refrigerant into the gas and the liquid phase. The separated liquid refrigerant flows through the evaporator, and the gas refrigerant is sucked into the compressor. In the above process, the ejector suppresses the occurrence of vortices during expansion and isentropically expands, thereby recovering the energy that has been lost due to the vortex generation. By converting the recovered kinetic energy to pressure energy in the mixing part, the suction pressure of the compressor is increased and efficiency is improved. Further, since the state of the refrigerant at the evaporator inlet is liquid phase, the performance of the evaporator can also be improved by reducing the pressure loss in the evaporator, resulting in an increase in the heat transfer coefficient, so that a significant improvement in COP is expected to be achieved.

Figure 2.13 illustrates a schematic diagram of the proposed CO_2 ultra-High and Low Temperature Heat Pump System (HLTHPS) with the ejector cycle, which shows the principle of the CO_2 refrigeration with dry ice sublimation together with generation of high temperature water supply (wet steam).

As seen in Figure 2.13, in the HPC, the process from 1–2 represents the compression process, in which the gas CO_2 is compressed to become higher in pressure and temperature. The compressed gas CO_2 is condensed into the liquid CO_2 in the condensing process of 2–3. The process of 3–4 represents the expansion process, in which the liquid CO_2 expands into

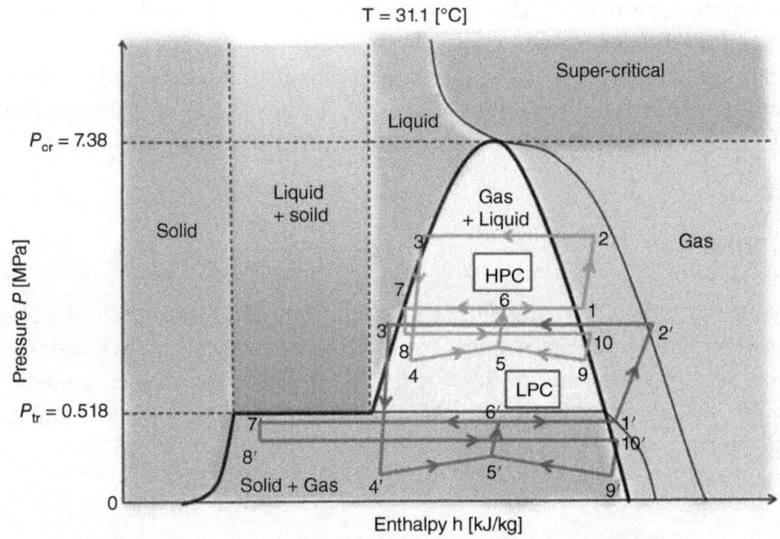

Figure 2.13 Experimental thermodynamic cycle of CO_2 ultra-low temperature cascade refrigeration system with ejector enhancement and principle of CO_2 refrigeration with dry ice sublimation.

the gas-liquid two-phase flow in the ejector. At point 5, low temperature CO_2 (point 4) and high temperature CO_2 (point 9) are entrained into the mixing chamber in the ejector. The mixture flows through the ejector diffuser where it recovers to the pressure at point 6. At this point, the mixture is separated into liquid CO_2 (point 7) and gas CO_2 (point 1). Separated gas CO_2 is expanded due to decreasing pressure at process 7–8. Then the CO_2 two-phase flow absorbs the heat and is turned into the gaseous phase in the evaporation process of 8–10. The whole refrigeration system is composed of two CO_2 refrigeration compression cycles and is arranged in cascade, namely the evaporation process of 8–10 in the HPC is coupled with the condensing process of 2′-3′ in the LPC through the brine channel. In the LPC, this cascade arrangement is capable of cooling the CO_2 to below approximately −15°C, passing through the condensing process of 2′–3′. The sufficiently cooled liquid CO_2 expands and dry ice (solid-gas two-phase flow) is formed in the expansion owing to the fact that the CO_2 exceeds the triplepoint in the CO_2 P-h diagram, in which the triple point of CO_2 is at the pressure of 0.518 MPa and temperature of −56.6°C. In the LPC, a sufficiently cooled mixture of dry ice and gas CO_2 is separated into dry ice slurry and gas CO_2 in the separation process (at point 6′). The dry ice slurry sublimates and absorbs a great deal of heat quantity in the evaporation/sublimation process of 8′–10′, in which the evaporator/sublimator can achieve a cryogenic cooling ability below the CO_2 triple-point temperature of −56.6°C.

From the above explanation, h_D and $h_{D'}$ in Eq. (2.26) become higher by installing the ejectors into the cascade heat pump system, and COP of the whole system with ejector COP_{eje} becomes higher than $\text{COP}_{\text{system}}$ in Eq. (2.24), so that a significant improvement in COP would be achieved:

$$\text{COP}_{\text{eje}} > \text{COP}_{\text{system}} \tag{2.28}$$

The use of ejectors in the thermodynamic cycle of CO_2 for the ultra-low temperature cascade refrigeration system, particularly the one with dry ice sublimation as demonstrated in this chapter, is new and still conceptual. However, the feasibility analysis indicates that the potential of the proposed system should be high enough so that the system with ejectors has to be looked into more deeply in view of both theoretical study (not only COP but also exergy analysis) and construction of a carefully designed experimental system.

Acknowledgments

A part of this chapter is referred to the PhD thesis of Chayadit Pumaneratkul [14]. The authors acknowledge the financial support from HighEFF under the FME scheme (Centre for Environment-friendly Energy Research, 257632/E20). The discussion with Mr. Zhao-Rui Peng and the proofreading by Mr. Guan-Bang Wang are also highly appreciated.

Nomenclature

c	specific heat, J kg^{-1} K^{-1}
E	exergy, J

h	specific enthalpy, J kg^{-1}
$\dot{m}$	mass flow rate, kg s^{-1}
P	pressure, Pa
Q	heat, J
R	gas constant, J mol^{-1} K^{-1}
s	specific entropy, J kg^{-1} K^{-1}
T	temperature, K
V	volume, m^3
W	work, J

Greek Letters

α	coefficient of equation of state
Δ	change
δ	infinitesimal segment
ε	exergy efficiency
η	efficiency
λ	thermal conductivity, W m^{-1} K^{-1}
μ	dynamic viscosity, Pa s
ρ	density, kg m^{-3}

Subscripts

c	critical
$cool$	cooling
$Carnot$	reversible
H	hot
h	high pressure
hp	heat pump
L	cold
l	low pressure
p	isobaric
ref	refrigerator
II	second-law

Abbreviations

CFC	Chlorofluorocarbon
COP	coefficient of performance
EOS	equation of state
GHG	greenhouse gas
GWP	Global Warming Potential

HCFC	Hydrochlorofluorocarbon
HFC	Hydrofluorocarbon
HLTHPS	High and Low Temperature Heat Pump System
HPC	high-pressure cycle
LPC	low-pressure cycle
ODP	Ozone Depletion Potential

References

1 Thurston, R.H. (1887). A history of the growth of the steam-engine (No. 24). Kegan Paul.

2 Mayumi, K. (1991). Temporary emancipation from land: from the industrial revolution to the present time. *Ecological Economics* 4 (1): 35–56.

3 Hu, Y., Naito, S., Kobayashi, N., and Hasatani, M. (2000). CO_2, NOx and SO2 emissions from the combustion of coal with high oxygen concentration gases. *Fuel* 79 (15): 1925–1932.

4 Perlwitz, J., Pawson, S., Fogt, R.L. et al. (2008). Impact of stratospheric ozone hole recovery on Antarctic climate. *Geophysical Research Letters* 35 (8): L08714.

5 Fahey, D.W. (2013). The Montreal Protocol protection of ozone and climate. *Theoretical Inquiries in Law* 14 (1): 21–42.

6 Thompson, D.W., Solomon, S., Kushner, P.J. et al. (2011). Signatures of the Antarctic ozone hole in Southern Hemisphere surface climate change. *Nature Geoscience* 4 (11): 741.

7 Norman, C., DeCanio, S., and Fan, L. (2008). The Montreal Protocol at 20: ongoing opportunities for integration with climate protection. *Global Environmental Change* 18 (2): 330–340.

8 Brack, D. (2017). *International Trade and the Montreal Protocol*. Routledge.

9 Lorentzen, G. (1995). The use of natural refrigerants: a complete solution to the CFC/HCFC predicament. *International Journal of Refrigeration* 18 (3): 190–197.

10 Velders, G.J., Fahey, D.W., Daniel, J.S. et al. (2009). The large contribution of projected HFC emissions to future climate forcing. *Proceedings of the National Academy of Sciences* 106 (27): 10949–10954.

11 Rogelj, J., Den Elzen, M., Höhne, N. et al. (2016). Paris Agreement climate proposals need a boost to keep warming well below 2 C. *Nature* 534 (7609): 631.

12 Ministry of the Environment Japan (2015). Law on regulation of management and rational use of fluorocarbons. http://www.env.go.jp/earth/ozone/cfc/law/kaisei_h27/index.html (accessed February 2018).

13 Ke, J., Han, B., George, M.W. et al. (2001). How does the critical point change during a chemical reaction in supercritical fluids? A study of the hydroformylation of propene in supercritical CO_2. *Journal of the American Chemical Society* 123: 3661–3670.

14 Pumaneratkul, C. (2018). Basic characteristics of rankine cycle with functional elements, using supercritical carbon dioxide. Ph.D. thesis. Doshisha University.

15 Zhang, X.R., Yamaguchi, H., Uneno, D. et al. (2006). Analysis of a novel solar energy-powered Rankine cycle for combined power and heat generation using supercritical carbon dioxide. *Renewable Energy* 31 (12): 1839–1831.

16 Cavallini, A. (2004). Properties of CO_2 as a refrigerant. In European seminar-CO_2 as a refrigerant: theoretical and design aspects.

17 Wakefield, C.B. (1996). Non ideal gas behavior (No. CONF-960220--). Edmond, OK (United States): University of Central Oklahoma.

18 Peng, D.Y. and Robinson, D.B. (1976). A new two-constant equation of state. *Industrial & Engineering Chemistry Fundamentals* 15 (1): 59–64.

19 Angus, S., Armstrong, B., and De Reuck, K.M. (1976). Carbon Dioxide: International Thermodynamic Tables of the Fluid State-3. Carbon Dioxide: International Thermodynamic Table of the Fluid State.

20 PROPATH Group. (1997). PROPATH: a program package for thermophysical properties.

21 Yunus, C.A. and Afshin, J.G. (2011). *Heat and Mass Transfer: Fundamentals and Applications*. McGraw-Hill Higher Education.

22 Çengel, Y.A. and Boles, M.A. (2015). *Thermodynamics: An Engineering Approach*, 8e. New York, NY: Mcgraw-Hill.

23 Dincer, I. and Rosen, M.A. (2007). *Exergy: Energy, Environment and Sustainable Development*, 1e. Oxford, UK: Elsevier Science.

24 Kizilkan, O., Sencan, A., and Kalogirou, S.A. (2007). Thermoeconomic optimization of a LiBr absorption refrigeration system. *Chemical Engineering and Processing: Process Intensification* 46 (12): 1376–1384.

25 Roller, D.H. and Thilorier, M. (1952). Thilorier and the first solidification of a" permanent" gas (1835). *Isis* 43 (2): 109–113.

26 Wurtz, C.A. (1876). Dictionnaire de chimie pure et appliquée. In Librairie Hachette et Cie. vol. 1: No. 1, 751.

27 Lorentzen, G. (1989): Method of operating a vapour compression cycle under trans- or supercritical conditions.

28 Elworthy, H.S. (1900). Apparatus for making pure carbonic acid. U.S. Patent 658, 627.

29 Gantz, C. (2015). *Refrigeration: A History*. McFarland.

30 Howells, H.R. (1959). Marine refrigeration. *Proceedings of the Institution of Mechanical Engineers* 173 (1): 46–58.

31 Almqvist, E. (2003). Development of the industrial gas business. In: *History of Industrial Gases*, 187–302. Boston, MA: Springer.

32 Bhatti, M.S. (1999). A historical look at chlorofluorocarbon refrigerants. *ASHRAE Transactions* 105: 1186.

33 Radermacher, R. and Kim, K. (1996). Domestic refrigerators: recent developments. *International Journal of Refrigeration* 19 (1): 61–69.

34 Anderssen, E., Bonneau, P., Sandaker, H., Jones, T. et al. (1999). Fluorocarbon evaporative cooling developments for the ATLAS pixel and semiconductor tracking detectors.

35 McLinden, M.O. and Didion, D.A., (1988). The search for alternative refrigerants – a molecular approach.

36 Scurlock, R.G. (1990). A matter of degrees: a brief history of cryogenics. *Cryogenics* 30 (6): 483–500.

37 Lorentzen, G. (1994). Revival of carbon dioxide as a refrigerant. *International Journal of Refrigeration* 17 (5): 292–301.

38 Okada, T. (2007). Development of CO_2 heat pump hot water system "ECO CUTE". *Mitsubishi Electric Advance* 120: 6–8.

39 Hashimoto, K. (2006). Technology and market development of CO_2 heat pump water heaters (Eco Cute) in Japan. *IEA Heat Pump Centre Newsletter* 24 (3): 12–16.

40 Nekså, P., Rekstad, H., Zakeri, G.R., and Schiefloe, P.A. (1998). CO_2-heat pump water heater: characteristics, system design and experimental results. *International Journal of Refrigeration* 21 (3): 172–179.

41 Sarkar, J., Bhattacharyya, S., and Gopal, M.R. (2004). Optimization of a transcritical CO_2 heat pump cycle for simultaneous cooling and heating applications. *International Journal of Refrigeration* 27 (8): 830–838.

42 Yamaguchi, H. and Zhang, X.R. (2009). A novel CO_2 refrigeration system achieved by CO_2 solid–gas two-phase fluid and its basic study on system performance. *International Journal of Refrigeration* 32 (7): 1683–1693.

43 Yamaguchi, H., Zhang, X.R., and Fujima, K. (2008). Basic study on new cryogenic refrigeration using CO_2 solid–gas two phase flow. *International Journal of Refrigeration* 31 (3): 404–410.

44 Garthwaite, G.A. (1997). Chilling and freezing of fish. In: *Fish Processing Technology* (ed. G.M. Hall), 93–118. Boston, MA: Springer.

45 Yamaguchi, H., Niu, X.D., Sekimoto, K., and Nekså, P. (2011). Investigation of dry ice blockage in an ultra-low temperature cascade refrigeration system using CO_2 as a working fluid. *International Journal of Refrigeration* 34 (2): 466–475.

46 Niu, X.D., Yamaguchi, H., Iwamoto, Y., and Nekså, P. (2010). Experimental study on a CO_2 solid–gas-flow-based ultra-low temperature cascade refrigeration system. *International Journal of Low Carbon Technologies* 6 (2): 93–99.

47 Yamasaki, H., Yamaguchi, H., Hattori, K., and Neksa, P. (2017). Experimental observation of CO_2 dry-ice behavior in an evaporator/sublimator. *Energy Procedia* 143: 375–380.

48 Haida, M., Banasiak, K., Smolka, J. et al. (2016). Experimental analysis of the R744 vapour compression rack equipped with the multi-ejector expansion work recovery module. *International Journal of Refrigeration* 64: 93–107.

49 Jin, Z., Hafner, A., Eikevik, T.M., and Nekså, P. (2019). Preliminary study on CO_2 transcritical ejector enhanced compressor refrigeration system for independent space cooling and dehumidification. *International Journal of Refrigeration*.

50 Takeuchi, H., Nishijima, H., Ikemoto, T. et al. (2009). The World's First Ejector Cycle® for DENSO's transport refrigerator. *DENSO Technical Review* 14: 65–73.

3

Fluid Dynamics and Heat Transfer of Supercritical Carbon Dioxide Cooling

Brian M. Fronk

School of Mechanical, Industrial and Manufacturing Engineering, Oregon State University, Corvallis, OR, USA

3.1 Supercritical Properties

Unlike conventional vapor compression heat pump systems, there is no condensation of the refrigerant on the high-pressure side of a transcritical carbon dioxide heat pump cycle. Rather, heat is rejected through a constant pressure, non-isothermal supercritical gas cooling process, as shown in Figure 3.1.

Here, carbon dioxide is initially at a temperature and pressure above the critical point ($T_{crit} = 30.98°C$ and $P_{crit} = 7.38$ MPa for CO_2). In this region, CO_2 properties are similar to those of a gas, with relatively low density, specific heat capacity and thermal conductivity. As the CO_2 is cooled at a constant pressure, it undergoes a transition from this "gas-like" state to a "liquid-like" state without the formation of discrete phases. As an example, at a reduced pressure of 1.5 (~11 MPa), the density increases by 4.6 times and specific heat by 1.7 times as CO_2 is cooled from 120 to 20°C. Importantly, these change in properties are highly non-linear with change in temperature, as shown in Figure 3.2 for different reduced pressures.

The figure shows steep gradients in density and viscosity, and spikes in the Prandtl number and specific heat temperature in the vicinity of the critical point. The figure also shows that as the reduced pressure increases, the temperature at which these spikes increases, and the band of temperature in which the properties rapidly change widens. The temperature at which the specific heat is maximum for a given supercritical pressure is referred to as the "pseudo-critical temperature." This terminology will be used in the remainder of this chapter. Importantly, these regions of rapid property change happen in temperatures and pressures that are regularly encountered in CO_2 heat pump systems.

The drastic variation of thermophysical properties can have a dramatic impact on the in-tube convective heat transfer behavior. Steep gradients in radial and axial density due to heating or cooling can result in buoyancy and flow acceleration effects which may enhance or deteriorate heat transfer. Furthermore, large increases in Prandtl number and specific heat capacity can locally enhance the heat transfer coefficient. These competing effects are poorly predicted by correlations and models developed for single-phase, constant property fluids. Thus, there is a need to have specific heat transfer and pressure drop models for the accurate design of transcritical CO_2 heat pump systems.

Transcritical CO_2 Heat Pump: Fundamentals and Applications,
First Edition. Xin-Rong Zhang and Hiroshi Yamaguchi.

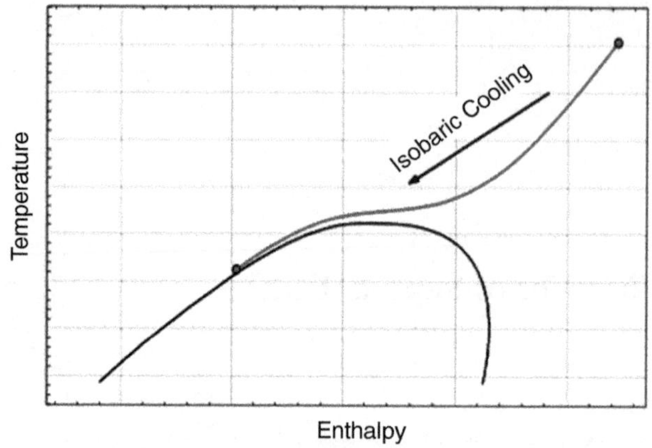

Figure 3.1 Temperature-enthalpy diagram with a constant pressure, non-isothermal gas cooling process shown.

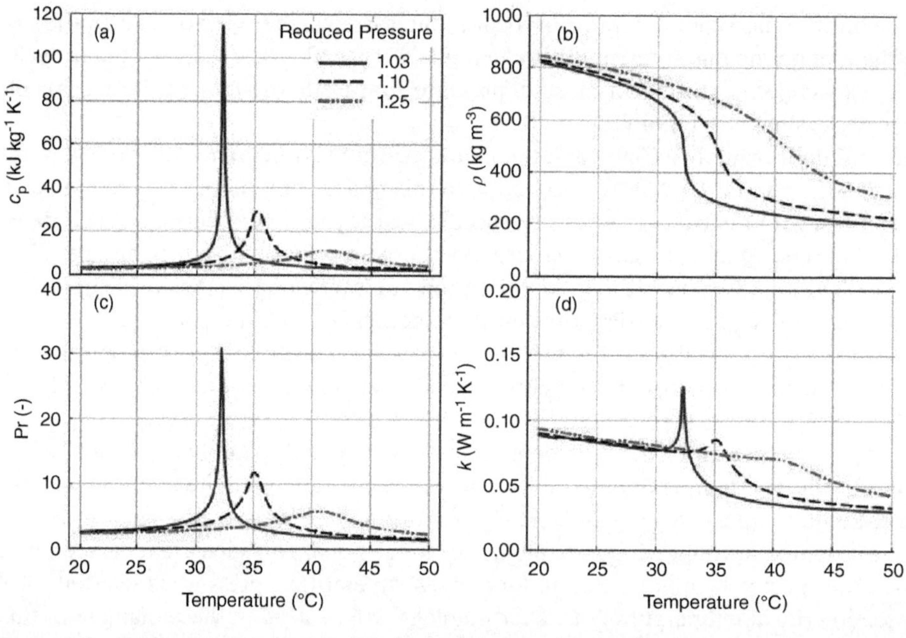

Figure 3.2 Supercritical carbon dioxide (a) specific heat, (b) density, (c) Prandtl number and (d) thermal conductivity as a function of temperature and reduced pressure.

3.2 Supercritical Heat Transfer Fluid Mechanics

Supercritical water and carbon dioxide were identified as potential nuclear reactor coolants dating back to the 1950s. This spurred significant research activity investigating the *heating* of supercritical fluids. Experiments with supercritical CO_2 [1–9] have focused on large diameter ($4.08 < D < 22.7$ mm), uniformly heated circular tubes at low to moderate heat fluxes ($0.05 < q'' < 330$ W cm^{-2}). At temperatures and pressures much greater than the critical point, sCO_2 behaves approximately as an ideal gas, with heat transfer reasonably well predicted by conventional convection models and correlations [10, 11]. In the pseudo-critical region, the underlying fluid mechanics and thermal transport are very different. A spike in specific heat capacity (Figure 3.2) and the related Prandtl number reduces conduction resistance within the turbulent boundary layer, enhancing heat transfer. Early researchers attempted to account for this and other variable thermophysical properties of supercritical fluids by defining a reference temperature for evaluating Reynolds and Prandtl number and/or by adding empirical bulk-to-wall property ratios to established constant property correlations [2, 12]. However, most of these empirical correlations were applicable for a limited range of applied heat and mass flux conditions, as shown in Hall *et al.* [13].

3.2.1 Buoyancy, Flow Acceleration and Oscillations in Near-Critical Flows

Mechanistic understanding of supercritical heat transfer has mostly focused on heating of supercritical fluids. However, the effects of buoyancy, flow acceleration and oscillations on heat transfer near the critical point can also be important for cooling. While a local spike in specific heat and Prandtl number tends to enhance heat transfer, in internal flows, transverse density gradients due to heating or cooling will induce buoyancy, which can suppress or further enhance heat transfer depending on orientation [14].

This motivated multiple experimental and computational [15–22] investigations to develop a more mechanistic understanding of near-critical heat transfer by considering buoyancy effects (during heating) at steady state. Wang *et al.* [20] used Laser Doppler Anemometry (LDA) to determine mean velocity profiles and turbulent flow statistics for both buoyancy-aided (upward) and buoyancy-opposed (downward) flow configurations. Interestingly, buoyancy aided flows initially show a heat transfer deterioration before recovering as the buoyancy parameter (Bo^*) increases (Figure 3.3). The LDA measurements revealed that even though the mean velocity for the buoyancy aided case increased in the vicinity of the wall, stream wise and cross stream velocity fluctuations were being suppressed. On the other hand, a reduction in the near wall velocity, for buoyancy opposed flows, enhances both stream wise and cross stream velocity fluctuations. Similar results were reported by Harris *et al.* [21] using particle image velocimetry, and are consistent with early visualization studies of Kline *et al.* [23], who visualized the structure of the turbulent boundary layer under the influence of both favorable and adverse pressure gradients. The recovery in heat transfer as Bo^* increases is due to a second transition of the boundary layer to a turbulent natural convection boundary layer as evidenced by the fact that the location for the maximum in stream wise velocity fluctuations is offset from the location of the maximum in the mean velocity [24, 25].

Investigations of uniformly heated horizontal flows [26–29] have shown that buoyancy forces are responsible for inducing secondary flow patterns in the channel cross section, leading to an accumulation of low density and low momentum fluid at the channel top surface and increasing the velocity of the fluid near the bottom wall of the channel. This degrades heat transfer at the top of the channel and enhances it at the bottom. Thus, it was established that buoyancy forces act to modify the stream wise and cross stream velocity fluctuations in near-critical, steady state flows, thereby increasing or reducing the Reynolds

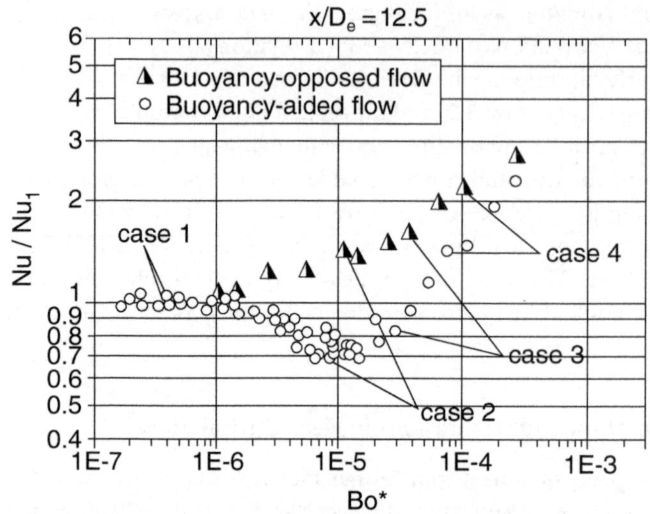

Figure 3.3 Effects of buoyancy on heat transfer during heating of supercritical fluid for two different flow configurations. Source: Reprinted from International Journal of Heat and Fluid Flow, 25(3), J. Wang, J. Li, and J.D. Jackson, 2004, with permission from Elsevier.

stress in flow cross section and potentially enhancing or degrading heat transfer. These effects are particularly pronounced when the temperature gradient between the bulk and wall is large enough such that the pseudocritical temperature occurs within the near-wall region (i.e. $T_b < T_{pc} < T_w$) [19, 22, 30–32].

At the same time, the axial decrease of density in heated near-critical flows causes bulk acceleration of the fluid, and can reduce turbulent thermal transport, as reported in the initial investigations of [33] for internal flows and [23] for external turbulent boundary layers. The favorable pressure gradient acts as a stabilizing influence and suppresses the ejection events from the near wall region [21, 23, 34, 35]. Under extreme conditions, the turbulent boundary layer can undergo a reverse transition to a laminar boundary layer, accompanied by a sharp reduction in the thermal transport capabilities [14, 36]. This phenomenon has been observed for all flow orientations.

Helmholtz (low frequency high amplitude), thermo-acoustic (high frequency low amplitude), and other flow oscillations [13, 37–45] have also been observed for heated flows near the critical point. Similar to buoyancy effects, oscillations have been found to most likely occur when $T_b < T_{pc} < T_w$ [13, 44]. Likely temperature profiles causing oscillations are shown schematically in Figure 3.4, adapted from Linne *et al.* [45]. For these conditions, a potential explanation for the start of oscillations is that a small change in near-wall temperature decreases the viscosity (which varies sharply with temperature near the pseudo-critical point) and thins the laminar sublayer, decreasing thermal resistance, which decreases wall temperature, causing a subsequent increase in near-wall viscosity, thickening of the laminar sublayer, and increase in near-wall temperature [46]. Other explanations are related to the rapid change in density and other physical properties near the pseudo-critical point. These oscillations have been observed to enhance heat transfer [47, 48], but can also cause local thermal cycling leading to fatigue failure and destructive

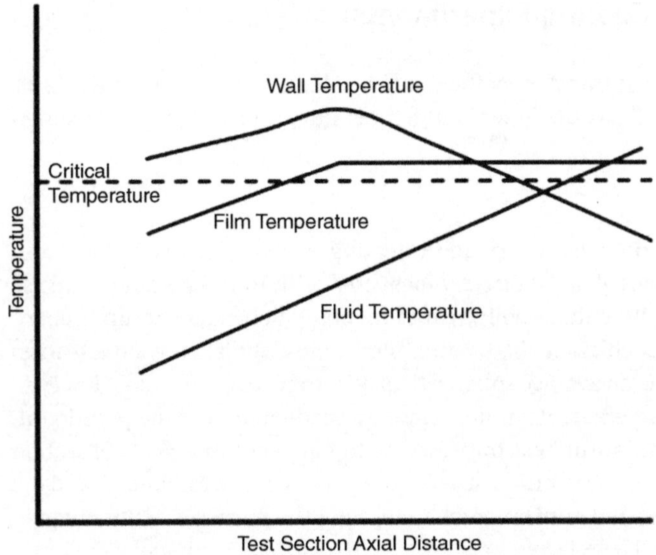

Figure 3.4 Temperature profiles with high probability of instabilities during heating of supercritical fluids. Source: Adapted from Linne *et al.* [45].

pressure pulsations. Two-phase explanations rely on a pseudo-boiling analogy [49, 50], arguing that low-density "gas" near the heated surface departs as a variable density volume, yielding behavior similar to the transition from nucleate to film boiling with increasing heat flux. While many investigators suggest that the pseudo-boiling analogy is not physically realistic [14, 51], consensus on the underlying mechanisms driving heat transfer deterioration/enhancement and flow instabilities remains elusive [52]. During cooling, energy is removed from the near wall boundary layer, generally avoiding the oscillations observed in heated supercritical flows.

3.2.2 Outline of Remainder of Chapter

In CO_2 heat pump applications, in-tube *cooling* of supercritical CO_2 is of interest. Here, energy is being removed, resulting in higher density, more thermally conductive regions near the wall, which tend to enhance heat transfer (i.e. greater than expected for a flow with constant properties) in the vicinity of the pseudo-critical point [7]. Furthermore, regions of heat transfer deterioration, flow pulsations or pseudo-boiling type behavior are not expected in cooling applications. Therefore, while heating has been more widely studied, caution must be exercised when attempting to extrapolate results to supercritical cooling.

In the remainder of this chapter, experimental techniques for characterizing supercritical cooling in macro and mini/micro channels will be introduced. This is followed by a discussion and comparison of supercritical CO_2 heat transfer correlations for heat transfer and pressure drop with and without lubricants.

3.3 Supercritical Gas Cooling Experiments

The local in-tube convective heat transfer coefficient (h) is defined according to Eq. (3.1). Here, q'' is the local heat flux, T_w is the inner wall temperature and T_b is the bulk supercritical fluid temperature.

$$q'' = h(T_b - T_w) \tag{3.1}$$

In supercritical heating experiments, electric Joule heating is often used to provide a constant input heat flux that enables q'' to be directly measured with low uncertainty. Precise control of the heat flux is more difficult in cooling experiments, which makes accurate determination of the heat transfer coefficient challenging. Thus, most studies rely on a Wilson plot-type approach to determine an average supercritical heat transfer coefficient. However, this approach can obscure large gradients in heat transfer coefficient near the pseudocritical point and hide local non-uniform heat transfer due to buoyancy effects. This section describes some of the experimental techniques used in CO_2 gas cooling experiments and the observed trends in "large" tubes and mini/microchannels. For the purposes of this chapter, tubes with diameter >3 mm will be referred to as large, and less than 3 mm will be referred to as mini/micro channels [53, 54].

3.3.1 Single-Tube Studies

Baskov *et al.* [7] conducted one of the first studies of cooling of supercritical carbon dioxide in a large vertical tube with an internal diameter of 4.12 mm cooled by a water-alcohol mixture. Experiments were conducted at reduced pressures from 1.08 to 1.63 and sCO_2 temperatures from 17 to 212°C. They used a heat flux meter to determine the local heat flux at multiple axial distances along the tube. A representative heat flux meter is shown in Figure 3.5. Here, fluid flows through the center tube, thermocouples are embedded in the thick tube wall and a uniform heat flux is applied at the outer wall surface.

This concept relies on the measurement of two temperatures in a solid material with known thermal conductivity separated by a known distance. The heat flux can then be

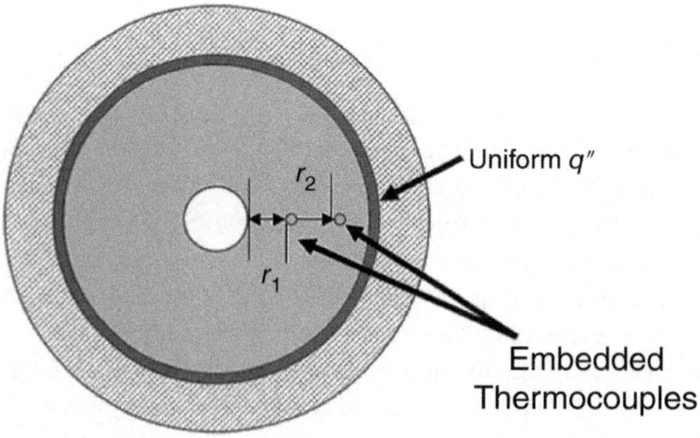

Figure 3.5 Cross section schematic of heat flux meter concept.

determined by applying the appropriate form of Fourier's law:

$$q'' = -k\frac{dT}{dx} \tag{3.2}$$

The coolant temperature was measured at each axial location, and the bulk sCO_2 temperature was calculated from the local enthalpy determined from an energy balance between each measurement location. Variations of this technique for determining bulk fluid temperature are used by many research groups. A challenge with this approach is that the radial distance between wall temperature measurements must be large enough that the temperature difference can be measured with low uncertainty. This necessitates a thick-walled tube, which increases the conductive thermal resistance and increases the uncertainty in the calculated convective sCO_2 heat transfer coefficient. Furthermore, increase in the tube wall thickness with a highly conductive material such as copper increases the potential for axial conduction, which can also increase experimental uncertainty.

Citing potentially high uncertainty of the thick-wall heat meter approach of Baskov *et al.* [7], Yoon *et al.* [55] deployed a segmented test section technique to evaluate quasi-local heat transfer of sCO_2 in a horizontal tube with inner diameter of 7.73 mm. A conceptual schematic of their approach is shown in Figure 3.6.

Here, sCO_2 flows through a center tube which is surrounded by eight subsections, each forming a tube-in-tube heat exchanger with a length of 470 mm. Water flows through each subsection to provide cooling to the test section. Water and supercritical CO_2 temperature and pressure are measured at the inlet and outlet of each subsection using immersed thermocouples. The outer wall temperature at top, bottom, and side positions are measured in each subsection. The subsection heat duty was then found from a water-side energy balance, and it was assumed the heat flux for each subsection was constant. Using this, the measured wall temperature and the average subsection sCO_2 temperature, the heat transfer coefficient could be found from Eq. (3.1).

Subsequent authors have refined this technique by exploring more, or shorter test sections to more effectively measure "local" heat transfer coefficient. In this approach, as the cooling water flow rate is increased or the subsection length decreases, the measured temperature difference decreases, which increases the uncertainty in the measured heat duty. To mitigate this, the cooling water flow rate could be decreased, which would result in a larger temperature difference for a given thermal input. However, here, the water-side convective thermal resistance would also increase, and the temperature difference between the outer wall and the sCO_2 would decrease, increasing uncertainty. Another way to mitigate the problem would be to increase the length of the subsection. With this approach, local variations in the heat transfer coefficient could not be quantified. Thus, design of an experiment of supercritical CO_2 cooling requires careful consideration of the competing sources of uncertainty to measure local heat transfer coefficients.

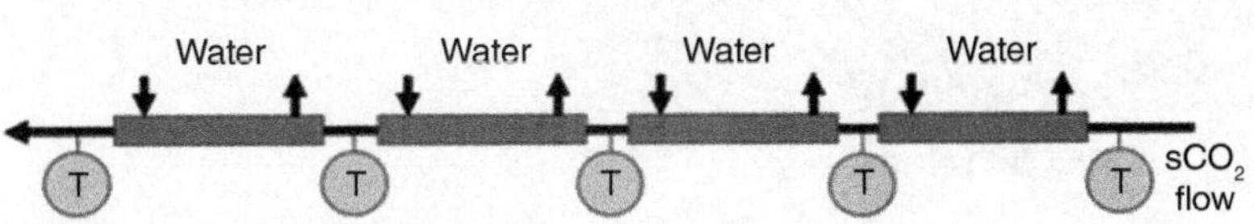

Figure 3.6 Schematic of quasi-local heat transfer coefficient measurement techniques.

3.3.2 Mini/Microchannel Studies

Mini- and microchannel geometries ($D_H < 3$ mm) have shown significant promise to safely contain the high pressures of supercritical carbon dioxide while still maintaining thin tube walls [56, 57]. The experimental challenges described above are even greater when considering mini- and microchannels. Notably, the heat duties decrease as the channel size goes down, making measurement more difficult. In addition, the heat transfer coefficients tend to increase with decreasing hydraulic diameter, decreasing the temperature difference between the wall and bulk fluid and increasing experimental uncertainty.

To mitigate the challenge of small heat duty, many investigators have conducted investigation of supercritical CO_2 cooling in multi-port mini/microchannel tubes. These studies also had the practical importance of evaluating a geometry that is commonly used in CO_2 heat pump equipment. Huai *et al.* [58] measured heat transfer in a 10 multi-port extruded aluminum tube with circular channels of diameter 1.31 mm. The test section was 500 mm long, 20 mm wide, and 2 mm thick. The sCO_2 was cooled by water flowing through copper blocks attached to the top and bottom of the test section, shown schematically in Figure 3.7.

The local heat flux was measured using 12 heat flux sensors placed in between the cooling blocks and microchannel tube, and the outer-wall temperature of the microchannel tube was measured with 24 K-type thermocouples. Here, the measured heat flux from each sensor was used to calculate the local enthalpy (and thus temperature) of the supercritical CO_2 with an energy balance. The measured wall temperature could then be used to determine the local heat transfer coefficient. In this approach, the measurement uncertainty is highly dependent on the heat flux sensor used, and the temperature difference between the wall and the supercritical CO_2.

3.3.3 Summary of Experimentally Observed Effects

Over the past 20 years, there have been numerous experimental studies on supercritical carbon dioxide gas cooling in various geometries. An exhaustive review of each study is beyond the scope of this chapter, but most studies are conducted following an experimental approach similar to those discussed above. Data from many of these investigations have been used to develop different correlations and models, discussed in the following sections. More detailed reviews of experimental studies can be found in [54, 59, 60].

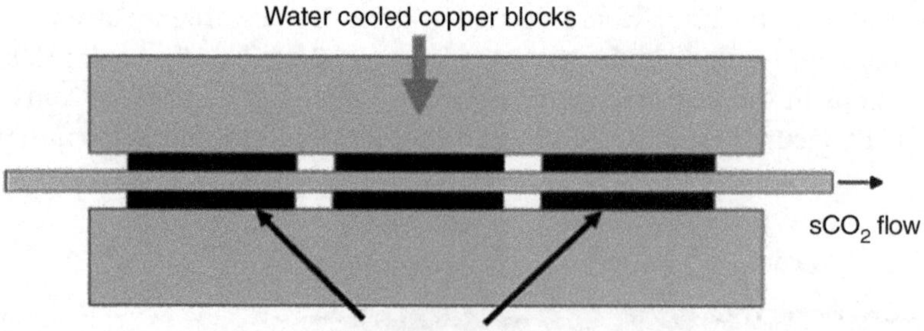

Figure 3.7 Schematic of a side view of the setup of Huai *et al.* showing a microchannel tube in between a set of water-cooled copper blocks and discrete heat flux sensors.

In general, most experimental studies on supercritical gas cooling show qualitative agreement regarding the effects of different parameters such as pressure, temperature, and mass flux on the convective heat transfer coefficient [54]. There is still some uncertainty on the relative importance of buoyancy in horizontal and vertical channels during supercritical cooling processes in conditions relevant to CO_2 heat pumps. These effects are summarized below.

Most important is the effects of temperature and pressure. Representative data from Oh and Son [53] for a horizontal, circular tube with inner diameter of 7.75 mm at different temperatures and pressures is shown in Figure 3.8. As the bulk supercritical carbon dioxide temperature nears the pseudo-critical point, there is a dramatic spike in the convective heat transfer coefficient due to the large increase in the Prandtl number at this point. As pressures increase, the magnitude of the spike decreases and occurs at higher temperatures, consistent with the thermophysical property trends shown in Figure 3.2, previously. At temperatures away from the pseudo-critical point, the heat transfer behavior approaches that of single-phase liquid or gas, and the effects of pressure are diminished. These trends are replicated in most published studies, and the spike in heat transfer at the pseudo-critical point is well established.

Similarly, there is good agreement in the literature on the effect of mass flux. Increasing the mass flux for a given pressure increases the Reynolds number, increasing the convective heat transfer coefficient. Figure 3.9 shows representative data from Oh and Son [53] for tube diameters of 4.55 and 7.75 mm.

The importance of buoyancy in *heated* supercritical carbon dioxide flows is well established. Large differences in the heat transfer coefficient in vertical upward and downward heating have been observed, as buoyancy effects either aid or deteriorate heat transfer [61]. The effects of buoyancy during cooling and in what conditions they are present are not as well established. Jiang *et al.* [62] showed experimentally a difference in local heat transfer of supercritical CO_2 cooled by water in upward versus downward flow in

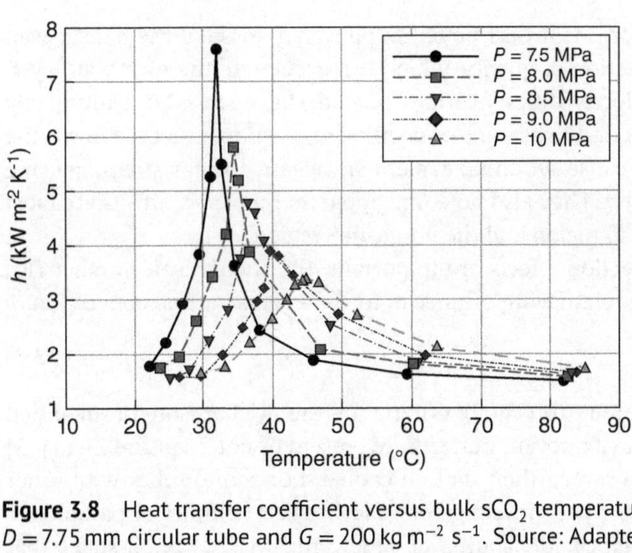

Figure 3.8 Heat transfer coefficient versus bulk sCO_2 temperature at varying pressures in a $D = 7.75$ mm circular tube and $G = 200$ kg m^{-2} s^{-1}. Source: Adapted from Oh and Son [53].

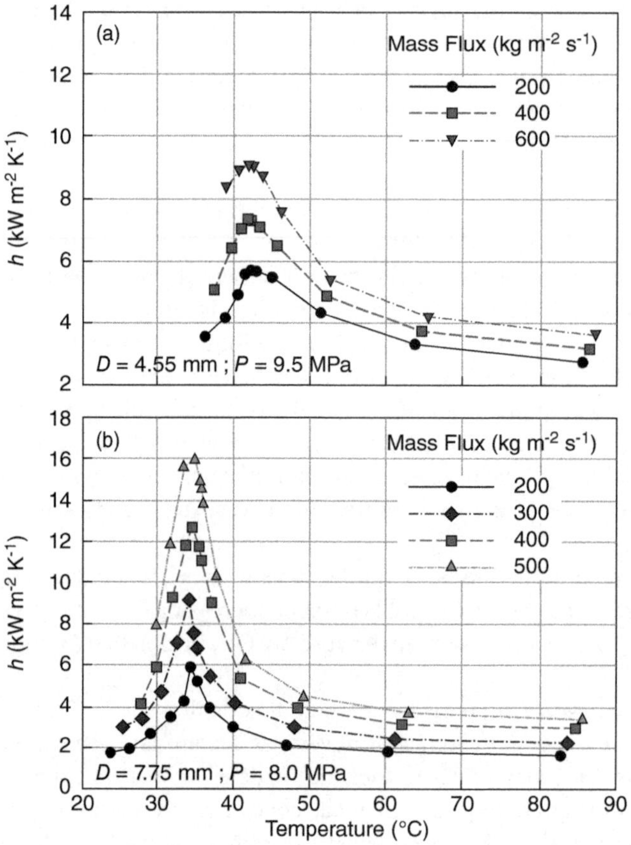

Figure 3.9 Heat transfer coefficient versus bulk sCO_2 temperature at varying mass flux in (a) $D = 4.55$ mm and (b) D = 7.75 mm circular tube. Source: Adapted from Oh and Son [53].

a 2 mm inner diameter tube. For downward flow, the buoyancy force of the denser wall layer is consistent with the direction of gravity, which can accelerate the near-wall layer, decreasing the wall-to-bulk velocity difference and related shear stress and ultimately reducing turbulence production and heat transfer coefficient. As flow proceeds down the tube, the buoyancy forces can cause an enhancement in negative shear stress, causing heat transfer coefficient to recover. They also note that apparent buoyancy effects are most pronounced in the pseudo-critical region and the liquid-like region.

To predict when mixed convection effects are important, the Richardson number (Ri, Eq. (3.3)) is used to quantify the relative importance of forced versus natural convection.

$$\mathrm{Ri} = \frac{\mathrm{Gr}}{\mathrm{Re}^2} \tag{3.3}$$

Many investigators use different threshold criteria of the Richardson or modified Richardson number to determine if buoyancy effects are present or not. Liao and Zhao [63] used a threshold of $\mathrm{Ri} < 10^{-3}$ to screen their data in cooled, horizontal tubes with inner diameter from 0.70 to 2.16 mm. Many investigators use the semi-empirical parameters of Jackson and Hall [64], developed for heating of supercritical CO_2, which states that buoyancy effects are important when Eq. (3.4) is satisfied:

$$\frac{\mathrm{Gr}}{\mathrm{Re}^{2.7}} > 10^{-5} \tag{3.4}$$

Still others have applied the transition criterion of Petukhov and Polyakov [65], in which a threshold Grashof number (Eq. (3.5)) is defined which corresponds to a value in which a 1% deviation in the Nusselt number from forced convection only is observed. The threshold Grashof number is compared to the Grashof number defined in terms of heat flux (Gr_q Eq. (3.6)). For values of Gr_q/Gr_{th} greater than 1, buoyancy effects are expected to be important.

$$Gr_{th} = 3 \times 10^{-5} \cdot \mathrm{Re}_b^{2.75} \overline{\mathrm{Pr}}^{(1/2)}[1 + 2.4\mathrm{Re}^{-1/8}(\mathrm{Pr}^{-2/3} - 1)]$$

where

$$\begin{aligned} \overline{\mathrm{Pr}} &= \frac{i_w - i_b}{T_w - T_b} \frac{\mu_b}{k_b} \\ Gr_q &= \frac{g\bar{\beta}q''D^4}{v_b^2 k_b} \end{aligned} \tag{3.5}$$

where

$$\bar{\beta} = \frac{1}{\rho_{film}} \frac{\rho_b - \rho_w}{T_w - \rho_b} \tag{3.6}$$

Caution must be exercised when attempting to utilize these and other criteria, or when comparing results between studies. Investigators use different combinations of properties defined at wall temperature, bulk temperature, average, and integrated values to evaluate Grashof and Reynolds numbers, which can cause confusion. At present, there is still no solid consensus on the threshold for onset of buoyancy effects during supercritical cooling of carbon dioxide [66]. This remains an area of active research.

3.4 Supercritical CO_2 Heat Transfer Correlations

The heat transfer phenomena described above are not seen in subcritical, constant property fluids, motivating the development of predictive correlations for the design of CO_2 gas coolers. Predicting supercritical heat transfer during *heating* has relied on empirical correlations using bulk-to-wall property corrections to a Dittus-Boelter type correlation: $Nu = C\mathrm{Re}^m\mathrm{Pr}^n$ [1, 51, 67–69]. However, these correlations often suffer from inconsistent reference temperatures for fluid properties and dimensionless parameters [10], and diverging predictions when compared to one another, particularly in the pseudocritical regime [11, 70].

Research on supercritical cooling has primarily focused on experimental studies of sCO_2 for heating, ventilation, air conditioning and refrigeration (HVAC&R) applications [54, 59, 71, 72]. As with supercritical heating, most correlations for supercritical cooling rely on empirical (from sCO_2 data) bulk-to-wall property corrections to account for supercritical effects [7, 63, 73–79]. The correlations developed generally for sCO_2 fail to extend to other fluids with significantly different properties [80, 81], and exhibit poor agreement in the pseudocritical regime, as will be discussed below.

To improve the understanding of supercritical CO_2 thermal hydraulics, numerical simulation of supercritical flow is a growing topic of interest [82–87]. However, it remains unclear if standard models for turbulence and turbulent heat transfer are applicable to the highly varying properties in the supercritical regime [8, 88, 89]. Much of the numerical work has been focused on laminar flow [90, 91]. These limited results do suggest the presence of non-uniform temperature and velocity profiles due to buoyancy effects [53, 90, 92–94]. This area remains one of active research and development. At present, there is no consistent, mechanistic model accounting for the underlying physical transport phenomena for supercritical cooling. Thus, designers should attempt to identify correlations and models that were developed under similar conditions to those expected in operation. This section introduces selected models that are relevant for design of sCO_2 heat pump gas coolers.

3.4.1 Constant Property Turbulent Correlations

Despite the unique heat transfer behavior near the critical point, often a turbulent, single-phase, constant property heat transfer correlation is sufficient for gas cooler design [95, 96]. This is particularly true for air-cooled gas coolers, where the air-side resistance is the dominant heat transfer resistance, and accurate prediction of the CO_2 heat transfer is less important. Common correlations are the Dittus-Boelter [97] turbulent correlation (Eq. (3.7)), or the Gnielinski [98] correlation (Eq. (3.8)).

$$\mathrm{Nu} = \frac{hD}{k} = 0.023 \cdot \mathrm{Re}^{0.8} \cdot \mathrm{Pr}^{\,n}$$

$$n = 0.4 \text{ for heating and } n = 0.3 \text{ for cooling} \tag{3.7}$$

$$\mathrm{Nu} = \frac{(f/8)\cdot(\mathrm{Re}-1000)\cdot \mathrm{Pr}}{1 + 12.7\cdot (f/8)^{1/2}\cdot(\mathrm{Pr}^{\,2/3}-1)} \tag{3.8}$$

In both correlations, the Reynolds and Prandtl numbers are typically evaluated at the bulk fluid temperature, although in some cases researchers have used the wall temperature, or some average of the wall and bulk. The Gnielinski [98] correlation is valid for turbulent flows with Reynolds ranging from 2300 to 5×10^6 and Prandtl number for 0.5 to 2000. The friction factor in Eq. (3.8) is calculated from the correlation of Filonenko [99], shown in Eq. (3.9). Generally, both the Dittus-Boelter and the Gnielinski [98] correlation will under predict the heat transfer coefficient near the pseudo-critical temperature.

$$f = (0.79 \cdot \ln(\mathrm{Re}) - 1.64)^{-2} \tag{3.9}$$

3.4.2 Krasnoschekov et al. (1970) Correlation

The correlation of Krasnoschekov *et al.* [100] was one of the earliest developed for supercritical cooling in horizontal flow, and attempted to account for the difference between bulk and wall temperatures through the use of property correction ratios.

$$\mathrm{Nu}_w = \frac{hD}{k_w} = \mathrm{Nu}_{o,w} \cdot \left(\frac{\rho_w}{\rho_b}\right)^n \cdot \left(\frac{\bar{c}_p}{c_{p,w}}\right)^m$$

where

$$\bar{c}_p = \frac{i_b - i_w}{T_b - T_w} \tag{3.10}$$

Here, $Nu_{o,w}$ is the Nusselt number calculated from the Petukhov and Kirillov [101] correlation (Eq. (3.11)) with the Reynolds and Prandtl number evaluated at the wall temperature and the friction factor from Eq. (3.9).

$$Nu_{o,w} = \frac{(f/8) \cdot Re_w \cdot Pr_w}{1.07 + 12.7 \cdot (f/8)^{1/2} \cdot (Pr_w^{2/3} - 1)} \tag{3.11}$$

The exponents n and m are provided graphically in the original reference as a function of pressure. The graphical nature does limit the ease of use of this correlation. However, the correlation shows that when the tube wall temperature is below the critical temperature of the fluid, the predicted heat transfer coefficient using the property corrections is seen to increase compared to that of a constant property single-phase fluid, as is observed experimentally. Many of the following correlations adopt a similar methodology for modifying single-phase correlations to account for the pseudo-critical region effects.

3.4.3 Ghajar and Asadi (1986) Correlation

Ghajar and Asadi [102] performed a study comparing existing empirical heat transfer correlations in the near-critical region. To eliminate errors from different property inputs used by the different investigators who proposed these correlations, they re-evaluated the numerical constants in the equations using the same physical property inputs. This was accomplished by curve-fitting the equations under evaluation to the experiment data, based on the best available property inputs. The forced convection correlations were then compared against a large bank of data of supercritical and near-critical carbon dioxide and steam. The heat flux for the carbon dioxide data ranged from 0.8 to 1100 W cm^{-2} and the mass flux from 260 to 25 000 kg m^{-2} s^{-1}. For water, the heat flux ranged from 11.6 to 2320 W cm^{-2} and the mass flux from 170 to 30 000 kg m^{-2} s^{-1}. The authors found that a correlation with the following form predicted the data the best:

$$Nu = a \cdot Re_b^b \cdot Pr_b^c \left(\frac{\rho_w}{\rho_b}\right)^d \cdot \left(\frac{\bar{c}_p}{c_{p,b}}\right)^n \tag{3.12}$$

The constant a and the exponents b, c, and d are curve-fitted constants equal to 0.025, 0.8, 0.417, and 0.32 for CO_2, respectively. The parameter n is determined from the criterion of Jackson and Fewster as follows:

For $T_b < T_w \leq T_{pc}$ and $T_w > T_b \geq 1.2T_{pc}$; $n = 0.4$

For $T_b \leq T_{pc} < T_w$; $n = 0.4 + 0.2 \cdot (T_w/T_{pc} - 1)$

For $T_{pc} \leq T_b \leq 1.2T_{pc}$ and $T_b < T_w$; $n = 0.4 + 0.2 \cdot (T_w/T_{pc} - 1) \cdot [1 - 5 \cdot (T_b/T_{pc} - 1)]$

where T_b, T_w, and T_{pc} are the bulk fluid temperature, the wall temperature and the critical temperature of the fluid, respectively.

3.4.4 Pitla et al. (2002) Correlation

Motivated by the resurgence of carbon dioxide as a refrigerant in heat pump applications, Pitla *et al.* [59, 75, 103, 104] conducted a series of experimental and numerical investigations on supercritical gas cooling. In their initial review article [59], they compared several correlations [7, 100] and concluded that there was disagreement among correlations in the pseudo-critical region, and that differences were apparent between cooling in vertical and horizontal tubes.

This motivated an experimental and numerical study [103, 104], and ultimately the proposal of a new correlation [75] for supercritical cooling of CO_2 at conditions of interest to the HVAC&R industry:

$$\mathrm{Nu} = \frac{hD}{k_b} = \left(\frac{\mathrm{Nu}_w + \mathrm{Nu}_b}{2}\right) \cdot \frac{k_w}{k_b} \tag{3.13}$$

Here, $\mathrm{Nu_w}$ and $\mathrm{Nu_b}$ are each calculated using the Gnielinski [98] correlation (Eq. (3.8)) using the wall and bulk properties, respectively. They found the best fit was obtained using the bulk fluid inlet velocity to calculated $\mathrm{Re_w}$, and local mean velocity to calculate $\mathrm{Re_b}$, regardless of the position within the tube. They found that 85% of their data (horizontal tube with ID = 4.72 mm) was predicted within ±20% by the proposed correlation. However, as noted by Oh and Son [53], the assumption that the heat transfer coefficient is an equal weighting of wall and bulk properties can cause an under prediction of the heat transfer coefficient in the pseudo-critical region.

3.4.5 Son and Park (2006) Correlation

Son and Park [74] conducted experiments in similar cooling conditions to Pitla *et al.* in a horizontal tube with ID = 7.75 mm, mass flux from 200 to 400 kg m^{-2} s^{-1}, and inlet pressures from 75 to 100 bar. From their data, they proposed a heat transfer correlation separated into regions above and below the pseudo-critical temperature:

$$\mathrm{Nu} = \frac{hD}{k_b} = \begin{cases} \mathrm{Re}_b^{0.55}\,\mathrm{Pr}_b^{0.23} \cdot \left(\dfrac{c_{p,b}}{c_{p,w}}\right)^{0.15} & \dfrac{T_b}{T_{pc}} > 1 \\ \mathrm{Re}_b^{0.35}\,\mathrm{Pr}_b^{1.9} \cdot \left(\dfrac{c_{p,b}}{c_{p,w}}\right)^{-3.4} \cdot \left(\dfrac{\rho_b}{\rho_w}\right)^{-1.6} & \dfrac{T_b}{T_{pc}} \le 1 \end{cases} \tag{3.14}$$

Below the pseudo-critical temperature their data was predicated with a mean deviation of 16.3%, and with a mean deviation of 17.6% above the pseudo-critical temperature. By comparison, the Pitla *et al.* [75] correlation had a mean deviation of 36.4% with their data.

3.4.6 Oh and Son (2010) Correlation

Oh and Son [53] conducted supercritical CO_2 cooling experiments in horizontal tubes with diameters of 4.55 and 7.75 mm, mass fluxes from 200 to 600 kg m^{-2} s^{-1}, and inlet pressures ranging from 75 to 100 bar, as discussed in Section 3.3.3. They compared their data with 10 correlations from the literature. Of these, the Pitla *et al.* [75] model showed the best agreement, although with under prediction in the pseudo-critical region. Thus, they introduced another new correlation similar in form to Son and Park, with exponents in Eq. (3.15) determined via least square curve fitting.

$$\mathrm{Nu} = \frac{hD}{k_b} = \begin{cases} 0.023\,\mathrm{Re}_b^{0.7}\,\mathrm{Pr}_b^{2.5} \cdot \left(\dfrac{c_{p,b}}{c_{p,w}}\right)^{-3.5} & \dfrac{T_b}{T_{pc}} > 1 \\ 0.023\,\mathrm{Re}_b^{0.6}\,\mathrm{Pr}_b^{3.2} \cdot \left(\dfrac{c_{p,b}}{c_{p,w}}\right)^{-4.6} \cdot \left(\dfrac{\rho_b}{\rho_w}\right)^{3.7} & \dfrac{T_b}{T_{pc}} \le 1 \end{cases} \tag{3.15}$$

3.4.7 Microchannel Correlations

The above correlations were developed for channels with hydraulic diameters larger than 3 mm. Very small channels are attractive for CO_2 heat pump gas cooler designs as they (i) can provide very high heat transfer coefficients, (ii) require relatively thin tube walls to safely contain the working pressure and (iii) minimize refrigerant charge. Thus, there have been several correlations developed specifically for mini/microchannel tubes.

Liao and Zhao [63] developed a correlation based on experiments from single horizontal microtubes ($0.5 < D < 2.16$ mm) at pressure from 74 to 120 bar and temperatures from 20 to 120°C. They speculated that the large property variations were causing buoyancy to have some influence on the heat transfer, even in the small diameter tubes. Thus, they proposed a correlation that modified the Dittus-Boelter correlation with property ratio corrections and the Richardson number:

$$\mathrm{Nu}_w = \frac{hD}{k_w} = 5.57\left(\frac{Gr}{\mathrm{Re}_b^2}\right)^{0.205}\left(\frac{\rho_b}{\rho_w}\right)^{0.437}\left(\frac{\bar{c}_p}{c_{p,w}}\right)^{0.411} \cdot \mathrm{Nu}_{db,w}$$

where

$$\mathrm{Nu}_{db,w} = 0.023 \cdot \mathrm{Re}_w^{0.8} \cdot \mathrm{Pr}_w^{0.3} \tag{3.16}$$

Dang and Hihara [77] measured quasi-local heat transfer coefficients in horizontal tubes with internal diameters of 1, 2, 4, and 6 mm, mass flux from 200 to 1200 kg m^{-2} s^{-1}, temperatures from 30 to 70°C, pressure from 80 to 100 bar, and cooling heat flux of 6–33 kW m^{-2}. For the 1 and 2 mm tubes, the mass flux was limited to 800–1200 kg m^{-2} s^{-1}. From their data, they concluded that at bulk temperatures less than T_{pc}, the heat transfer coefficient was less sensitive to changes in diameter or heat flux, while at temperatures higher than T_{pc} there was a dependence on diameter and heat flux and an average property in the radial direction was required. With this as a basis, they modified the Gnielinski [98] correlation:

$$\mathrm{Nu}_f = \frac{hD}{k_f} = \frac{(f_f/8)\cdot(\mathrm{Re}_b - 1000)\cdot \mathrm{Pr}}{1 + 12.7\cdot(f_f/8)^{1/2}\cdot(\mathrm{Pr}^{2/3} - 1)}$$

where

$$f_f = [1.82\log_{10}(\mathrm{Re}_f) - 1.64]^{-2}$$

$$\mathrm{Pr} = \begin{cases} c_{p,b}\mu_b/k_b & \text{for } c_{p,b} \ge \bar{c}_p \\ \bar{c}_p\mu_b/k_b & \text{for } c_{p,b} < \bar{c}_p \text{ and } \mu_b/k_b \ge \mu_f/k_f \\ \bar{c}_p\mu_f/k_f & \text{for } c_{p,b} < \bar{c}_p \text{ and } \mu_b/k_b < \mu_f/k_f \end{cases} \tag{3.17}$$

Here, the subscript f refers to properties evaluated at the film temperature ($T_f = [T_b + T_w]/2$), and the mean specific heat is defined as in Eq. (3.10). They found the correlation predicted 93% of their 458 data points within ±20%, and that the pressure drop correlation in Eq. (3.17) predicted the 1 and 2 mm data well (only diameter for which data were available).

Huai *et al.* [58] investigated heat transfer of supercritical CO_2 in a multiport tube (channel diameter = 1.31 mm). The tube geometry is consistent with what might be expected in air-cooled, CO_2 heat pump gas coolers. They found their data was over predicted by the Liao and Zhao correlation, which they theorized may be due to the correlation being developed from single tube rather than multiport tube data. Thus, they introduced a new correlation, shown below.

$$Nu = \frac{hD}{k_b} = 2.2186 \cdot 10^{-2} \cdot \mathrm{Re}_b^{0.8}\,\mathrm{Pr}_b^{0.3}\left(\frac{\rho_b}{\rho_w}\right)^{-1.4652}\left(\frac{\bar{c}_p}{c_{p,w}}\right)^{0.0832} \tag{3.18}$$

3.4.8 Comparison of Correlations

As shown above, there has been a proliferation of correlations to predict heat transfer of supercritical carbon dioxide during cooling. Typically, these correlations use a single-phase correlation as the base, and modify it with wall-to-bulk property ratios and empirical exponents and constants. This has led to many correlations very similar in form but with different constants due to the data for which they were developed. As an example, the predictions from six of the correlations discussed are compared to the constant property Gnielinski [98] correlation in Figure 3.10 as a function of temperature and at two different reduced pressures. The y-axis shows the prediction of each correlation divided by the predicted value of the Gnielinski [98] correlation at those conditions. The mass flux is assumed to be $G = 400\,\mathrm{kg\,m^{-2}\,s^{-1}}$, the tube diameter is 6 mm and the cooling heat flux is assumed constant at $5\,\mathrm{kW\,m^{-2}}$. Figure 3.10 shows deviation from one correlation to another, despite often being developed from very similar datasets. This disagreement is particularly large in the pseudo-critical region. Thus, these highly empirical correlations cannot be generally applied, and the designer should use care when selecting an appropriate correlation. The results also suggest that a more mechanistic understanding of supercritical CO_2 gas cooling that can be generalized remains elusive.

3.5 Supercritical CO_2 Pressure Drop

Accurate prediction of pressure drop is also required for design of CO_2 heat pump equipment. As CO_2 remains a single phase fluid during the gas cooler process, the complexities of predicting two-phase pressure drop are avoided. Thus, CO_2 gas cooling pressure drop generally follows similar trends as for a single phase fluid, and the pressure drop can be evaluated as follows:

$$\Delta P = \rho \cdot f \cdot V^2 \frac{L}{2D} \tag{3.19}$$

For a fixed mass flux, pressure drop will increase with decreasing tube diameter as the velocity increases. Similarly, pressure drop will increase for a given tube diameter as mass

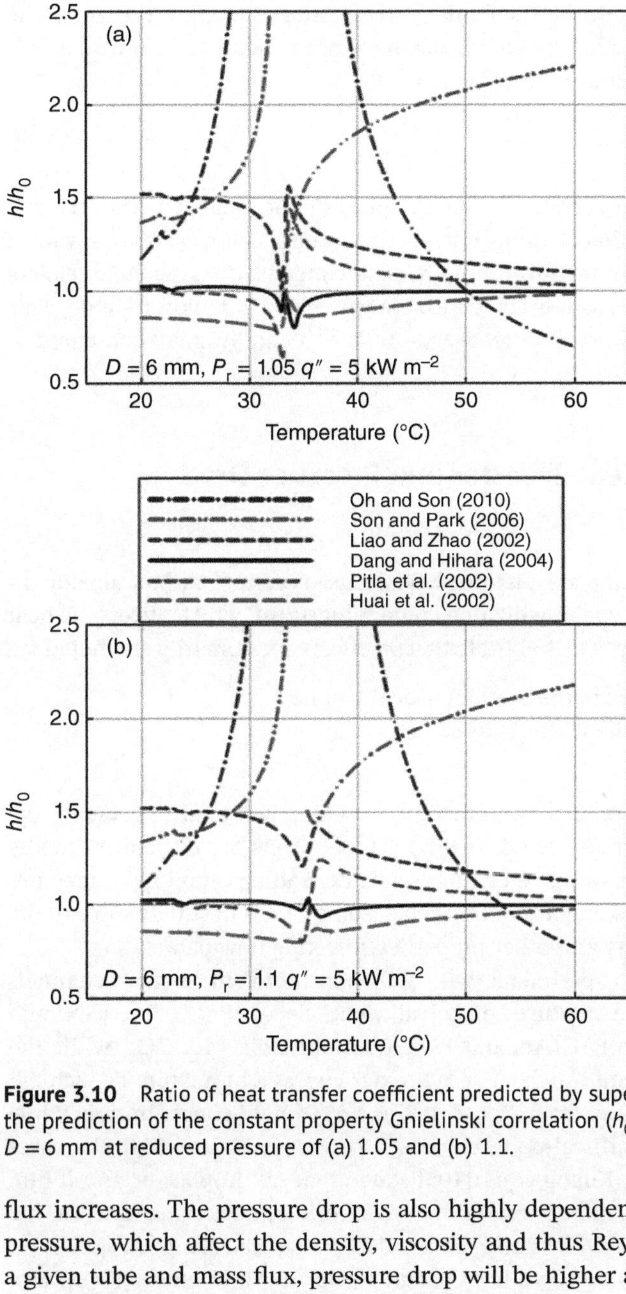

Figure 3.10 Ratio of heat transfer coefficient predicted by supercritical cooling correlation (h) and the prediction of the constant property Gnielinski correlation (h_0) for $G = 400\ kg\ m^{-2}\ s^{-1}$ and $D = 6$ mm at reduced pressure of (a) 1.05 and (b) 1.1.

flux increases. The pressure drop is also highly dependent on temperature and operating pressure, which affect the density, viscosity and thus Reynolds number and velocity. For a given tube and mass flux, pressure drop will be higher at temperatures greater than the pseudo-critical temperature where the fluid behaves as a low density gas. Thus, a segmented or discretized approach is necessary to approximate the local friction factor as the CO_2 temperature changes along a tube.

For most applications of interest for sCO_2 heat pumps, the flow is turbulent, and the Darcy friction factor (f) in Eq. (3.19) is a function of the Reynolds number and wall roughness. Common expressions for the single phase, turbulent friction factor include the Blasius equation (Eq. (3.20)), or the Filonenko [99] equation, shown in Eq. (3.9). The use of standard, single-phase friction factor models have been found to be adequate for predicting CO_2 pressure drop during gas cooling [54, 60].

$$f = \frac{0.316}{\mathrm{Re}_b^{1/4}} \tag{3.20}$$

When selecting a friction factor correlation, the designer should check whether the calculated friction factor is the Fanning or Darcy form of the friction factor, as the Darcy form is four times greater than the Fanning form. Finally, as in condensation, the bulk velocity of the supercritical CO_2 decreases as it is cooled from a gas-like to a liquid-like state. This decrease in momentum causes a pressure recovery, but it is generally small compared to the frictional pressure loss.

3.6 Supercritical CO_2 Heat Transfer and Pressure Drop with Lubricants

The above experimental studies and correlations have focused on pure carbon dioxide. In a CO_2 heat pump, CO_2 in the gas cooler will often contain lubricant, which affects the heat transfer and pressure drop. Cheng *et al.* [54] state that oil effects are primarily attributed to:

1. Modification of fluid properties in the pseudo-critical regime
2. Distortion of temperature gradient effects due to oil
3. Presence of two-phase oils

As with condensation, the presence of oil during gas cooling will tend to decrease the heat transfer coefficient and increase pressure drop. These effects are difficult to model generally, as CO_2 and oils can be miscible or immiscible depending on oil type, pressure and temperature. Further, for oils that are not miscible, some CO_2 will still dissolve in the oil, which changes the oil viscosity and other properties, affecting flow parameters.

Kuang *et al.* [105] conducted experiments with CO_2 and lubricant in microchannels (D = 0.86 mm). They considered mixtures of polyalkylene glycol (PAG), PAG/AN, and polyolester glycol (POE) oil. The PAG/AN and PAG are immiscible with CO_2, while the POE is miscible. Experiments were run from zero up to 5% by weight mixture of each oil. They found pressure drop increased from 20% to 49% at a 5% weight concentration, while the heat transfer at the pseudo-critical point decreased between 31% and 58%. For cases where oil/CO_2 were immiscible, Kuang *et al.* [105] stated that oil droplets or an oil film can form on inner tube surfaces, increasing thermal resistance and frictional drag on the bulk flow. Similar results were found by Yun *et al.* [106] and Zhao and Jiang [107].

Dang *et al.* [72] conducted a study with CO_2 and PAG, polyvinyl ether (PVE), and PAG-PVE copolymer (ECP) oil. They measured heat transfer, pressure drop, and visualized the flow. They found that there was little evidence of oil films/droplets at temperatures below the pseudo-critical temperature in the case of the oil with highest solubility (PVE), and that the heat transfer coefficient did not change appreciably with oil concentration. However, at temperatures at or greater than the pseudo-critical temperature they found all three oils caused a decrease in heat transfer coefficient. As CO_2 temperature increases, the density drops rapidly, while the oil density difference is less sensitive to changes in temperature. Thus, the oil to CO_2 density and viscosity difference increases, increasing interfacial roughness and presence of oil droplets [107].

3.6.1 CO_2/Lubricant Pressure Drop Correlations

At present, there is no general correlation that can predict pressure drop of CO_2/lubricant mixtures over a wide range of geometries, operating conditions, oil types and concentrations. As CO_2 heat pump technology advances, this remains an area of active research. Zhao and Jiang [107] presented empirical correlations for CO_2 + POE (miscible) and CO_2 + PAG oils based on data from horizontal 1- and 2 mm tubes. For CO_2 + POE, they proposed the following:

$$\frac{\Delta P_{CO_2-oil}}{\Delta P_{CO_2}} = 0.816\left(\frac{\rho_{oil}}{\rho_{CO_2}}\right)^{0.246}\left(\frac{x\cdot\mu_{oil}}{\mu_{CO_2}}\right)^{0.008} \tag{3.21}$$

Here, x is the oil concentration by weight percent, and ΔP_{CO_2} is the pressure drop of pure CO_2 as predicted by Petukhov's correlation (Eq. (3.22)). The constant and exponents in Eq. (3.21) were determined from a least square fit method. They stated that the correlation is valid for small tubes (D = 1–2 mm), for pressures from 80 to 110 bar, mass fluxes from 400 to 1200 kg m^{-2} s^{-1}, temperatures from 20 to 90°C and POE oil weight concentrations less than 2%.

$$\frac{f}{f_0} = \left(\frac{\mu_w}{\mu_b}\right)^{0.24}$$

where

$$f_0 = [1.82\log_{10}(\text{Re}_b) - 1.64]^{-2} \tag{3.22}$$

PAG oil is immiscible, and thus the pressure drop model is different and is a function of the oil weight concentration and the solubility of CO_2 (mass %) in the oil. They propose different correlations for oil weight percentages less than and greater than 1%:

$$\frac{\Delta P_{CO_2-oil}}{\Delta P_{CO_2}} = 0.029\left(\frac{\rho_{oil}}{\rho_{CO_2}}\right)^{0.280}\left(\frac{x\cdot\mu_{oil}}{\omega\cdot\mu_{CO_2}}\right)^{0.775} \quad \text{for } x \le 1\%$$

$$\frac{\Delta P_{CO_2-oil}}{\Delta P_{CO_2}} = 0.608\left(\frac{\rho_{oil}}{\rho_{CO_2}}\right)^{-0.002}\left(\frac{x\cdot\mu_{oil}}{\omega\cdot\mu_{CO_2}}\right)^{0.254} \quad \text{for } x > 1\% \tag{3.23}$$

Here, ΔP_{CO_2} is the pressure drop of pure CO_2 as predicted by the Filonenko [99] equation (Eq. (3.9)), and ω is the solubility of CO_2 in mass percent in the oil. In their paper, they provided an alternative formulation of Eq. (3.23) which better predicts the data, but yields physically unrealistic predictions (pressure drop ratio decreases as viscosity ratio increases) for oil concentrations less than 1%. Care should be applied when applying either the POE or PAG correlation, as they are more indicative of qualitative behavior than quantitative behavior.

3.6.2 CO_2/Lubricant Heat Transfer Correlations

Like the pressure drop correlations, different approaches are warranted when the CO_2/oil mixture is miscible or immiscible. For CO_2 and POE mixtures, Zhao *et al.* [108] proposed a heat transfer correlation similar in form to their pressure drop correlations above. Experiments were conducted in D = 1.98 and 4.14 mm tubes during cooling of supercritical CO_2 with oil concentrations from 0 to 2 (wt%), temperatures from 20 to 100°C, pressures from 80 to 110 bar and mass flux from 400 to 1200 kg m^{-2} s^{-1}. For the no oil case, their data was well predicted by the Dang and Hihara [77] correlation, described in Section 3.4.7.

$$\frac{h_{CO_2-oil}}{h_{CO_2}} = 0.764\left(\frac{\rho_{oil}}{\rho_{CO_2}}\right)^{0.530}\left(\frac{x \cdot \mu_{oil}}{\mu_{CO_2}}\right)^{-0.227} \quad \text{for } T_b/T_{pc} > 1$$
$$\frac{h_{CO_2-oil}}{h_{CO_2}} = 1.186\left(\frac{\rho_{oil}}{\rho_{CO_2}}\right)^{-0.236}\left(\frac{x \cdot \mu_{oil}}{\mu_{CO_2}}\right)^{-0.114} \quad \text{for } T_b/T_{pc} \leq 1 \tag{3.24}$$

Here, h_{CO_2} is the no oil prediction of Dang and Hihara (Eq. (3.17)). The correlation predicted 90% of their data within ±20%.

Jung and Yun [109] developed a modeling approach for CO_2 + PAG (immiscible) that considered the flow pattern. For oil concentration less than 1%, a homogenous model was used, as there was no observation of an oil film and oil droplets were approximately evenly distributed in the flow from the data of Dang *et al.* [110]. With this assumption, the correlation of Gnielinski [98] was used, where all properties were evaluated by average mixture properties according to mixing rules presented in the original reference. For higher oil fractions (~5% by weight), the flow patterns became more complicated and a separated flow model was used considering oil droplet entrainment, and properties of the liquid phase considering CO_2 solubility. In all of these CO_2/lubricant models, accurate prediction of the combined system thermophysical properties is essential for obtaining reasonable predictions.

3.7 Summary and Need for Additional Research

Supercritical heat transfer differs from single-phase subcritical heat transfer due to a rapid property variation in the vicinity of the pseudo-critical point. For cooling, this leads to enhanced heat transfer near the pseudo-critical temperature that is poorly predicted by single-phase correlations. This has led to the development of numerous empirical correlations to predict supercritical cooling heat transfer of CO_2. The form of these correlations has been a single-phase correlation modified with property correction ratios and other dimensionless groups, weighted with curve fit exponents. While these correlations often predict the data from which they are developed well, they show deviation when compared to one another and with other data. Thus, it can be concluded that there is no general, physically based correlation that can predict a wide range of data well, particularly in the pseudo-critical region. Developing a model like this remains the subject of active research. However, away from the pseudo-critical point the correlations are in better agreement and the heat transfer behavior more closely resembles that of a single-phase fluid.

Thus, the designer of CO_2 heat pump gas coolers must weigh how important accurate prediction of the heat transfer in the pseudo-critical region is, as it occurs over a relatively narrow temperature band. Many investigators have shown that single-phase correlations are sufficient for design, particularly when the thermal resistance of the CO_2 is much lower than the cooling fluid, such as in air-cooled gas coolers. Furthermore, conventional correlation for single-phase, turbulent pressure drop have been shown to predict CO_2 pressure drop adequately.

One of the areas of supercritical CO_2 thermal hydraulics important for heat pumps that requires additional investigation is the effect of lubricants on system thermophysical properties, pressure drop, and heat transfer coefficient. There are clear negative impacts of entrained lubricant, and some initial correlations and modeling approaches have been introduced. However, these approaches require further development to be generally applicable and accurate over a wide range of conditions.

Nomenclature

c_p	specific heat, kJ kg^{-1} K^{-1}
Bo*	buoyancy parameter, $Gr_q/(Re^{3.425} Pr^{0.8})$
C	constant, (–)
D	diameter, m
f	friction factor, (–)
G	mass flux, kg m^{-2} s^{-1}
Gr	Grashof number, (–)
Gr_q	Grashof number based on wall heat flux, $g\beta q'' D^4/(\nu^2 k)$
Gr_{th}	threshold Grashof number defined in Eq. (3.5), (–)
h	heat transfer coefficient, kW m^{-2} K^{-1}
i	specific enthalpy, kJ kg^{-1}
k	thermal conductivity, W m^{-1} K^{-1}
L	length, m
Nu	Nusselt number, (–)
P	pressure, MPa
Pr	Prandtl number (–)
q"	heat flux, W m^{-2}
Re	Reynolds number, (–)
Ri	Richardson number, (–)
T	temperature, °C
V	velocity, m s^{-1}
x	oil concentration, (wt%)

Greek Symbols

β	coefficient of thermal expansion, K^{-1}
ρ	density, kg m^{-3}

ν	kinematic viscosity, m^{-2}, s^{-1}
μ	dynamic viscosity, kg m^{-1} s^{-1}
ω	solubility of CO_2 in oil, (mass %)

Subscripts

b	bulk
crit	critical point
CO_2	carbon dioxide property
db	related to Dittus-Boelter correlation
f, film	related to film temperature
oil	lubricating oil property
pc	pseudocritical
r	reduced property
w	wall

References

1 Krasonshchekov, E.A. and Protopopov, V.S. (1966). Experimental study of heat exchange in carbon dioxide in the supercritical range at high temperature drops. *Teplofizika Vysokikh Temperatur* 4: 389–398.

2 Bringer, R.P. and Smith, J.M. (1957). Heat transfer in the critical region. *AIChE Journal* 3: 49–55. https://doi.org/10.1002/aic.690030110.

3 Petukhov, B.S., Krasnoschekov, E.A., and Protopopov, V.S. (1961). An investigation of heat transfer to fluids flowing in pipes under supercritical conditions. *1961 International Heat Transfer Conferences ASME*, Boulder, CO, USA, pp. 569–578.

4 Tanaka, H., Nishiwaki, N., Hirata, M. (1967). Turbulent heat transfer to supercritical carbon dioxide, *JSME Semi-International Symposium*, JSME, Tokyo, Japan. pp. 127–134.

5 Bourke, P.J., Pulling, D.J., Gill, L.E., and Denton, W.H. (1970). Forced convective heat transfer to turbulent CO_2 in the supercritical region. *International Journal of Heat and Mass Transfer* 13: 1339–1348. https://doi.org/10.1016/0017-9310(70)90074-8.

6 Shiralkar, B. and Griffith, P. (1970). The effect of swirl, inlet conditions, flow direction, and tube diameter on the heat transfer to fluids at supercritical pressure. *Journal of Heat Transfer* 92: 465. https://doi.org/10.1115/1.3449690.

7 Baskov, V.L. (1977). Heat transfer with turbulent flow of a liquid at supercritical pressure in tubes under cooling conditions. *High Temperature* 15: 81–86.

8 Kurganov, V.A. and Kaptilnyi, A.G. (1993). Flow structure and turbulent transport of a supercritical pressure fluid in a vertical heated tube under the conditions of mixed convection: experimental data. *International Journal of Heat and Mass Transfer* 36: 3383–3392.

9 Fewster, J., Jackson, J.D. (2004). Experiments on supercritical pressure convective heat transfer having relevance to SPWR. *2004 International Congress on Advances in Nuclear Power Plants – ICAPP'04*, Pittsburgh, PA, USA, pp. 537–551.

10 Cheng, X. and Schulenberg, T. (2001). Heat transfer at supercritical pressures – literature review and application to an HPLWR, Karlsruhe, Germany.

11 Pioro, I.L., Khartabil, H.F., and Duffey, R.B. (2004). Heat transfer to supercritical fluids flowing in channels – empirical correlations (survey). *Nuclear Engineering and Design* 230: 69–91. https://doi.org/10.1016/j.nucengdes.2003.10.010.

12 Deissler, R.G. (1954). Heat transfer and fluid friction for fully developed turbulent flow of air and supercritical water with variable fluid properties. *Transactions of the ASME* 76: 73–85.

13 Hall, W., Jackson, J., and Watson, A. (1968). Forced convection heat transfer to fluids at supercritical pressure. *Proceedings of the Institute of Mechanical Engineers* 182.

14 Jackson, J.D. (2013). Fluid flow and convective heat transfer to fluids at supercritical pressure. *Nuclear Engineering and Design* 264: 24–40. https://doi.org/10.1016/j.nucengdes.2012.09.040.

15 Jackson, J.D. (2017). Models of heat transfer to fluids at supercritical pressure with influences of buoyancy and acceleration. *Applied Thermal Engineering* 124: 1481–1491. https://doi.org/10.1016/j.applthermaleng.2017.03.146.

16 Bourassa, C. and Thomas, F.O. (2009). An experimental investigation of a highly accelerated turbulent boundary layer. *Journal of Fluid Mechanics* 634: 359. https://doi.org/10.1017/S0022112009007289.

17 Jackson, J.D., Cotton, M.A., and Axcell, B.P. (1989). Studies of mixed convection in vertical tubes. *International Journal of Heat and Fluid Flow* 10: 2–15. https://doi.org/10.1016/0142-727X(89)90049-0.

18 Jackson, J.D. (2015). Screening and correlating data on heat transfer to fluids at supercritical pressure. *Journal of Nuclear Engineering and Radiation Science* 2: 011001. https://doi.org/10.1115/1.4031378.

19 Bae, Y.Y., Kim, H.Y., and Kang, D.J. (2010). Forced and mixed convection heat transfer to supercritical CO_2 vertically flowing in a uniformly-heated circular tube. *Experimental Thermal and Fluid Science* 34: 1295–1308. https://doi.org/10.1016/j.expthermflusci.2010.06.001.

20 Wang, J., Li, J., and Jackson, J.D. (2004). A study of the influence of buoyancy on turbulent flow in a vertical plane passage. *International Journal of Heat and Fluid Flow* 25: 420–430. https://doi.org/10.1016/j.ijheatfluidflow.2004.02.008.

21 Harris, J.R., Lance, B.W., and Smith, B.L. (2015). Experimental validation data for computational fluid dynamics of forced convection on a vertical flat plate. *Journal of Fluids Engineering* 138: 011401. https://doi.org/10.1115/1.4031007.

22 Kim, D.E. and Kim, M.H. (2010). Experimental study of the effects of flow acceleration and buoyancy on heat transfer in a supercritical fluid flow in a circular tube. *Nuclear Engineering and Design* 240: 3336–3349. https://doi.org/10.1016/j.nucengdes.2010.07.002.

23 Kline, S.J., Reynolds, W.C., Schraub, F.A., and Runstadler, P.W. (1967). The structure of turbulent boundary layers. *Journal of Fluid Mechanics* 30: 741–773. https://doi.org/10.1017/S0022112067001740.

24 Tsuji, T. and Nagano, Y. (1989). Velocity and temperature measurements in a natural convection boundary layer along a vertical flat plate. *Experimental Thermal and Fluid Science* 2: 208–215. https://doi.org/10.1016/0894-1777(89)90035-6.

25 Tsuji, T. and Nagano, Y. (1988). Turbulence measurements in a natural convection boundary layer along a vertical flat plate. *International Journal of Heat and Mass Transfer* 31: 2101–2111. https://doi.org/10.1016/0017-9310(88)90120-2.

26 Pidaparti, S.R., McFarland, J.A., Mikhaeil, M.M. et al. (2015). Investigation of buoyancy effects on heat transfer characteristics of supercritical carbon dioxide in heating mode. *Journal of Nuclear Engineering and Radiation Science* 1: 1–10. https://doi.org/10.1115/1.4029592.

27 Bazargan, M., Fraser, D., and Chatoorgan, V. (2005). Effect of buoyancy on heat transfer in supercritical water flow in a horizontal round tube. *Journal of Heat Transfer* 127: 897–902. https://doi.org/10.1115/1.1929787.

28 Wang, J., Guan, Z., Gurgenci, H. et al. (2018). Computational investigations of heat transfer to supercritical CO_2 in a large horizontal tube. *Energy Conversion and Management* 157: 536–548. https://doi.org/10.1016/j.enconman.2017.12.046.

29 Adebiyi, G.A. and Hall, W.B. (1976). Experimental investigation of heat transfer to supercritical pressure carbon dioxide in a horizontal pipe. *International Journal of Heat and Mass Transfer* 19: 715–720.

30 Kim, H.Y., Kim, H., Song, J.H. et al. (2007). Heat transfer test in a vertical tubes using CO_2 at supercritical pressures. *Journal of Nuclear Science and Technology* 44: 285–293. https://doi.org/10.1080/18811248.2007.9711284.

31 Kim, H., Kim, H.Y., Song, J.H., and Bae, Y.Y. (2008). Heat transfer to supercritical pressure carbon dioxide flowing upward through tubes and a narrow annulus passage. *Progress in Nuclear Energy* 50: 518–525. https://doi.org/10.1016/j.pnucene.2007.11.065.

32 Zahlan, H., Groeneveld, D., and Tavoularis, S. (2015). Measurements of convective heat transfer to vertical upward flows of CO_2 in circular tubes at near-critical and supercritical pressures. *Nuclear Engineering and Design* 289: 92–107. https://doi.org/10.1016/j.nucengdes.2015.04.013.

33 Perkins, K.R. (1975). *Turbulence Structure in Gas Flows Relaminarizing by Heating*. University of Arizona http://hdl.handle.net/10150/565327.

34 Kays, W.M., Crawford, M.E., and Weigand, B. (2004). *Convective Heat and Mass Transfer*, 4e. Boston: McGraw-Hill Higher Education.

35 Davidson, P. (2015). *Turbulence: An Introduction for Scientists and Engineers*, 2e. Oxford University Press.

36 Jackson, J.D. (2012). Heat transfer deterioration in tubes caused by bulk flow acceleration due to thermal and frictional influences. *2012 International Congress on Advances in Nuclear Power Plants – ICAPP'12*, ANS, pp. 223–230.

37 Hunt, S. and Heister, S. (2017). Thermoacoustic oscillations in multipath heated fuel circuits. *Journal of Heat Transfer* 139: 091801. https://doi.org/10.1115/1.4036597.

38 Linne, D., Meyer, M., Edwards, T. (1997). Evaluation of heat transfer and thermal stability of supercritical JP-7 fuel. *33rd Joint Propulsion Conference and Exhibit, American Institute of Aeronautics and Astronautics, Reston*, Virigina. https://doi.org/10.2514/6.1997-3041

39 Wang, H., Zhou, J., Pan, Y., and Wang, N. (2016). Experimental investigation on the characteristics of thermo-acoustic instability in hydrocarbon fuel at supercritical pressures. *Acta Astronautica* 121: 29–38. https://doi.org/10.1016/j.actaastro.2015.12.047.

40 Li, H., Qin, J., Jiang, Y. et al. (2019). Experimental study on the thermodynamic characteristics of the high temperature hydrocarbon fuel in the cooling channel of the hypersonic vehicle. *Acta Astronautica* 155: 63–79. https://doi.org/10.1016/j.actaastro.2018.11.021.

41 Zuber, N. (1966). An analysis of thermally induced flow oscillations in the near-critical and super-critical thermodynamic region, Hunstville, AL. https://ntrs.nasa.gov/archive/nasa/casi.ntrs.nasa.gov/19670004205.pdf (accessed 4 September 2020).

42 Krasnoshchekov, E.A. and Protopopov, V.S. (1966). Experimental study of heat exchange in carbon dioxide in the supercritical region at high temperature drops. *High Temperature* 4: 375–382.

43 Stewart, E., Stewart, P., and Watson, A. (1973). Thermo-acoustic oscillations in forced convection heat transfer to supercritical pressure water. *International Journal of Heat and Mass Transfer* 16: 257–270.

44 Kurganov, V.A., Zeigarnik, Y.A., and Maslakova, I.V. (2012). Heat transfer and hydraulic resistance of supercritical-pressure coolants. Part I: specifics of thermophysical properties of supercritical pressure fluids and turbulent heat transfer under heating conditions in round tubes (state of the art). *International Journal of Heat and Mass Transfer* 55: 3061–3075. https://doi.org/10.1016/j.ijheatmasstransfer.2012.01.031.

45 Linne, D., Meyer, M., Braun, D., Keller, D. (2000). Investigation of instabilities and heat transfer phenomena in supercritical fuels at high heat flux and temperatures. *36th AIAA/ASME/SAE/ASEE Joint Propulsion Conference and Exhibit, American Institute of Aeronautics and Astronautics, Reston*, Virigina. https://doi.org/10.2514/6.2000-3128.

46 Hines, W.S. (1962). Pressure oscillations associated with heat transfer to hydrocarbon fluids at supercritical pressures and temperatures. *ARS Journal* 32: 361–366. https://doi.org/10.2514/8.6020.

47 Petukhov, B.S. (1968). Heat transfer in a single-phase medium under supercritical conditions (survey). *High Temperature* 6: 696–709.

48 Sevastyanov, V.V., Sinitsin, A.T., and Yakaitis, F.L. (1980). Study of heat exchanger process in supercritical region of parameters under conditions of high-frequency pressure oscillations. *High Temperature* 18: 433–439.

49 Ackerman, J.W. (1970). Pseudoboiling heat transfer to supercritical pressure water in smooth and ribbed tubes. *Journal of Heat Transfer* 92: 490. https://doi.org/10.1115/1.3449698.

50 Pioro, I. and Mokry, S. (2011). Heat transfer to fluids at supercritical pressures. In: *Heat Transfer: Theoretical Analysis, Experimental Investigations and Industrial Systems* (ed. A. Belmiloudi), 481–504. InTech https://doi.org/10.5772/1756.

51 Swenson, H.S., Carver, J.R., and Kakarala, C.R. (1965). Heat transfer to supercritical water in smooth-bore tubes. *Journal of Heat Transfer* 87: 477–484.

52 Wang, J., Li, H., Yu, S., and Chen, T. (2011). Investigation on the characteristics and mechanisms of unusual heat transfer of supercritical pressure water in vertically-upward tubes. *International Journal of Heat and Mass Transfer* 54: 1950–1958. https://doi.org/10.1016/j.ijheatmasstransfer.2011.01.008.

53 Oh, H.K. and Son, C.H. (2010). New correlation to predict the heat transfer coefficient in-tube cooling of supercritical CO_2 in horizontal macro-tubes. *Experimental Thermal and Fluid Science* 34: 1230–1241. https://doi.org/10.1016/j.expthermflusci.2010.05.002.

54 Cheng, L., Ribatski, G., and Thome, J.R. (2008). Analysis of supercritical CO_2 cooling in macro- and micro-channels. *International Journal of Refrigeration* 31: 1301–1316. https://doi.org/10.1016/j.ijrefrig.2008.01.010.

55 Yoon, S.H., Kim, J.H., Hwang, J.W. et al. (2003). Heat transfer and pressure drop characteristics during the in-tube cooling process of carbon dioxide in the supercritical region. *International Journal of Refrigeration* 26: 857–864. https://doi.org/10.1016/S0140-7007(03)00096-3.

56 Fronk, B.M. and Garimella, S. (2011). Water-coupled carbon dioxide microchannel gas cooler for heat pump water heaters: part I – experiments. *International Journal of Refrigeration* 34: 7–16. https://doi.org/10.1016/j.ijrefrig.2010.05.004.

57 Hyder, M.B. and Fronk, B.M. (2018). Simulation of thermal hydraulic performance of multiple parallel micropin arrays for concentrating solar thermal applications with supercritical carbon dioxide. *Solar Energy* 164: 327–338. https://doi.org/10.1016/j.solener.2018.02.035.

58 Huai, X.L., Koyama, S., and Zhao, T.S. (2005). An experimental study of flow and heat transfer of supercritical carbon dioxide in multi-port mini channels under cooling conditions. *Chemical Engineering Science* 60: 3337–3345. https://doi.org/10.1016/j.ces.2005.02.039.

59 Pitla, S., Robinson, D., Groll, E., and Ramadhyani, S. (1998). Heat transfer from supercritical carbon dioxide in tube flow: a critical review. *HVAC&R Research* 4: 281–301. https://doi.org/10.1080/10789669.1998.10391405.

60 Rao, N.T., Oumer, A.N., and Jamaludin, U.K. (2016). State-of-the-art on flow and heat transfer characteristics of supercritical CO_2 in various channels. *Journal of Supercritical Fluids* 116: 132–147. https://doi.org/10.1016/j.supflu.2016.05.028.

61 Pioro, I.L. and Duffey, R.B. (2007). *Heat Transfer and Hydraulic Resistance at Supercritical Pressures in Power Engineering Applications*, 1e. ASME.

62 Jiang, P.X., Zhao, C.R., Shi, R.F. et al. (2009). Experimental and numerical study of convection heat transfer of CO_2 at super-critical pressures during cooling in small vertical tube. *International Journal of Heat and Mass Transfer* 52: 4748–4756. https://doi.org/10.1016/j.ijheatmasstransfer.2009.06.014.

63 Liao, S.M. and Zhao, T.S. (2002). Measurements of heat transfer coefficients from supercritical carbon dioxide flowing in horizontal mini/micro channels. *Journal of Heat Transfer* 124: 413. https://doi.org/10.1115/1.1423906.

64 Jackson, J.D. and Hall, W.B. (1979). Forced convection heat transfer to fluids at supercritical pressure. In: *Turbulent Forced Convection in Channels and Bundles: Theory and Applications to Heat Exchangers and Nuclear Reactors* (eds. B. Spalding and S. Kakac), 563–611. Hemisphere Publishing Corporation.

65 Petukhov, B. and Polyakov, A. (1988). Heat Transfer in Turbulent Mixed Convection. Hemisphere Publishing Corporation.

66 Wang, K.Z., Xu, X.X., Liu, C. et al. (2017). Experimental and numerical investigation on heat transfer characteristics of supercritical CO_2 in the cooled helically coiled tube. *International Journal of Heat and Mass Transfer* 108: 1645–1655. https://doi.org/10.1016/j.ijheatmasstransfer.2017.01.004.

67 Yamagata, K., Nishikawa, K., Hasegawa, S. et al. (1972). Forced convection heat transfer to supercritical water flowing in tubes. *International Journal of Heat and Mass Transfer* 15: 2575–2593.

68 Kirillov, P.L., Yur'ev, Y.S., and Bobkov, V.P. (1990). *Handbook of Thermal-Hydraulics Calculations* (in Russian). Moscow, Russia: Energoatomizdat Publishing House.

69 Jackson, J.D. (2002). Consideration of the heat transfer properties of supercritical water in connection with the cooling of advanced nuclear reactors. *13th Pacific Basin Nuclear Conference*, Shenzhen City, China.

70 Pioro, I., Mokry, S., and Draper, S. (2011). Specifics of thermophysical properties and forced-convective heat transfer at critical and supercritical pressures. *Reviews in Chemical Engineering* 27: 191–214. https://doi.org/10.1515/REVCE.2011.501.

71 Groll, E.A. and Kim, J.H. (2007). Review article: review of recent advances toward transcritical CO_2 cycle technology. *HVAC&R Research* 13: 499–520.

72 Dang, C. and Hihara, E. (2012). Predicting the cooling heat transfer coefficient of supercritical CO_2 with a small amount of entrained lubricating oil by using the neural network method. *International Journal of Refrigeration* 35: 1130–1138. https://doi.org/10.1016/j.ijrefrig.2012.01.022.

73 Petrov, N.E. and Popov, V.N. (1989). Heat transfer and turbulent flow of cooled helium with supercritical state parameters in a tube. *High Temperature* 27: 396–403.

74 Son, C.-H. and Park, S.-J. (2006). An experimental study on heat transfer and pressure drop characteristics of carbon dioxide during gas cooling process in a horizontal tube. *International Journal of Refrigeration* 29: 539–546. https://doi.org/10.1016/j.ijrefrig.2005.10.010.

75 Pitla, S.S., Groll, E.A., and Ramadhyani, S. (2002). New correlation to predict the heat transfer coefficient during in-tube cooling of turbulent supercritical CO_2. *International Journal of Refrigeration* 25: 887–895.

76 Huai, X. and Koyama, S. (2007). An experimental study of carbon dioxide condensation in mini channels. *Journal of Thermal Science* 13: 358–365.

77 Dang, C. and Hihara, E. (2004). In-tube cooling heat transfer of supercritical carbon dioxide. Part 1. Experimental measurement. *International Journal of Refrigeration* 27: 736–747. https://doi.org/10.1016/j.ijrefrig.2004.04.018.

78 Kuang, G., Ohadi, M., and Dessiatoun, S. (2008). Semi-empirical correlation of gas cooling heat transfer of supercritical carbon dioxide in microchannels. *HVAC&R Research* 14: 861–870. https://doi.org/10.1080/10789669.2008.10391044.

79 Liu, Z.-B., He, Y.-L., Yang, Y.-F., and Fei, J.-Y. (2014). Experimental study on heat transfer and pressure drop of supercritical CO_2 cooled in a large tube. *Applied Thermal Engineering* 70: 307–315. https://doi.org/10.1016/j.applthermaleng.2014.05.024.

80 Zhao, C.R. and Jiang, P.X. (2011). Experimental study of in-tube cooling heat transfer and pressure drop characteristics of R134a at supercritical pressures. *Experimental Thermal and Fluid Science* 35: 1293–1303. https://doi.org/10.1016/j.expthermflusci.2011.04.017.

81 Garimella, S., Mitra, B., Andresen, U.C. et al. (2015). Heat transfer and pressure drop during supercritical cooling of HFC refrigerant blends. *International Journal of Heat and Mass Transfer* 91: 477–493. https://doi.org/10.1016/j.ijheatmasstransfer.2015.07.024.

82 Valueva, E.P. and Kulagin, E.N. (2012). A numerical study of heat transfer and turbulent flow of carbon dioxide in a tube at supercritical pressure. *Thermal Engineering* 59: 40–47. https://doi.org/10.1134/S0040601512010132.

83 Li, X., Huai, X., Cai, J. et al. (2011). Convective heat transfer characteristics of China RP-3 aviation kerosene at supercritical pressure. *Applied Thermal Engineering* 31: 2360–2366. https://doi.org/10.1016/j.applthermaleng.2011.03.036.

84 Sharabi, M. and Ambrosini, W. (2009). Discussion of heat transfer phenomena in fluids at supercritical pressure with the aid of CFD models. *Annals of Nuclear Energy* 36: 60–71. https://doi.org/10.1016/j.anucene.2008.10.006.

85 Zhang, P., Huang, Y., Shen, B., and Wang, R.Z. (2011). Flow and heat transfer characteristics of supercritical nitrogen in a vertical mini-tube. *International Journal of Thermal Sciences* 50: 287–295. https://doi.org/10.1016/j.ijthermalsci.2010.06.014.

86 Jiang, P.X., Zhang, Y., Zhao, C.R., and Shi, R.F. (2008). Convection heat transfer of CO_2 at supercritical pressures in a vertical mini tube at relatively low Reynolds numbers. *Experimental Thermal and Fluid Science* 32: 1628–1637. https://doi.org/10.1016/j.expthermflusci.2008.05.006.

87 Meng, H.Y., Wang, S.Z., Zhou, L. et al. (2014). Influence of inlet mass flow rate on heat transfer of supercritical liquefied natural gas in horizontal tubes. *Advances in Materials Research* 960–961: 433–437. https://doi.org/10.4028/www.scientific.net/AMR.960-961.433.

88 Yoo, J.Y. (2012). The turbulent flows of supercritical fluids with heat transfer. *Annual Review of Fluid Mechanics* 45: 495–525. https://doi.org/10.1146/annurev-fluid-120710-101234.

89 Dang, C. and Hihara, E. (2004). In-tube cooling heat transfer of supercritical carbon dioxide. Part 2. Comparison of numerical calculation with different turbulence models. *International Journal of Refrigeration* 27: 748–760. https://doi.org/10.1016/j.ijrefrig.2004.04.017.

90 Cao, X.L., Rao, Z.H., and Liao, S.M. (2011). Laminar convective heat transfer of supercritical CO_2 in horizontal miniature circular and triangular tubes. *Applied Thermal Engineering* 31: 2374–2384. https://doi.org/10.1016/j.applthermaleng.2011.03.038.

91 Dang, C. and Hihara, E. (2010). Numerical study on in-tube laminar heat transfer of supercritical fluids. *Applied Thermal Engineering* 30: 1567–1573. https://doi.org/10.1016/j.applthermaleng.2010.03.010.

92 Du, Z., Lin, W., and Gu, A. (2012). Numerical study on supercritical CH4/N2 cooling in a horizontal tube. *ASME 2012 Heat Transfer Summer Conference*, ASME, pp. 1003–1012.

93 Du, Z., Lin, W., and Gu, A. (2010). Numerical investigation of cooling heat transfer to supercritical CO_2 in a horizontal circular tube. *Journal of Supercritical Fluids* 55: 116–121. https://doi.org/10.1016/j.supflu.2010.05.023.

94 Du, Z., Lin, W., and Gu, J. (2012). Numerical investigation for heat transfer of supercritical CO_2 cooled in a vertical circular tube. *Heat Transfer Engineering* 33: 905–911. https://doi.org/10.1080/01457632.2012.654452.

95 Fronk, B.M. and Garimella, S. (2011). Water-coupled carbon dioxide microchannel gas cooler for heat pump water heaters: part II – model development and validation. *International Journal of Refrigeration* 34: 17–28. https://doi.org/10.1016/j.ijrefrig.2010.05.012.

96 Goodman, C., Fronk, B.M., and Garimella, S. (2011). Transcritical carbon dioxide microchannel heat pump water heaters: part II – system simulation and optimization. *International Journal of Refrigeration* 34: 870–880. https://doi.org/10.1016/j.ijrefrig.2010.12.001.

97 Dittus, F.W. and Boelter, L.M.K. (1930). *Heat Transfer in Automobile Radiators of the Tubular Type*. Berkeley: University of California.

98 Gnielinski, V. (1976). New equation for heat and mass transfer in turbulent pipe and channel flow. *International Journal of Chemical Engineering* 16: 354–368.

99 Filonenko, G. (1954). Hydraulic resistance in pipes. *Teploenergetika* 1: 40–44.

100 Krasnoshchekov, E.A., Kuraeva, I.V., and Protopopov, V.S. (1970). Local heat transfer of carbon dioxide at supercritical pressure under cooling conditions. *Teplofizika Vysokikh Temperatur* 7: 922–930.

101 Petukhov, B.S. and Kirillov, V.V. (1958). On heat exchange at turbulent flow of liquid in pipes. *Teploenergetika* 4: 63–68.

102 Ghajar, A.J. and Asadi, A. (1986). Improved forced convective heat-transfer correlations for liquids in the near-critical region. *AIAA Journal* 24: 2030–2037.

103 Pitla, S., Groll, E., and Ramadhyani, S. (2001). Convective heat transfer from in-tube cooling of turbulent supercritical carbon dioxide: part 2 – experimental data and numerical predictions. *HVAC&R Research* 7: 367–382. https://doi.org/10.1080/10789669.2001.10391281.

104 Pitla, S., Groll, E., and Ramadhyani, S. (2001). Convective heat transfer from in-tube flow of turbulent supercritical carbon dioxide: part 1 – numerical analysis. *HVAC&R Research* 7: 345–366. https://doi.org/10.1080/10789669.2001.10391280.

105 Kuang, G., Ohadi, M., and Zhao, Y. (2003). Experimental study of miscible and immiscible oil effects on heat transfer coefficients and pressure drop in microchannel gas cooling of supercritical CO_2, *ASME SUmmer Heat Transfer Conference ASME*, Las Vegas, NV, pp. 671–675. https://doi.org/10.1115/HT2003-47473.

106 Yun, R., Hwang, Y., and Radermacher, R. (2007). Convective gas cooling heat transfer and pressure drop characteristics of supercritical CO_2/oil mixture in a minichannel tube. *International Journal of Heat and Mass Transfer* 50: 4796–4804. https://doi.org/10.1016/j.ijheatmasstransfer.2007.03.018.

107 Zhao, C.R. and Jiang, P.X. (2014). Predictions of in-tube cooling pressure drops for CO_2 mixed with lubricating oil at supercritical pressures. *Applied Thermal Engineering* 73: 527–536. https://doi.org/10.1016/j.applthermaleng.2014.07.072.

108 Zhao, C.R., Jiang, P.X., and Zhang, Y.W. (2011). Flow and convection heat transfer characteristics of CO_2 mixed with lubricating oil at super-critical pressures in small tube during cooling. *International Journal of Refrigeration* 34: 29–39. https://doi.org/10.1016/j.ijrefrig.2010.06.013.

109 Jung, J.Y. and Yun, R. (2013). Prediction of gas cooling heat transfer coefficients for CO_2-oil mixtures. *International Journal of Refrigeration* 36: 129–135. https://doi.org/10.1016/j.ijrefrig.2012.09.005.

110 Dang, C., Iino, K., and Hihara, E. (2008). Study on two-phase flow pattern of supercritical carbon dioxide with entrained PAG-type lubricating oil in a gas cooler. *International Journal of Refrigeration* 31: 1265–1272. https://doi.org/10.1016/j.ijrefrig.2008.01.014.

4

Boiling Flow and Heat Transfer of CO_2 in an Evaporator

Haruhiko Yamasaki

Department of Mechanical Engineering, Osaka Prefecture University, Sakai, Japan

4.1 Introduction

In the cooling process, fluids absorb a great deal of heat from a surrounding surface with evaporation and discharge heat to another surrounding surface with condensation. In order for efficient cooling, it is necessary that the fluid which has a large latent heat supplies and evaporates itself to a cooling space. Therefore, it is essential in the cooling process that the liquid substance should evaporate in to a gaseous phase and return to the liquid phase at the end of the process. Here, refrigeration plays an important role. That is, in general, refrigeration temperature can be decided by the saturation temperature of the refrigerant. The saturation temperature of the refrigerant is the thermo-physical property and obeys the vapor pressure curve. For example, in the case of water, it can be evaporated at 100°C under atmospheric pressure. However, water evaporates at 0.01°C at its triple point (6.1×10^2 Pa). It can be said that, when water is used as a refrigerant, it can be cooled to 0.01°C. In other words, by using refrigerants at low triple point, lower temperature can be realized. In general refrigeration systems, Freon-based refrigerant (Chlorofluorocarbons [CFCs] and Hydrochlorofluorocarbons [HCFCs]) and alternative Freon-based refrigerant (Hydrofluorocarbon [HFCs]) have been used for many decades [1–4]. However, they are being replaced by natural working fluid due to its high global warming effect [5]. From the view point of preventing global warming and protecting the ozone layer, natural refrigerants have been the subject of much attention [6, 7]. The representative natural refrigerants are shown in Table 4.1.

As shown in Table 4.1, to date, there are various natural refrigerants available. Ammonia as a refrigerant has so far been applied to heat pump and refrigeration systems in various industrial fields. Propane and Isobutane refrigerants have been tried for application in domestic refrigeration, and CO_2, as a refrigerant, has been applied to water heaters in recent years. The aforementioned refrigerants have had many successes in industrial, business and household use [9]. However, there are still many refrigerants which are not technically satisfying and safe as a reliable refrigerant. For example, Ammonia has toxicity and is flammable, and is still not suitable for various technical applications for refrigeration [10]. Many of the hydrocarbon-based refrigerants are also flammable which results in the lack

Transcritical CO_2 Heat Pump: Fundamentals and Applications,
First Edition. Xin-Rong Zhang and Hiroshi Yamaguchi.

Table 4.1 Features of natural refrigerants.

	Chemical formula	Molecular weight (kg kmol^{-1})	Boiling point (at atmosphere)	Latent heat of vaporization (kJ kg^{-1} at 25°C)	Safety
R290	C_3H_8	44.1	−42.09	335.3	Flammable
R600	C_4H_{10}	58.1	−0.55	360.9	Flammable
R600a	$CH(CH_3)_3$	58.1	−11.67	329.1	Flammable
R744	CO_2	44.01	−78.4	119.6	High pressure
R717	NH_3	17.03	−33.33	570.2	Toxicity Flammable
R718	H_2O	18	0.01	2442	—

R290: Propane; R600: Butane; R600a: Isobutane; R744: Carbon dioxide; R717: Ammonia; R718: Water (The properties are referred by Refprop ver. 9.1 [8]).

of requisite protection in flammable equipment. Water, when used as a refrigerant, is very safe, but it is not suitable for vapor compression refrigerator systems due to its extremely low vapor pressure and extremely low density. Among natural refrigerants, CO_2 is non-toxic and non-flammable, and has low global warming potential and zero ozone layer depletion as well. CO_2 can be expected to be utilized to save energy due to its high liquid density and high operational pressure. From the above-mentioned advantages, CO_2 is considered as a good refrigerant to replace conventional Freon and alternative Freon [6, 11–14].

In a conventional heat pump system using fluorocarbon or ammonia as a refrigerant, since the condensation temperature is constant in the condenser, an irreversible loss occurs in the heat exchanger between the heating medium and the refrigerant, so that in effect the heat exchange efficiency decreases. On the other hand, in a hot water heater using CO_2, although it is necessary to have higher operational pressure, the temperature changes continuously in the heat exchanger between the heating medium and the refrigerant due to the state of CO_2 being operated to a supercritical state of 7.38 MPa and 31.1°C. It has been a great success in high-temperature hot water supply using CO_2 as working fluid with high efficiency and reliability [15–17].

As for industrial refrigeration technology, CO_2 may not be suitable for medium- or large-scale refrigeration facilities with single refrigeration cycles, since the CO_2 cycle forms a cycle in a high pressure state. Therefore, CO_2/NH_3 cascade refrigeration systems have been in use in practice [18–21]. The schematic diagram of the CO_2/NH_3 cascade heat pump system is shown in Figure 4.1. This system consists of the NH_3 refrigeration cycle as a high temperature cycle and the CO_2 refrigeration cycle as a low temperature cycle, which are connected by a cascade condenser. The condensation temperature can be achieved at 30–50°C in the high temperature cycle whereas the condensation temperature can be at −50°C in the low temperature cycle [19]. When pressure is lowered below the triple point of −56.6°C at 0.518 MPa in the low temperature cycle in an expansion process, dry ice forms and blocks the evaporator, resulting in blockage with large heat loss, and inducing lower system efficiency. For the above reason, CO_2/NH_3 cascade heat pump

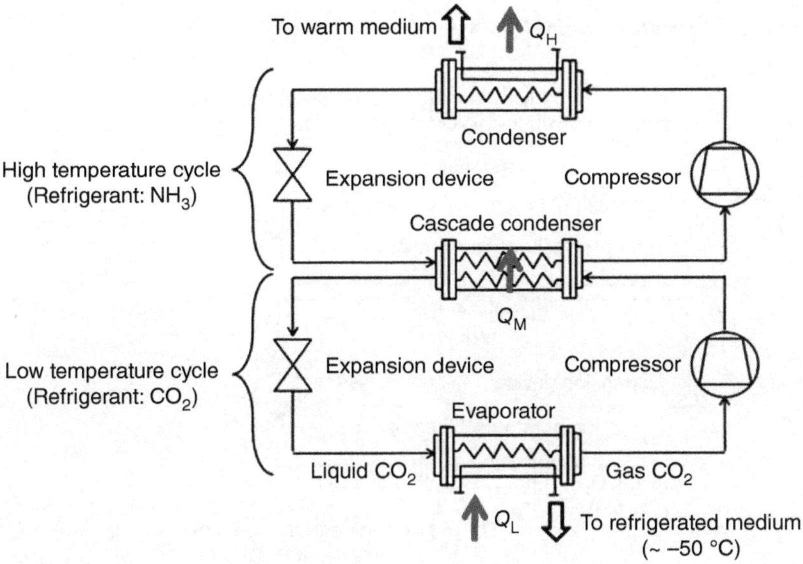

Figure 4.1 Schematic diagram of a CO_2/NH_3 cascade refrigeration system.

systems operate for cooling in the range of —-50°C in lower temperature cycles, not below −56°C where dry ice may be produced in the cooling process.

For the cryogenic temperature range below −50°C which is often used for the storage of big fish such as tuna, cooling semiconductors, storing medical materials, etc., it is difficult to utilize the refrigeration cycle with natural refrigerant as shown in Table 4.1. Although HCFC22 has been used as a refrigerant for cryogenic refrigeration systems until recently [22], it has been decided not to be manufactured after 2020 by the Montreal Protocol. In addition, HFC is also regulated by the Kyoto Protocol (1997). In addition to the ultra-low refrigeration system using the air cycle, a CO_2/CO_2 cascade heat pump system has been proposed for achieving cryogenic temperatures [23, 24]. The schematic diagram of a CO_2/CO_2 cascade refrigeration system is shown in Figure 4.2 as a reference. The system consists of two CO_2 heat pump systems in high and low temperature sides, which are connected by a cascade condenser. The condensation temperature can be around 100–130°C in the high temperature side and the condensation temperature of −60 to −70°C in the low temperature side. In the CO_2/NH_3 cascade heat pump system, the liquid CO_2 absorbs heat in the evaporation process as mentioned above. On the other hand, in the CO_2/CO_2 cascade heat pump system, the solid CO_2 (as dry ice) absorbs heat in a sublimation process. Although it is necessary to design the evaporator to avoid the blockage phenomena of dry ice, the CO_2/CO_2 refrigeration system has great potential to realize cryogenic temperatures with high heat transfer efficiency with dry ice sublimation.

In this chapter, liquid CO_2 with a boiling heat transfer below 0°C in a representative CO_2/NH_3 cascade heat pump system, and the dry ice sublimation heat transfer below the triple point of CO_2 in the CO_2/CO_2 cascade heat pump system technology are introduced and discussed to some extent.

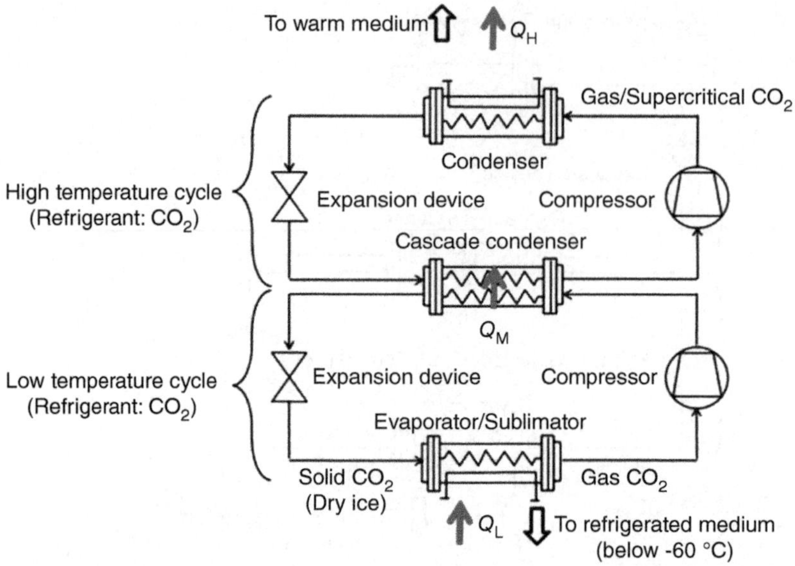

Figure 4.2 Schematic diagram of a CO_2/CO_2 cascade refrigeration system.

4.2 Boiling Heat Transfer of Liquid CO_2 in an Evaporator

Since carbon dioxide as a refrigerant has a low critical point at high pressure, CO_2 evaporates with much higher reduced pressure than other refrigerants. Based on the unique thermo-physical properties, heat transfer with associated phenomena is quite different from other known refrigerants. In particular, CO_2 as a refrigerant associated with its high vapor density, low viscosity and low surface tension coefficient has been expected to process excellent heat transfer properties. The variations of the surface tension coefficient and viscosity at the saturation temperature in the evaporation process (−60 to 30°C) are shown in Figures 4.3 and 4.4, respectively. For comparison with other conventional refrigerants, their properties are also shown in the figures. The physical properties are obtained and calculated from the database PROPATH [25]. As shown in Figures 4.3 and 4.4, the surface tension of CO_2 is lower than other refrigerants. For example, at 0°C, the surface tension of CO_2 is 16.6%, 44.3%, and 39.5% to that of the refrigerant R717, R290, and R134a, respectively. In addition, the viscosity of saturated liquid CO_2 at 0°C is 59.0% and 36.8% to that of the refrigerants R717 and R1334a, respectively. The lower surface tension and viscosity of CO_2 at low temperature may cause dryout at the early stage of boiling heat transfer. On the other hand, low surface tension and low viscosity promote the boiling bubble to form and detach, resulting in higher boiling heat transfer. Figure 4.5 depicts the liquid-vapor density ratio at the saturation temperature in the evaporation process (−50 to 0°C). The density ratio of liquid and vapor is 9.5 at 0°C (the saturation pressure of 3.48 MPa), whereas it is 29.0 at −30°C (saturation pressure of 1.43 MPa). That is to say, the void fraction increases with temperature decreasing at the same vapor quality. At the same temperature, the density ratio of saturated liquid to saturated vapor of CO_2 is smallest among other refrigerants. For example, at 0°C,

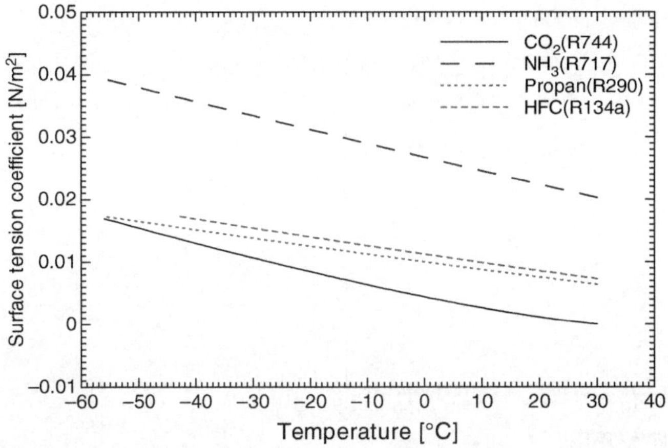

Figure 4.3 Relation between surface tension coefficient of refrigerants and evaporation temperature.

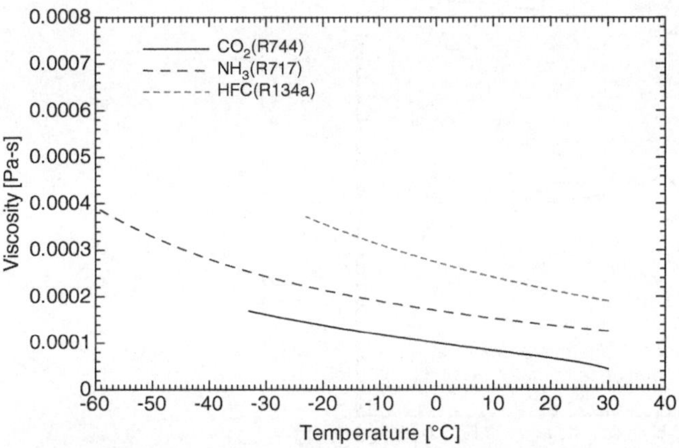

Figure 4.4 Relation between viscosity of saturated liquid of refrigerants and evaporation temperature.

the density ratio of CO_2 is only 5.1%, 10.6%, 7.0%, and 18.4% to that of the refrigerants R717, R134a, R600, and R290, respectively (Figure 4.6).

Analyzing boiling flow and boiling heat transfer in boiling two-phase flow patterns is very important since the boiling heat transfer is determined by boiling bubble and liquid-vapor flow patterns. Compared to conventional refrigerants, the experimental visualizations of boiling CO_2 flow are limited due to its high saturation pressure. Yun and Kim [26] showed the flow pattern at an evaporation temperature of 5.3°C in a rectangular head 2 mm × 16 mm, where the range of mass flux and heat flux are 217–1000 kg m^{-2} s^{-1} and 2–250 kw m^{-2}, respectively. Pettersen [27] showed the flow pattern at an evaporation temperature of 20°C in a 0.98 mm I.D. tube and heat flux of 13 kw m^{-2}. Their visualization results are shown in Figure 4.7. As the mass flux increases, the flow transition of intermittent-bubbly flow occurs at a lower vapor quality due to a lower surface tension. Also

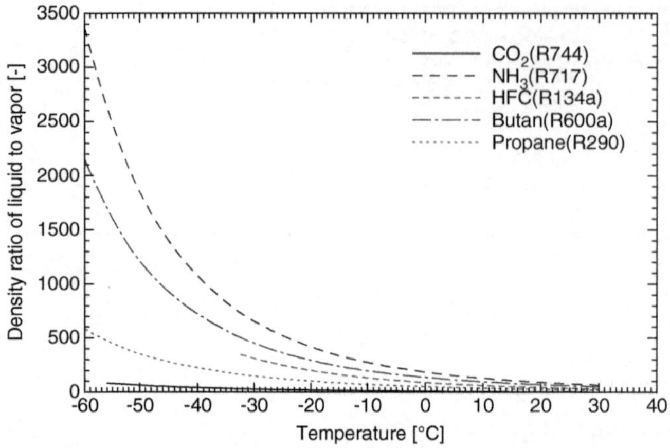

Figure 4.5 Relation between density ratio of saturated liquid to saturated vapor and evaporation temperature.

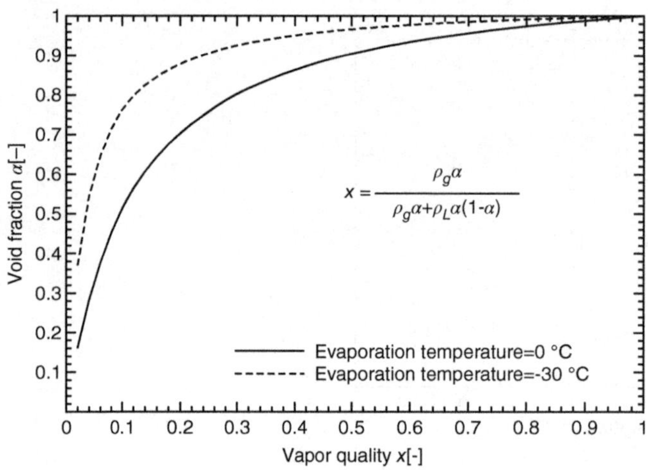

Figure 4.6 Relation between vapor quality and void fraction at evaporation temperature of −30 and 0°C.

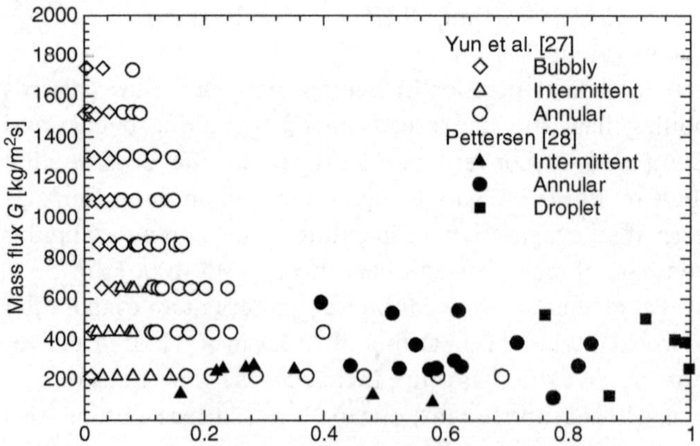

Figure 4.7 Flow region map with respect to mass flux G and vapor quality x.

as shown in Figure 4.7, when mass flux is larger than 868 kg m^{-2} s^{-1}, transit flow bubbly is directed to annular flow without having the intermittent flow. This can be caused by different heating conditions. Observations by Pettersen showed that intermittent flow occurred at low mass flow at vapor quality range of intermittent – annular flow pattern at higher mass flux. Then, droplet flow occurs at higher vapor quality. By comparing both results, the annular flow region shows good agreement with major flow pattern at high mass flux.

For the prediction of CO_2 flow pattern, the method proposed by Thome and EI Hajal [28] is often used. The proposed method is an updated version of flow pattern by Kattan *et al.* [29], which is based on a flow map for conventional refrigerants such as R134a, R123, R402a, R404a, and R502. R. Yun and Y. Kim [26] and Cheng *et al.* [30] have developed a new flow pattern map for CO_2. The prediction results put forward by R. Yun and Y. Kim [26] and the experimental result given by Pettersen [27] is shown in Figure 4.7. There is a reasonable agreement of the trend of intermittent to annular flow as shown in Eq. (4.1):

$$j_g/\alpha = C_0 j + f(v_\infty, \alpha) \tag{4.1}$$

where j and j_g are mean volumetric flux and superficial vapor velocity respectively, C_o is distribution parameter and $f\,(v_\infty, \alpha)$ indicates relative bubble velocity as a function of bubble rising velocity v_∞ and void fraction α. The CO_2 two-phase flow pattern maps demonstrate that the flow pattern visualization of CO_2 is limited significantly. It can be thought that the visualization at low evaporation temperature is required for better understanding of boiling heat transfer in an evaporator.

Before taking into account the results of heat transfer of CO_2 at lower saturation temperature, typical results of CO_2 heat transfer at different tube diameters are shown in Figure 4.8. Yun *et al.* [31] experimentally investigated CO_2 heat transfer in a stainless-steel tube of 6.0 mm I.D. tube with a heated length of 1.4 m. The decreasing trend of heat transfer coefficient with vapor quality is a general trend. In the quality range of 0.2–0.5, the heat transfer coefficient is independent from mass flux. This is probably due to a dominance of nucleate boiling. Thermophysical properties of CO_2 such as a lower surface tension, a lower viscosity and a lower density ration of liquid and vapor caused the dominance of nucleate boiling at lower vapor quality. The drop in heat transfer coefficient with vapor quality may be caused

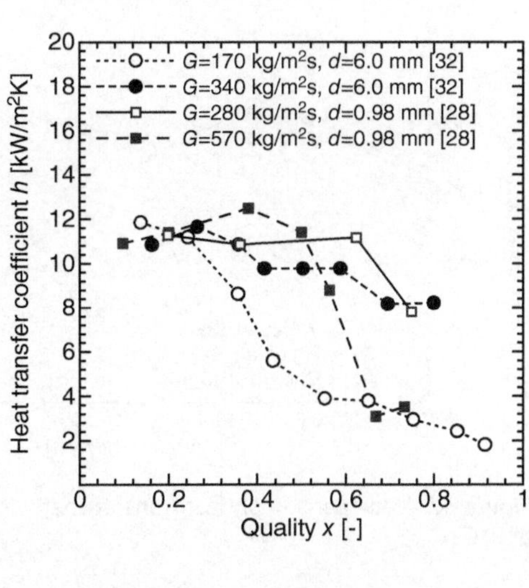

Figure 4.8 A comparison between the boiling heat transfer of CO_2 in macro channel [31] and micro channel [27] at $q = 10\,\text{kW m}^{-2}$ and $T_{sat} = 10\,°\text{C}$.

by partial dryout of liquid film. Pettersen [27] experimentally investigated CO_2 heat transfer in a quartz glass microchannel tube of 0.98 mm I.D. tube with 0.5m length. The results of Pettersen [27] show the same trend of that by Yun *et al.* [31]. However, the drop in the heat transfer coefficient with increasing vapor quality shift to the lower side of vapor quality as shown in Figure 4.8. This may be caused by nucleate boiling which is the dominant mechanism for evaporation in micro-scale with a small convection boiling contribution. For that reason, in the high mass flux region, the liquid film flowing along the heated tube should be evaporated and disappear, and dryout then occurs at low vapor quality.

Referring to Figure 4.6, the state of flow properties are verified, where the relationship between the void fraction and the quality is calculated by the equation of state for the liquid-vapor density difference, assuming the homogeneous flow with the slip ratio of 1. It is noted that the surface tension and the viscosity for the actual relation between the void ratio and the quality are both taken into account in the verification. In general, in the annular flow region, the boiling heat transfer increases with the increase of mass flow rate, since convective boiling heat transfer becomes dominant. In addition to the density difference of liquid and vapor, CO_2 has a smaller surface tension as well as associated viscosity change in comparison with conventional refrigerants, so that the film exfoliation can be assumed to be promoted, where the dryout might occur at relatively low quality. In effect, when CO_2 is used, it is considered for the heat exchanging process to have a high heat transfer coefficient at low saturation temperature. Owing to the reasons described above, there is a very strong relationship between the quality and the boiling phenomenon. Various researchers have explained the relationship between quality and boiling heat transfer as described below [32–38].

Figures 4.9 and 4.10 show the relationship between the quality and the heat transfer coefficient around the saturation temperature of 0 and −30°C in the past representative works. Table 4.2 shows the experimental data sources with each associated condition. As can be seen in Figure 4.9 of representative works, the heat transfer coefficient decreases as the quality increases. This is a rather peculiar phenomenon of CO_2 that is unlike the tendency

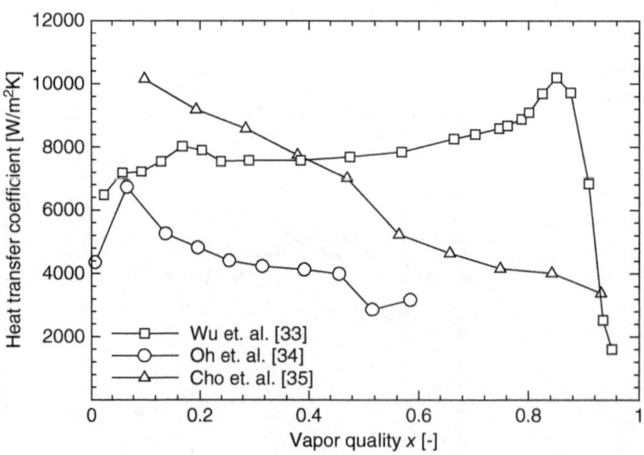

Figure 4.9 Comparison of heat transfer coefficient vs vapor quality at saturation temperature of 0°C.

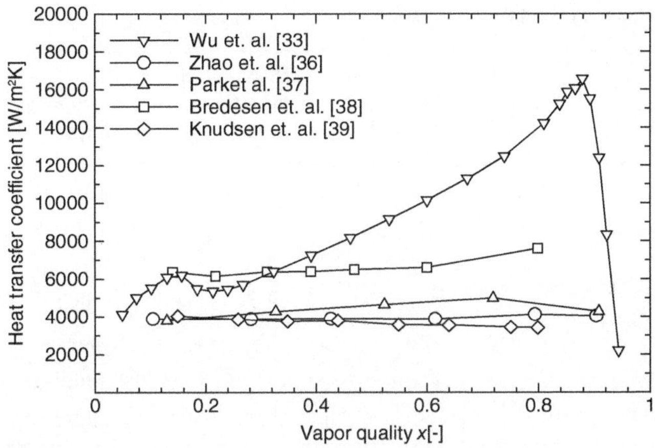

Figure 4.10 Comparison of heat transfer coefficient vs vapor quality at saturation temperature of −30°C.

of the heat transfer coefficient in conventional refrigerants. It is believed in general that nuclear boiling heat transfer is larger than convective boiling heat transfer due to the low surface tension coefficient and low viscosity of CO_2. In the case of a low surface tension coefficient, the wettability of the wall surface is largely reduced, the liquid film is easily peeled off from the wall, and the nucleation boiling may be persisted and rather highly promoted simultaneously. As shown in Figure 4.10, it is further found that, when the saturation temperature is −30°C, the value of heat transfer coefficient is relatively high compared to the heat transfer coefficient at 0°C. The heat transfer coefficient does not increase at vapor quality less than 0.3 because of the boiling in flow region transition from intermittent to annular. On the contrary, when the vapor quality is larger than 0.3, the heat transfer coefficient becomes larger than that at saturation temperature 0°C. Reduction of the saturation temperature usually leads to high frictional pressure drop at the gas-liquid interface due to high liquid density of CO_2 with low vapor density. Thus, it can be understood that the high heat transfer coefficient increases by reducing the saturation temperature of CO_2. However, although it has been pointed out that the pressure drop at the vapor-liquid interface may cause dryout at an early stage, there would not be detailed reports on the dry out available so far.

Figures 4.9 and 4.10 are the collection of various experimental data on heat fluxes based on mass flow rates. Besides being summarized in Table 4.2, various experiments have been conducted, such as by changing the tube diameter, saturation temperature, etc. In recent years, in order to suppress the dryout condition at low quality range, a study has been conducted to increase heat transfer and suppress dryout by using micro-fins in the heat transfer tube [34, 39]. This is a good example, as it is of major importance for dryout condition to be avoided.

In order to enhance the energy conversion efficiency of CO_2 evaporator for actual manufacturing systems, it is necessary to understand the vapor-liquid two-phase flow of CO_2, especially the pressure drop and two-phase flow pattern. As mentioned above, due to the small density ratio, the velocity difference of that between two-phase is less than that of

Table 4.2 Experimental studies on CO_2 boiling heat transfer.

Reference source	Saturation temperature (°C)	Heat flux ($kW\,m^{-2}$)	Mass flux ($kg\,m^{-3}\,s^{-1}$)	Diameter of evaporator tube (mm)	Circular single tube type
Wu *et al.* [32]	−30, 0	7.5	300	1.42	Stainless
Oh *et al.* [33]	0	10	200	7.75	Stainless
Cho *et al.* [34]	0	6	424	0.9	—
Zhao *et al.* [35]	−29.9	14.8	204.3	4.75	Stainless
Park *et al.* [36]	−30	5	200	6.1	Copper
Bredesen *et al.* [37]	−25	6	200	7	Aluminum
Knudsen *et al.* [38]	−28	8	80	10.08	Stainless

conventional refrigerants, which results in a small pressure drop of CO_2. Oh *et al.* [40] carried out an experimental investigation of CO_2 pressure drop at a low saturation temperature region from 0 and 20°C using 4.57 mm inner diameter horizontal stainless-steel tube. The heat flux was varied from 10 to 40 $kW\,m^{-2}$, mass fluxes ranging from 200 to 1000 $kg\,m^{-2}\,s^{-1}$. The results showed that the pressure drop became lower when the saturation temperature of CO_2 is higher. This is caused by the effect of the density and lower viscosity. When the saturation temperature is higher, the vapor velocity decreases due to the decrease in the liquid to vapor density ratio as shown in Figure 4.6. The decreasing trends were shown in various studies [34, 36, 41, 42].

As seen in the above experimental results tabulated in Table 4.2, the heat transfer of CO_2 largely depends on individual works where the material and shape of the heat transfer tube makes substantial difference, due to the high pressure drop, the low density ratio, and the low surface tension coefficient. At this stage, the difficulty of estimating a general correlation equation of the boiling heat transfer at this condition range can be understood. In recent years, with the development of databases of physical properties of CO_2, some effective correlation equations (although with limited applicability) have been proposed as displayed below. Thome *et al.* [28] considered the CO_2 database and proposed a correlation equation of CO_2 boiling heat transfer based on the correlation equation of Kattan *et al.* [43] as shown in Eq. (4.2)

$$h = [(h_{nb})^n + (h_{ce})^n]^{\frac{1}{n}} \tag{4.2}$$

where h_{nb} is nucleate evaporation contribution and h_{ce} is convective evaporation contribution. Thome *et al.* [28] have summarized the following equation using experimental results in a temperature range of −25 to 25°C.

$$h = \left(\frac{\theta_{dry}}{2\pi}\right) h_v + \left(\frac{2\pi - \theta_{dry}}{2\pi}\right) h_{wet} \tag{4.3}$$

$$h_v = 0.023\mathrm{Re}_g^{0.8}\,\mathrm{Pr}_g^{0.4}\frac{\lambda_g}{D} \tag{4.4}$$

$$h_{wet} = [(S \cdot h_{nb.CO_2})^3 + (h_{cb})^3]^{1/3} \tag{4.5}$$

$$h_{cb} = 0.0133[4G(1-x)\delta]/[\mu_l(1-\alpha)]^{0.69}\,\mathrm{Pr}_g^{0.4}\lambda_l/\delta \tag{4.6}$$

$$\alpha = \frac{x}{\rho_g}\left\{[1+0.12(1-x)]\left(\frac{x}{\rho_g}+\frac{1-x}{\rho_l}\right)+\frac{1.18(1-x)}{G}\left[\frac{g\sigma(\rho_l-\rho_g)}{\rho_l^2}\right]^{1/4}\right\}^{-1} \tag{4.7}$$

$$\delta = [\pi D(1-\alpha)]/[2(2\pi-\theta_{dry})] \tag{4.8}$$

$$S = (1-x)^{1/2}/(0.121\mathrm{Re}_\delta^{0.225}) \tag{4.9}$$

$$h_{nb.CO_2} = 0.71h_{nb} + 3970 \tag{4.10}$$

$$h_{nb} = 55\mathrm{Pr}^{\,0.12}(-0.4343\ln \mathrm{Pr})^{-0.55}M^{-0.5}q^{0.67} \tag{4.11}$$

where Re is Reynolds number, Pr is Prandtl number, λ is thermal conductivity, D is inner diameter, G is mass flow rate, x is quality, α is void fraction and M is molecular mass. In addition, θ_{dry} is dry angle which is considered at dry out condition. Equation (4.5) is a modification of Eq. (4.2) which considers the nucleate boiling heat transfer coefficient with the boiling suppression factor of CO_2. The convectional boiling heat transfer coefficient of Eq. (4.10) is formulated under film flow region to which the nuclear boiling heat transfer coefficient of Eq. (4.11) is modified in the Cooper [44] correlation, which excludes surface roughness correction term. Fang *et al.* [45] have proposed a correlation which used a much more experimental database with fitting function as follows:

$$h = 00061(S+F)\mathrm{Re}_l Fa^{0.11}\,\mathrm{Pr}_l^{\,0.4}/[\ln(1.024\mu_{l,f}/\mu_{l,w})]\frac{\lambda_l}{D} \tag{4.12}$$

$$S = 41000Bo^{1.13} - 0.275 \tag{4.13}$$

$$F = [x/(1-x)]^a(\rho_l/\rho_g)^{0.4} \tag{4.14}$$

$$a = \begin{cases} 0.48 + 0.00524(\mathrm{Re}_l Fa^{0.11})^{0.85} - 5.9\times10^{-6}(\mathrm{Re}_l Fa^{0.11})^{1.85} & \mathrm{Re}_l Fa^{0.11} < 600 \\ 0.87 & 600 \le \mathrm{Re}_l Fa^{0.11} \le 6000 \\ 160.8/(\mathrm{Re}_l Fa^{0.11})^{0.6} & 6000 > \mathrm{Re}_l Fa^{0.11} \end{cases} \tag{4.15}$$

where $\mu_{l,f}$ and $\mu_{l,w}$ are the liquid viscosities at the fluid temperature and the inner wall temperature, respectively, and Fa is the dimensionless number as defined by Eq. (4.16)

$$Fa = \frac{(\rho_l-\rho_g)\sigma}{G^2L} \tag{4.16}$$

where L is the characteristic length. Since the correlation equations of Fang *et al.* [45] are calculated from many experimental databases, satisfactory estimates of the correlation can be obtained even in the low temperature range below 0°C. Regarding other practical correlation equations, a correlation equation of boiling heat transfer of CO_2 based on the correlation equation of Chen *et al.* [46] is presented by Eq. (4.17) as follows.

$$h = S\cdot h_{nb} + F\cdot h_{cn} \tag{4.17}$$

The correlation equations proposed so far have an average absolute error in range of 10–30% [47]. One of the causes is due to the variation of experimental data. In order to

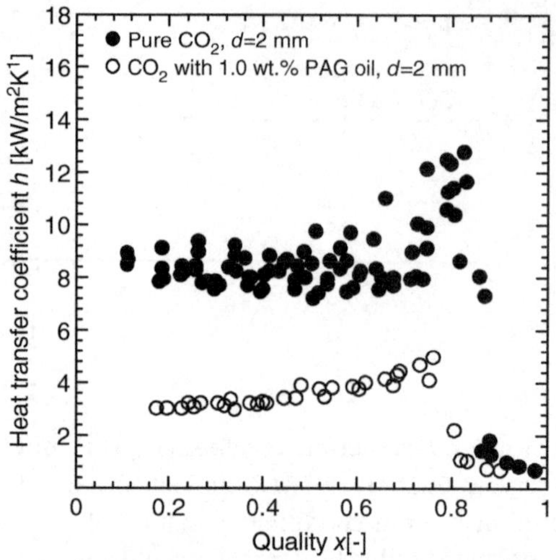

Figure 4.11 Effect of the oil on boiling heat transfer of CO_2 at $q = 18\,kW\,m^{-2}\,k^{-1}$ and $G = 720\,kg\,m^{-2}\,s^{-1}$ [48].

propose a more accurate correlation in practical future use, a greater number of accurate experimental databases should be supplied, such as heat transfer tube shape and surface roughness.

In an actual heat pump system, lubricating oil is essential to the compressor for sealing. When the lubricating oil flows into an evaporator, it usually causes some unexpected negative effects. Therefore, the heat transfer coefficient and pressure drop for CO_2-oil mixture are required to design an evaporator. Dang *et al.* [48] carried out experiments on the flow boiling of pure CO_2 and CO_2-polyalkylene glycol (PAG) mixtures in a smooth stainless steel tube (type 316) with an I.D. of 2 mm and length of 1.5 m. Figure 4.11, based on heat transfer data from their original paper, shows the comparison of the conduction of oil. With the presence of oil at concentration of 1 wt%, in the pre-dry out region, the heat transfer coefficient dramatically decreases since nucleate boiling is suppressed by the thermal resistance of the oil-rich sublayer. When the inner tube wall is covered with the oil layer, CO_2 could not reach the heated surface where bubble generation is blocked. The dryout quality and post dry out heat transfer are not influenced by the presence of oil in the experimental condition.

Pehlivanoglu *et al.* [49] have experimentally investigated the boiling heat transfer CO_2-oil mixture at a low saturation temperature region of −15 and −30°C using 6.1 and 9.6 mm inner diameter horizontal copper tube. The heat flux was varied from 2 to 15 kW m^2, mass fluxes ranging from 100 to 400 kg m^{-2} s^{-1}. The results also show the same trend with/without the presence of the oil at high saturation temperatures of CO_2 [50, 51]. The heat transfer coefficients of saturation temperature at −15°C are higher than that at −30°C since the oil affects the heat transfer coefficient. In addition, it may be considered that, when the saturation temperature decreases below −10°C, the density of CO_2 becomes higher than that of oil. There haven't been detailed reports on the flow behavior of CO_2-oil mixture at low saturation temperature so far.

4.3 Sublimation Heat Transfer of Dry Ice-Gas CO_2 in an Evaporator/Sublimator

In CO_2 heat pump systems, when the system operates in conditions where the temperature is below triple point, it has been found that dry ice blockage may occur in the evaporator which causes system operation failure. However, there are advantages that can be easily realized in a refrigeration system which could attain ultra-low temperature ranges by utilizing dry ice sublimation heat transfer. Compared with a traditional vapor-compression refrigeration cycle, a flow with sublimation can achieve stable operation with higher heat recovery capacity due to relatively high latent heat in sublimation. Owing to the fact that, in recent years, CO_2/CO_2 refrigeration systems have been gaining much attention in the field of ultra-low temperature refrigeration Huang *et al* [52] have proposed a refrigeration system using dry ice by carrying out numerical prediction. The numerical analysis of estimation of system characteristics shows that the coefficient of performance (COP) can be 50% higher than that of a conventional CO_2 refrigeration system due to higher sublimation latent heat. In the analysis, a Lagrangian particle-trajectory model together with a Nusselt-type model are presented [53] for the sedimentation and sublimation process during throttling from a high pressure CO_2 into the atmosphere, in order to obtain suitable parameters to get longer duration of the deposition and shorter duration of sublimation. An experimental work [54, 55] of simulating CO_2 flow through the safety valve has been carried out to verify the influence of upstream vapor quality, the valve opening on CO_2 freezing, and blockage in the valve and downstream line. Recently, a CO_2 ultra-low temperature cascade refrigeration system with dry ice sublimation has been proposed and introduced by Zhang *et al* [56]. The designed system has provided users with cryogenic cooling capacity below the CO_2 triple-point temperature of −56.6°C by expanding the liquid CO_2 into dry ice. They have also investigated the sublimation heat transfer of solid-gaseous CO_2 flow in an evaporator/sublimator using the CO_2/CO_2 cascade refrigeration system. A schematic diagram of the evaporator/sublimator of proposed ultra-low temperature CO_2 in the cascade refrigeration system is depicted in Figure 4.12. The evaporator/sublimator is a copper horizontal circular tube, which has an inner diameter of 0.04 mm and outer diameter of 0.45 mm. The length of the evaporator/sublimator is 5.0 m. As for heat sources, eight silicone gum-type heaters (200 V–300 W × 8) are twisted around the evaporator/sublimator.

Experimental results obtained in measuring outside wall temperatures and local Nusselt numbers under input of constant heat flux are shown in Figure 4.13a and b, respectively. In Figure 4.13a, each result is identified by input heat flux of 1592, 2122, and 2653 W m^{-2}. The local Nusselt number and local heat transfer coefficient h_x are calculated by the following formula:

$$Nu = h_x D/\lambda \tag{4.18}$$

$$h_x = q/(T_w - T_{in}) \tag{4.19}$$

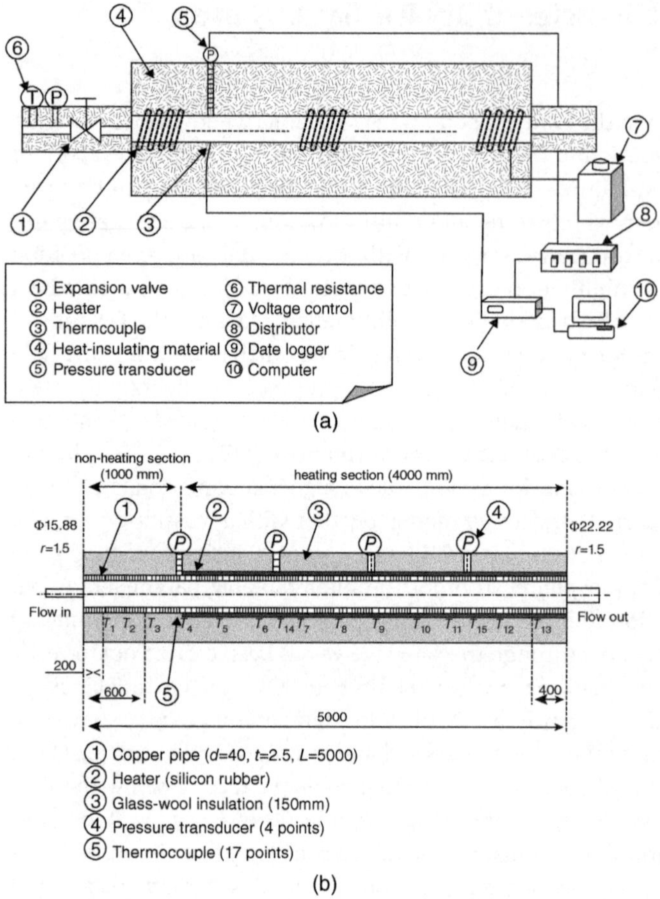

Figure 4.12 Schematic of the evaporator/sublimator in the ultra-low temperature CO_2 in the cascade refrigeration system. (a) Diagram of the test section and data acquisition and (b) detail of the test section and its measurement [56].

where, D is the internal diameter, λ is the thermal conductivity of gas CO_2, q is the heat flux, T_w is the inner wall temperature of the evaporator/sublimator and T_{in} is the CO_2 temperature, which can be regarded as the saturation temperature corresponding to the measured pressures in the evaporator/sublimator. From the Figure 4.13a, it is seen that the wall temperature reaches below −50°C by dry ice sublimation heat transfer before $x = 2$ m. And it can be further seen from $x = 2$ m that the wall temperature increases up to −35°C rapidly. It should be stressed that this is caused by the dry ice sublimating in the region of $x = 2.0$ m and that most of the dry ice particles have been sublimated with absorbing a great deal of heat. After $x = 2.0$ m, the CO_2 flow mainly become a gaseous state, so that the CO_2 temperature increases obviously when the evaporator/sublimator is heated. From Figure 4.13b, in the range of $x = 0$–2 m, where sublimation of dry ice takes place, the Nusselt number slightly increases along with the horizontal length x due to the fact that dry ice sublimation behavior may make the CO_2 flow field different. It is thought that the sublimation makes the thermal boundary layer thinner, and the solid particle sublimation makes the flow form

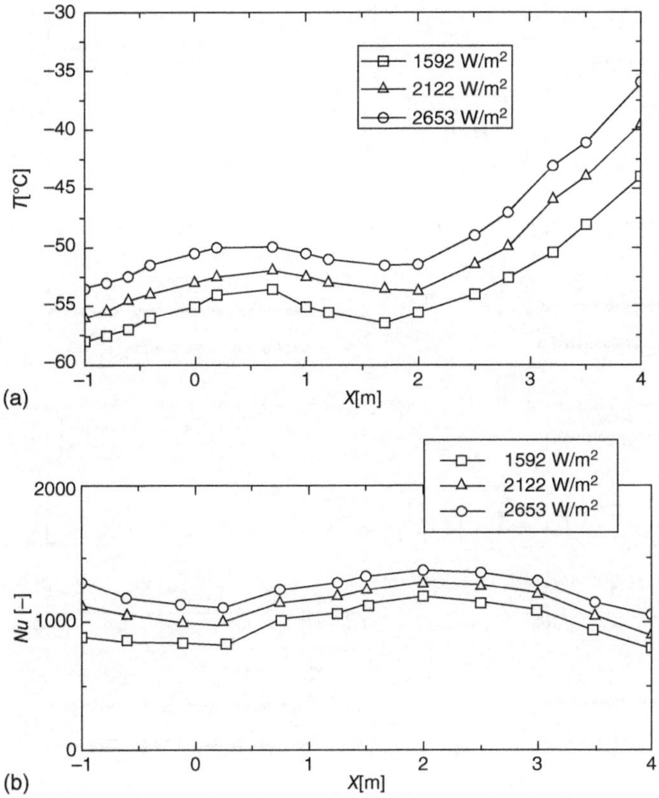

Figure 4.13 Variations of the measured data of the outside wall temperature and local Nusselt number along the horizontal length x of the evaporator/sublimator under the different heat fluxes, for the case of the condensation temperature of −25°C and the opening in the expansion valve of 15 mm. (a) Temperatures and (b) Nusselt number [56].

stronger turbulence. After the sublimation region, it was observed that the Nusselt number slightly decreases along with the evaporator/sublimator length x. This is caused by the development of a thermal boundary layer by CO_2 gaseous flow. From the above results, the sublimation heat transfer of dry ice seems to enhance the heat transfer level of the CO_2 solid-gas two-phase flow more than that of gas flow convection heat transfer.

Yamaguchi *et al* [57] also investigated temperature distribution in the evaporator/sublimator with time progress in order to understand the mechanism of blockage phenomena with dry ice sublimation. The temperature and pressure distribution in the evaporator/sublimator are shown in Figure 4.14. The illustrations of dry ice behavior are also displayed in Figure 4.14. The results are shown for conditions of condensation temperature of −25°C and input heat flux of 2122 W m^{-2}. From Figure 4.14, it is found that the sedimentation of dry ice first appears in the inlet region of the evaporator/sublimator, and then gradually moves downstream as time elapses. Before dry ice sublimation takes place, it can be clearly observed that temperature and pressure dramatically increase due to the solid-gaseous two-phase flow being entrapped in the evaporator/sublimator. The large temperature variation near the outlet of the observed evaporator is mainly due to the

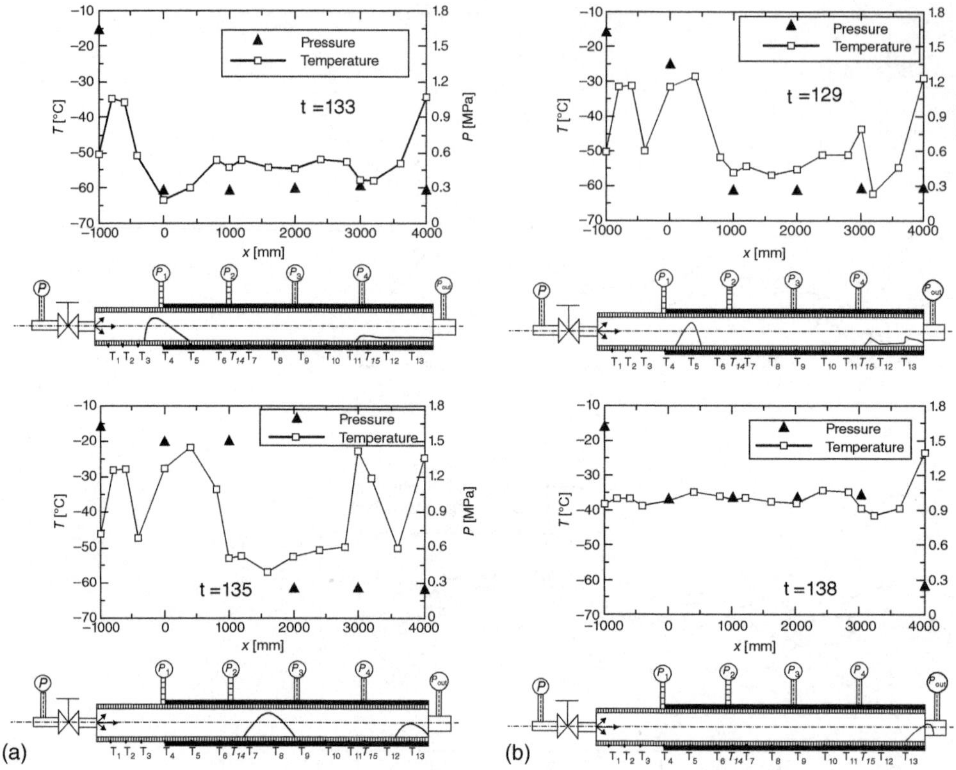

Figure 4.14 Variations of local pressure and wall temperature with the dry ice sedimentation inside the evaporator/sublimator at different times. (a) Expansion valve opening in 15 mm. (b) Expansion valve opening in 10 mm [57].

flow rapidly shrinking in the tube connected to the compressor. Dry ice blockage occurs in the small inlet tube after the expansion process at low mass flow rates, low condensation temperature and low heating power input. Also, it is thought that dry ice blockage occurs in the small inlet tube right after the expansion valve at low mass flow rates, and low condensation temperature with low heating power input. Based on the investigation, the dry ice blockage may be eliminated by adding greater heat input or increasing the opening of the expansion valve.

In order to prevent dry ice blockage in the evaporator/sublimator, it is also possible to alter the inlet shape of an evaporator/sublimator by engineering modification. The visualization results of dry ice behavior in the configuration of the sudden expansion channel and the modified tapered expansion channel are displayed in Figure 4.15. The visualization test set-up and detail structure are shown in Ref. [58].

As shown in Figure 4.15b, it can clearly be seen that the particle distribution is almost uniform along the inner wall in the case of the tapered channel. On the other hand, separation vortex is observed in the case of the sudden expansion channel (Figure 4.15a). It is thought that this vortex enhances the coalescence of the dry ice particles and forms larger dry ice particles compared with the tapered channel. As the results of the observation, in the

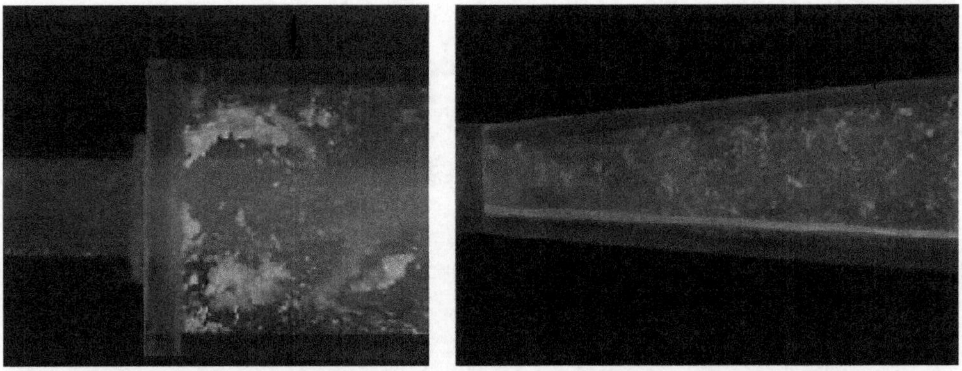

(a) sudden expansion channel (b) tapered expansion channel

Figure 4.15 Pictures of dry ice flow in the inlet of evaporator/sublimator by visualization set-up by high speed camera [58]. (a) Sudden expansion channel (b) tapered expansion channel.

case of the tapered channel, the average particle size was 1.68 mm, which is smaller than that of 2.02 mm in the case of the sudden expansion channel. It is thought from the result that the dry ice blockage, which may occur in practical CO_2/CO_2 refrigeration systems, can be satisfactorily eliminated by changing the inlet shape of the evaporator/sublimator from sudden expansion to tapered expansion channel. Iwamoto *et al* [59] have evaluated the sublimation heat transfer by changing the evaporator/sublimator inlet channel from a sudden expansion channel to a tapered expansion channel. Figure 4.16 shows the local heat transfer coefficient of CO_2 flow at various condensation temperatures along the evaporator/sublimator, in the case of the heat flux input of relatively high value 1910 W m^{-2} with the expansion valve opening of 15 mm.

As shown in Figure 4.16, the local heat transfer coefficient decreases when the condensation temperature decreases. In addition, the local heat transfer coefficient decreases in the range of 0–3000 mm, and then increases in the range of 3000–4000 mm. In order to discuss

Figure 4.16 Local heat transfer coefficient of CO_2 flow at various condensation temperatures along the evaporator [59].

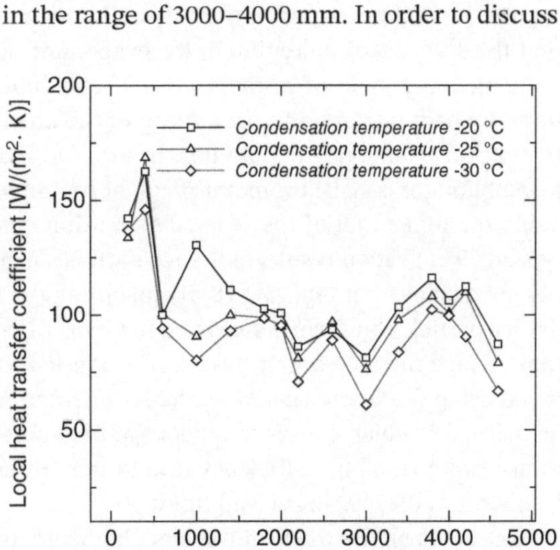

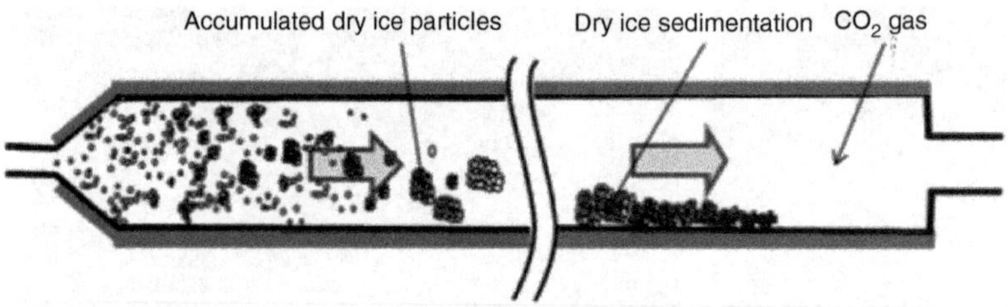

Figure 4.17 Illustration of dry ice particle behavior and dry ice sedimentation inside the evaporator/sublimator [59].

the variation of the local heat transfer coefficient along the evaporator/sublimator length shown in Figure 4.16, Figure 4.17 illustrates behaviors of dry ice particles and dry ice sedimentation inside the evaporator/sublimator. From the visualization test, it is considered that the dry ice particles uniformly distribute near the inlet of the tapered channel. The dry ice particles then flow toward the downstream with coalescence and collide with each other to become larger size particles, and then form dry ice sedimentation on the bottom of the evaporator/sublimator. The dry ice sedimentation is forced to flow away along the bottom, absorbing a great deal of heat, and changing to the gaseous phase. In the region of 0–3000 mm, on the other hand, the local heat transfer coefficient decreases along the evaporator/sublimator length, due to developing the thermal boundary layer. After $x = 3000$ mm, the local heat transfer coefficient increases because of the CO_2 sedimentation moving along the bottom of the evaporator/sublimator by sublimating and absorbing a great deal of heat, as shown in Figure 4.17. Resultantly, the heat transfer coefficient increases in the region of 3000–4000 mm due to the sedimentation sublimation. After the sublimation, the dry ice sedimentations change into the single gaseous phase, which means that the local heat transfer coefficient decreases along the evaporator length, owing to the thermal boundary developing. With replacing the evaporator/sublimator from the sudden expansion to the tapered expansion channel, it is found that sublimation heat transfer increases owing to the fact that the dry ice sedimentation in the evaporator/sublimator is greatly eased.

Along to the series of works, Yamasaki *et al* [60] have achieved the enhancement of sublimation heat transfer and preventing of the blockage phenomena by inducing a swirling flow in the evaporator/sublimator. In order to induce the swirling flow into the evaporator/sublimator, a swirl promoter made of stainless thin wire (of 1 mm diameter) is bonded along the inner wall of the tapered expansion channel of the evaporator/sublimator. The typical visualization results of dry ice particles using the swirl promoter and tapered channel are displayed in Figure 4.18. By installing the swirl promoter, it was observed that the dry ice particles uniformly dispersed in whole pipe cross-section by swirling motion, which may induce the increase of heat absorption in CO_2 solid-gas two-phase flow. Conversely, when using the tapered channel, the sedimentation occurs with settling down of larger dry ice particles, which causes the blockage phenomena. For the above reason, the swirl promoter can give higher efficiency due to the presence of dispersed dry ice particles in the large part of the evaporator/sublimator.

Figure 4.19 shows the heat transfer characteristics of the solid-gas two-phase flow inside the evaporator/sublimator, in the case of condensation temperatures of −20°C, heat flux

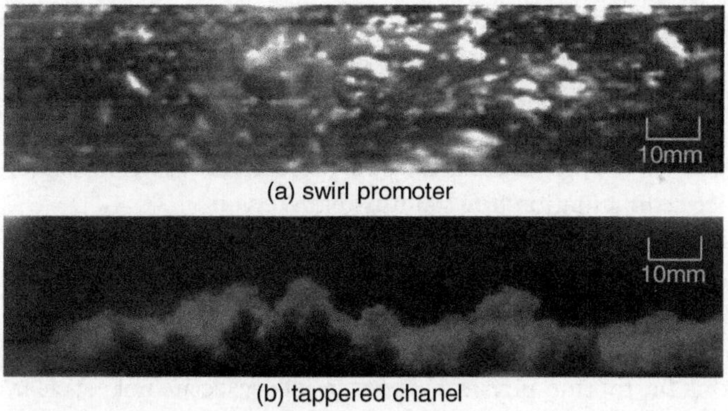

Figure 4.18 Visualization results of dry ice particles in an evaporator/sublimator (a) swirl promoter (b) tapered channel.

Figure 4.19 Heat transfer coefficient in cases with and without the swirl promoter.

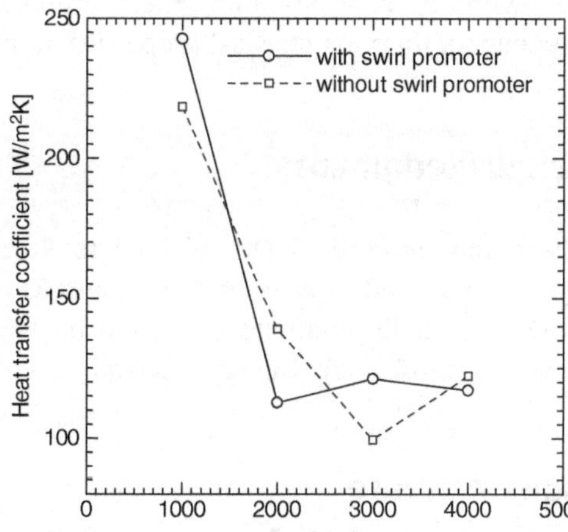

input of further higher 2904 W m^{-2} and the expansion valve opening of 25 mm. As shown in Figure 4.19, the local heat transfer coefficient at $x = 1000$ mm in the case with a swirl promoter is higher than that in the case without a swirl promoter. This is caused by inducing a stronger swirling flow that causes a large amount of dry ice particle to disperse along the inner wall of the pipe by absorbing a great deal of heat. In the range of $x = 1000$–2000 mm, both of the heat transfer coefficients show decreases, indicating the development of thermal boundary layers of gaseous phase along the wall. In the case with a swirl promoter, it is understood that the heat transfer coefficient is increased by active sublimation heat transport. Without a swirl promoter, on the other hand, the heat transfer coefficient tends to decrease at $x = 3000$ mm and then increase at $x = 4000$ mm. These trends can be explained by the large agglomeration of dry ice formed in the evaporator/sublimator, where the heat transfer is degraded as the major heat transported by heat conduction. Furthermore, when the sedimentation of dry ice that fills the bottom wall of the pipe occurs, it is thought that a vapor layer may be formed between the inner wall and the sedimentation of dry ice, leading to the heat transfer coefficient further deteriorating. At $x = 4000$ mm as shown in

Figure 4.19 in the case of a swirl promoter, the solid-gas two-phase changes into the single gaseous phase, which leads to the local heat transfer coefficient decreasing along the evaporator/sublimator length, owing to the thick thermal boundary developing, and resulting in an increase of the heat transfer coefficient. In the case without the swirl promoter, the heat transfer coefficient increases at $x = 4000$. This is caused by the increase in the heat transfer coefficient when residual dry ice sedimentation flows at upstream region.

Sublimation phenomena of solid-gaseous two-phase flow of CO_2 in an evaporator induce strong turbulence between the heated surface and dry ice since the difference in density between the vapor and dry ice is extremely high. In future work, it is necessary to consider the non-equilibrium effect in the complex flow phenomena taking place in the dry ice sublimation. It should be said that the thermo-physical properties of dry ice are not yet clear. Especially, the database of the thermal conductivity of dry ice is still not fully verified. In order to develop an ultra-low temperature refrigeration system, more studies should be conducted to obtain more knowledge on the heat transfer characteristics of the CO_2 solid-gas flow with dry ice sublimation, taking into account non-equilibrium physics with precise data base of thermos-physical properties in the extreme region.

Acknowledgments

The author is particularly indebted to Prof. Hiroshi Yamaguchi, Doshisha University, Kyoto, Japan and Prof. Peter Nekså, NTNU and SINTEF, Norway for their support. This work is partially financial supported by HighEFF under the FME-scheme (Centre for Environment-friendly Energy Research, 257632/E20).

Nomenclature

x	vapor quality (–)
j	volumetric flux, $m\,s^{-1}$
v	velocity, $m\,s^{-1}$
h	heat transfer coefficient, $k\,W\,m^{-2}\,K^{-1}$
G	mass flux, $kg\,m^{-2}\,s^{-1}$
Re	Reynolds number (–)
Pr	Prandtl number (–)
g	acceleration of gravity, $m\,s^{-2}$
S	boiling suppression factor (–)
Nu	Nusselt number (–)
T	temperature, K
COP	coefficient of performance

Greek Symbols

α	void fraction (–)
θ_{dry}	dry angle of tube, rad
λ	thermal conductivity, $W\,m^{-1}\,K^{-1}$
ρ	density, $kg\,m^{-3}$
δ	liquid film thickness, m
σ	surface tension coefficient, $kg\,s^{-2}$

Subscripts

g	gas phase
L	liquid phase
nb	nucleate evaporation contribution
ce	convective evaporation contribution
sat	saturation
v	vapor
wet	wet wall
cb	convection boiling
w	inner wall

References

1 Perarson, S.F. (2003). Refrigerants past, present and future. The international congress of refrigeration in Washington DC, USA. http://www.iifiir.org/en/doc/1055pdf (accessed February 2018).

2 Tsai, W.T. (2005). An overview of environmental hazards and exposure risk of hydrofluorocarbons (HFCs). *Chemosphere* 61 (11): 1539–1547.

3 Norman, C., DeCanio, S., and Fan, L. (2008). The Montreal Protocol at 20: ongoing opportunities for integration with climate protection. *Global Environmental Change* 18 (2): 330–340.

4 Harby, K. (2017). Hydrocarbons and their mixtures as alternatives to environmental unfriendly halogenated refrigerants: an updated overview. *Renewable and Sustainable Energy Reviews* 73: 1247–1264.

5 Velders, G.J.M., Fahey, D.W., Daniel, J.S. et al. (2009). The large contribution of projected HFC emissions to future climate forcing. *Proceedings of the National Academy of Sciences* 106 (27): 10949–10954.

6 Lorentzen, G. (1995). The use of natural refrigerants: a complete solution to the CFC/HCFC predicament. *International Journal of Refrigeration* 18 (3): 190–197.

7 Bamigbetan, O., Eikevik, T.M., Nekså, P., and Bantle, M. (2017). Review of vapour compression heat pumps for high temperature heating using natural working fluid. *International Journal of Refrigeration* 80: 197–211.

8 Lemmon, E.W., Huber, M.L., and McLinden, M.O. (2013). NIST standard reference database 23: Reference fluid and thermodynamic and transport propeties -REFPROP ver. 9.1. National Institute of Standards and Technology.

9 Babiloni, A.M., Esbri, J.N., Cervera, A.B. et al. (2015). Commercial refrigeration – an overview of current status. *International Journal of Refrigeration* 57: 186–196.

10 Pearson, A. (2008). Refrigeration with ammonia. *International Journal of Refrigeration* 31 (4): 545–551.

11 Lorentzen, G. and Pettersen, J. (1993). A new effect and environmentally benign system for car air-conditioning. *International Journal of Refrigeration* 16 (1): 4–12.

12 Lorentzen, G. (1994). Revival of carbon dioxide as a refrigerant. *International Journal of Refrigeration* 17 (5): 292–301.

13 Nekså, P. (2002). CO_2 heat pump systems. *International Journal of Refrigeration* 25 (4): 421–427.

14 Gullo, P., Hafner, A., and Banasiak, K. (2018). Transcritical R744 refrigeration systems for supermarket applications: current status and future perspectives. *International Journal of Refrigeration* 93: 269–310.

15 Okada, T. (2007). Development of CO_2 heat pump hot water system "ECO CUTE". *Mitsubishi Electric Advance* 120: 6–8.

16 Hashimoto, K. (2006). Technology and market development of CO_2 heat pump water heaters (EcoCute) in Japan. *IEA Heat Pump Centre Newsletter* 24 (3): 12–16.

17 Nekså, P., Rekstad, H., Zakeri, G.R., and Schiefloe, P.A. (1998). CO_2-heat pump water heater: characteristics, system design and experimental results. *International Journal of Refrigeration* 21 (3): 172–179.

18 Dopazo, J.A., Seara, J.F., Sieres, J., and Uhía, F.J. (2009). Theoretical analysis of a CO_2-NH_3 cascade refrigeration system for cooling applications at low temperature. *Applied Thermal Engineering* 29: 1577–1583.

19 Dopazo, J.A. and Seara, J.F. (2011). Experimental evaluation of a cascade refrigeration system prototype with CO_2 and NH_3 for freezing process applications. *International Journal of Refrigeration* 34: 257–267.

20 Aminyavari, M., Najafi, B., Shirazi, A., and Rinaldi, F. (2014). Exergetic, economic and environmental (3E) analyses, and multi-objective optimization of a CO_2/NH_3 cascade refrigeration system. *Applied Thermal Engineering* 65 (1–2): 42–50.

21 Rezayan, O. and Behbahaninia, A. (2011). Thermoeconomic optimization and exergy analysis of CO_2/NH_3 cascade refrigeration systems. *Energy* 36 (2): 888–895.

22 Devotta, S., Waghmare, A.V., Sawant, N.N., and Domkundwar, B.M. (2001). Alternatives to HCFC-22 for air conditioners. *Applied Thermal Engineering* 21 (6): 703–715.

23 Yamaguchi, H., Zhang, X.R., and Fujima, K. (2008). Basic study on new cryogenic refrigeration using CO_2 solid–gas two phase flow. *International Journal of Refrigeration* 31 (3): 404–410.

24 Yamaguchi, H. and Zhang, X.R. (2009). A novel CO_2 refrigeration system achieved by CO_2 solid–gas two-phase fluid and its basic study on system performance. *International Journal of Refrigeration* 32 (7): 1683–1693.

25 PROPATH Group (1997). PROPATH: a program package for thermophysical properties.

26 Yun, R. and Kim, Y. (2004). Flow regimes for horizontal two-phase flow of CO_2 in a heated narrow rectangular channel. *International Journal of Multiphase Flow* 30: 1259–1270.

27 Pettersen, J. (2004). Flow vaporization of CO_2 in microchannel tubes. *Experimental Thermal and Fluid Science* 28: 111–121.

28 Thome, J.R. and Hajal, J.E. (2004). Flow boiling heat transfer to carbon dioxide: general prediction method. *International Journal of Refrigeration* 27 (3): 294–301.

29 Kattan, N., Thome, J.R., and Favrat, D. (1998). Flow boiling in horizontal tubes: part1 – development of an adiabatic two-phase flow pattern map. *Journal of Heat Transfer* 120: 140–147.

30 Cheng, L., Ribatski, G., and Thome, J.R. (2008). New prediction methods for CO_2 evaporation inside tubes: part I – a two-phase flow pattern map and a flow pattern based phenomenological model for two-phase flow frictional pressure drops. *International Journal of Heat and Mass Transfer* 51 (1–2): 111–124.

31 Yun, R., Kim, Y., Kim, M.S., and Choi, Y. (2003). Boiling heat transfer and dryout phenomenon of CO_2 in a horizontal smooth tube. *International Journal of Heat and Mass Transfer* 46: 2353–2361.

32 Wu, J., Koettig, T., Franke, C. et al. (2011). Investigation of heat transfer and pressure drop of CO_2 two-phase flow in a horizontal minichannel. *International Journal of Heat and Mass Transfer* 54 (9–10): 2154–2162.

33 Oh, H.K., Ku, H.G., Roh, G.S. et al. (2008). Flow boiling heat transfer characteristics of carbon dioxide in a horizontal tube. *Applied Thermal Engineering* 28 (8–9): 1022–1030.

34 Cho, J.M. and Kim, M.S. (2007). Experimental studies of the evaporative heat transfer and pressure drop of CO_2 in smooth and micro-fin tubes of the diameters of 5 and 9.52 mm. *International Journal of Refrigeration* 30 (6): 986–994.

35 Zhao, X. and Bansal, P.K. (2007). Flow boiling heat transfer characteristics of CO_2 at low temperatures. *International Journal of Refrigeration* 30 (6): 937–945.

36 Park, C. and Hrnjak, P.S. (2006). CO_2 and R410A flow boiling heat transfer, pressure drop, and flow pattern at low temperature in a horizontal smooth tube. *International Journal of Refrigeration* 30 (1): 166–178.

37 Bredesen A.M., Harfner, A.H., Pettersen, J., et al. (1997). Heat transfer and pressure drop for in-tube evaporation of CO_2. *Proceedings of IIR Conference on Heat Transfer Issues in Natural Refrigerants*. USA: University of Maryland, pp. 35–49.

38 Knudsen, H.J. and Jessen, P.H. (1997). Heat transfer coefficient for boiling carbon dioxide. *Compression System with Natural Working Fluids, Sittard*, IEA Heat Pump Centre. The Netherlands, pp. 113–122.

39 Dang, C., Haraguchi, N., and Hihara, E. (2010). Flow boiling heat transfer of carbon dioxide inside a small-sized microfin tube. *International Journal of Refrigeration* 33 (4): 655–663.

40 Oh, H.K. and Son, C.H. (2011). Flow boiling heat transfer and pressure drop characteristics of CO_2 in horizontal tube of 4.57-mm inner diameter. *Applied Thermal Engineering* 31: 163–172.

41 Yoon, S.H., Cho, E.S., Hwang, Y.W. et al. (2004). Characteristics of evaporative heat transfer and pressure drop of carbon dioxide and correlation development. *International Journal of Refrigeration* 27: 111–119.
42 Choi, K.I., Pamitran, A.S., and Oh, J.T. (2007). Two-phase flow heat transfer of CO_2 vaporization in smooth horizontal minichannels. *International Jounral of Refrigeration* 30 (5): 767–777.
43 Kattan, N., Thome, J.R., and Favrat, D. (1998). Flow boiling in horizontal tubes: part 3 – development of a new heat transfer model based on flow pattern. *Journal of Heat Transfer* 120 (1): 156–165.
44 Cooper, M.G. (1984). Heat flow rates in saturated nucleate pool boiling – a wide ranging examination using reduced properties. *Advanced Heat Transfer* 16: 157–239.
45 Fang, X. (2013). A new correlation of flow boiling heat transfer coefficients for carbon dioxide. *International Journal of Heat and Mass Transfer* 64: 802–807.
46 Chen, J.C. (1966). Correlation for boiling heat transfer to saturated fluids in convective flow. *Industrial & Engineering Chemistry Process Design and Development* 5 (3): 332–339.
47 Fang, X., Zhou, Z., and Li, D. (2013). Review of correlations of flow boiling heat transfer coefficients for carbon dioxide. *International Journal of Refrigeration* 36 (8): 2017–2039.
48 Dang, C., Haraguchi, N., Yamada, T. et al. (2013). Effect of lubricating oil on flow heat transfer of carbon dioxide. *International Journal of Refrigeration* 36: 136–144.
49 Pehlivanoglu, N., Kim, S., and Hrnjak, P.S. (2010). Effect of oil on heat transfer and pressure drop of R744 in 6.1 mm horizontal smooth tube. *Proceeding of International Refrigeration and Air Conditioning Coference at Purdue*, Paper 1125.
50 Gao, L. and Honda, T. (2006). Experiments on flow boiling heat transfer of pure CO_2 and CO_2-oil mixtures in horizontal smooth and micro-fin tubes. *Proceeding of International Refrigeration and Air Conditioning Coference at Purdue*, Paper 770.
51 Zhao, Y., Molki, M., Ohadi, M.M. et al. (2002). Flow boiling of CO_2 with miscible oil in microchannels. *ASHRAE Transactions* 108 (1): 135–144.
52 Huang, D.P., Ding, G.L., and Quack, H. (2008). New refrigeration system using CO_2 vapor-solid as refrigerant. *Frontiers of Energy and Power Engineering in China* 2 (4): 494–498.
53 Huang, D.P., Ding, G.L., and Quack, H. (2007). Theoretical analysis of deposition and melting process during throttling high pressure CO_2 into atmosphere. *Applied Thermal Engineering* 27 (8–9): 1295–1302.
54 Huang, D.P., Ding, G.L., and Quack, H. (2007). Experimental observation and empirical estimation of formation of solid carbon dioxide in safety valves for refrigerating system. *HVAC&R Research* 13 (1): 77–92.
55 Huang, D.P., Quack, H., and Ding, G.L. (2007). Experimental study of throttling of carbon dioxide refrigerant to atmospheric pressure. *Applied Thermal Engineering* 27 (11–12): 1911–1922.
56 Zhang, X.R. and Yamaguchi, H. (2011). An experimental study on heat transfer of CO_2 solid-gas two phase flow with dry ice sublimation. *International Journal of Thermal Sciences* 50 (11): 2228–2234.
57 Yamaguchi, H., Niu, X.D., Sekimoto, K., and Nekså, P. (2011). Investigation of dry ice blockage in an ultra-low temperature cascade refrigeration system using CO_2 as a working fluid. *International Journal of Refrigeration* 34 (2): 466–475.

58 Yamaguchi, H., Iwamoto, Y., Ozaki, S., and Neksa, P. (2014). Experimental observation of sedimentation phenomena of CO_2 dry ice in model channel. *Proceedings of 11th IIR on Natural Refrigerations*, Hangzhou, China. 3 p.

59 Iwamoto, Y., Yamasaki, H., Niu, X.D. et al. (2015). Experimental study on system performance of ultra-low temperature cascade refrigeration system using carbon dioxide with tapered evaporator/sublimator. *Proceedings of ICR 2015, Yokohama*, Japan. Paper ID 943. 8 p.

60 Yamasaki, H., Yamaguchi, H., Hattori, K., and Neksa, P. (2017). Experimental observation of CO_2 dry-ice behavior in an evaporator/sublimator. *Energy Procedia* 143: 375–380.

5

Theoretical Analysis of the CO_2 Expansion Process

Ammar M. Bahman[1], Riley B. Barta[2], Eckhard A. Groll[2] and Davide Ziviani[2]

[1] *Mechanical Engineering Department, College of Engineering and Petroleum, Kuwait University, Kuwait City, Kuwait*
[2] *Ray W. Herrick Laboratories, School of Mechanical Engineering, Purdue University, West Lafayette, IN, USA*

5.1 Introduction

Improving the energy efficiency of thermal systems such as heat pumps or refrigeration systems is one of the main ways to compensate for the increasing global energy demand. Vapor compression systems utilized for space heating and cooling of buildings are main contributors to the worldwide energy consumption. A large majority of heat pumps available in the market operate with a conventional vapor compression cycle. Thus, a number of opportunities exist to improve the efficiency of the cycle and, ultimately, reduce the energy consumption. Many different sources of inefficiencies reduce the performance of vapor compression cycles. The four main components of a vapor compression cycle are the evaporator, compressor, condenser, and expansion valve. The compressor, evaporator, and condenser are the main contributors to decreased efficiency. Some examples of losses in the compressor are leakage, friction associated with mechanical elements, and heat transfer mechanisms. Heat exchanger losses can often be attributed to non-ideal component sizing and design. The expansion process is another source of losses in heat pump systems. In most standard systems today, the expansion process occurs through the orifice of a thermostatic expansion valve (TXV) or an electronic expansion valve (EXV). This expansion process is often regarded as a free or passive expansion process because it does not harvest the energy potential of the high-pressure refrigerant at the inlet as the energy is dissipated in the form of heat due to friction. In the past, attempting to harvest the lost energy from the expansion process in terms of useful power has typically been neglected due to the comparatively small quantity available. In the case of residential heat pump systems, the average total available power from the expansion process ranges from 100 to 300 W, whereas the total system power consumption is on the scale of 4–5 kW with 2–3 tons of refrigeration capacity.

The expansion work recovery becomes more important when carbon dioxide (CO_2 or R-744) is employed as the working fluid in vapor compression cycles. Although the use of CO_2 was known since the early twentieth century for marine applications, ammonia (NH_3), chlorofluorocarbons (CFCs), hydrochlorofluorocarbons (HCFCs), and later, hydrofluorocarbons (HFCs) were preferred due to more favorable characteristics such as

Transcritical CO_2 Heat Pump: Fundamentals and Applications,
First Edition. Xin-Rong Zhang and Hiroshi Yamaguchi.

higher critical temperatures and lower operating pressures. However, in the early 1990s, environmental concerns led to a revival of natural refrigerants. In particular, CO_2 has been extensively investigated as an alternative refrigerant in different applications including automotive air-conditioning, industrial refrigeration, military environmental control units (ECUs), and heat pump water heaters [1]. Because of its critical point characteristics (304.1 K and 7377.3 kPa), many refrigeration and air-conditioning applications require a transcritical CO_2 cycle, which are often less efficient compared to conventional vapor compression cycles. Therefore, cycle enhancements and expansion work recovery are key aspects to be considered and analyzed in order to improve the efficiency of baseline transcritical CO_2 cycles.

This chapter focuses on the theoretical and practical aspects of the CO_2 expansion process and provides insights into the ongoing research on this topic. In Sections 5.2 and 5.2.1, the basic transcritical CO_2 cycle is analyzed to understand the intrinsic irreversibilities of the cycle compared to conventional vapor-compression systems. In Section 5.2.2, a simplified thermodynamic model is employed to understand the impact of introducing an expansion work recovery device in the system to replace the throttling valve. The expansion process can also be effectively exploited by utilizing the high kinetic energy of the flow by means of an ejector-expansion device. In Section 5.3, the fundamentals of transcritical ejector-expansion CO_2 cycles as well as the expansion process through the ejector are outlined. In the last part of this chapter, Section 5.4, examples of expansion devices installed in transcritical CO_2 cycles are discussed.

5.2 Thermodynamic Analysis of the Expansion Process in Transcritical CO_2 Cycles

CO_2 is a natural substance that is plentiful in the Earth's atmosphere. Although CO_2 contributes to greenhouse effects, utilizing this natural refrigerant may result in possible long-term positive environmental impact since it is not as harmful to the world as more common fluorocarbon-based manufactured refrigerants. Furthermore, new generations of working fluids such as hydrofluoro-olefins (HFO) and hydrochlorofluoro-olefins (HCFO) are only temporary solutions to the long-term need of refrigerants.

When considering the properties of CO_2 [2], the critical temperature is generally lower than typical values of the heat sinks in air-conditioning systems. This aspect forces the heat rejection portion of the cycle to reach transcritical operation, which leads to higher pressure values in the systems compared to conventional refrigerants [3].

The intrinsic thermodynamic efficiency of a basic transcritical CO_2 cycle is lower than conventional vapor-compression cycles even in the case that better heat exchanger approach temperatures can be achieved due to superior heat transfer characteristics of CO_2 [4]. For these reasons, several cycle designs and enhancements have been considered and studied to demonstrate the advantages of transcritical CO_2 cycles for different areas, including residential, automobile air-conditioning, and industrial applications [5].

5.2.1 Thermodynamic Losses

There are several reasons for modifying the basic transcritical cycle, including improvement of energy efficiency, increase of capacity for a given system and component size, and

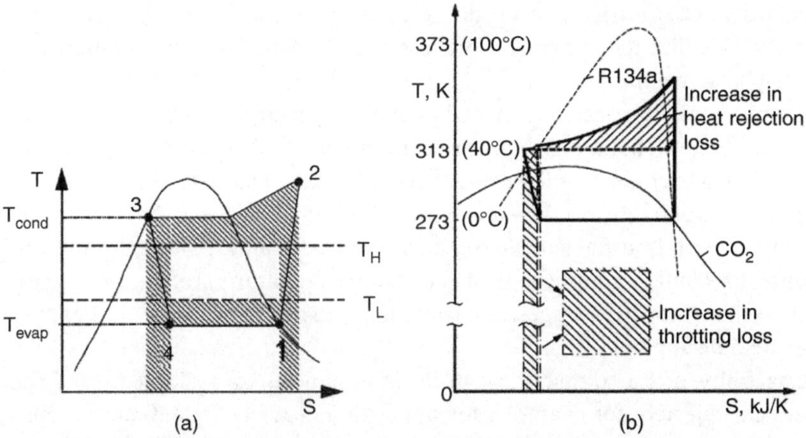

Figure 5.1 (a) Irreversibilities of a conventional Evans-Perkins cycle; (b) comparison of thermodynamic cycles for R-134a and CO_2 in T-s diagrams, showing additional thermodynamic losses for the CO_2 cycle when assuming equal evaporating temperature and equal minimum heat rejection temperature. Source: Adapted from Kim *et al.* [2].

adaptation of the heat rejection temperature profile to given requirements, e.g. in a heating system.

By considering the conventional Evans-Perkins cycle shown in the schematic of Figure 5.1a, different sources of irreversibilities can be identified with respect to the corresponding Carnot cycle, which is defined between T_H and T_L:

- compressor losses
- throttling losses
- desuperheating losses
- heat exchange losses
- pressure drops in the heat exchangers and distribution piping.

The conventional Evans-Perkins cycle can be directly compared with the transcritical CO_2 cycle, as represented in Figure 5.1b. To be consistent, the evaporating temperature and minimum heat rejection temperature are assumed to be given. By analyzing the two cycles, it can be seen that the transcritical cycle presents higher thermodynamic losses. In fact, due to the higher average temperature of heat rejection and the larger throttling loss, the theoretical cycle input work for CO_2 increases compared to a conventional refrigerant (in this case R-134a). Furthermore, the throttling losses are a function of the temperatures before and after the throttling device as well as the working fluid properties. Because of the high liquid specific heat and low evaporation enthalpy of CO_2 near the critical point, the decrease in capacity becomes significant and the compressor power increases. Nevertheless, the minimum heat rejection temperature is usually lower in the transcritical CO_2 cycle for a given heat sink inlet temperature and heat exchanger size [2]. Additionally, the operation of the CO_2 cycle results in higher evaporating temperatures for a given load, heat source temperature, and heat exchanger size.

Many researchers have analyzed the performance of the transcritical CO_2 refrigeration cycle in order to identify opportunities to improve the system energy efficiency. By performing a Second Law analysis, Robinson and Groll [6] found that the isenthalpic expansion process in a transcritical CO_2 refrigeration cycle is a major contributor to the cycle irreversibility due to the fact that the expansion process takes a path from the supercritical region into the two-phase region.

The thermodynamic cycle losses can be reduced by exploring a large number of modifications to the cycle architecture, including staging of compression and expansion, splitting of flows, use of internal heat exchange, and work-generating expansion instead of throttling. Lorentzen [3] outlined several advanced heat pump cycles and circuits for CO_2, including two-stage cycles, cycles with internal subcooling and cycles with an expander. In order to reduce the throttling loss and to adapt the heat rejection temperature profile, cycles with two or more compression/throttling stages, internal heat transfer, subcooling, and expansion work recovery can be applied.

The economic viability of the transcritical cycle is enhanced by making use of the high-temperature heat rejected, for example, for domestic hot water in stationary applications and for reheating/defogging in mobile applications. Theoretically, the same options are available in subcritical systems, but the relatively small amount of recoverable high-temperature heat has meant that it is usually wasted. The potential payoff is generally greater in CO_2 systems. Therefore, in transcritical heat pumps many more options exist for reversing flow between heating and cooling modes and for meeting simultaneous loads. Placement of the reversing valves is further complicated by the existence of the internal heat exchanger, where decisions must be made about preferences for counter- vs. parallel flow in heating mode.

5.2.2 Effect of Expansion Process

As previously discussed (refer to Figure 5.1b), the throttling losses in a transcritical CO_2 cycle are larger compared to conventional refrigerants. To reduce the expansion losses, internal heat exchange and expansion work recovery can be considered to improve the system coefficient of performance (COP).

In order to better understand the effects of both technical solutions, two different transcritical CO_2 cycles are compared. The schematic of the two cycles is shown in Figure 5.2. In particular, the main difference between the cycles is the expansion process. In fact, one cycle employs a throttling valve and, in the other, the expansion process occurs through a work recovery device that contributes to reduce the total input work to the cycle. The thermodynamic cycles are shown in Figure 5.3a. The cycle *1-2-4-5h-1* represents the transcritical CO_2 cycle with an isenthalpic or adiabatic expansion process (i.e. ideal expansion valve). Whereas, the cycle *1-2-4-5t-1* has the expansion through a valve (*4-5h*) replaced by an expander to extract useful work (*4-5t*). Both cycles have also been analyzed with different degrees of internal heat recovery where part of the energy from the heat rejection process (*3-4*) is transferred to the working fluid exiting the evaporator prior to the compression process (*6-1*). Figure 5.3b shows how the internal heat exchange is integrated into the thermodynamic cycles. Theoretically, the internal heat exchange should ensure a superheated state of the CO_2 prior to entering the compressor as well as reduce the total input power of the compressor. Both the isentropic compression process (*1-2s*) and the isentropic expansion process (*4-5s*) are indicated by dashed lines on the $T - s$ diagrams of Figures 5.3a, and b. By referring to the thermodynamic state numbering of Figure 5.3b, a simplified

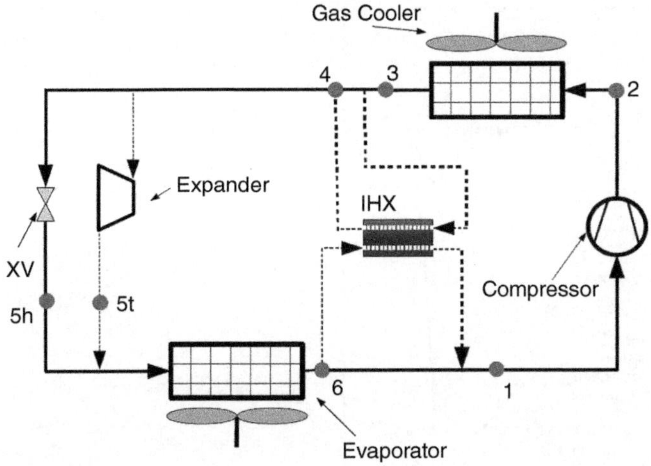

Figure 5.2 Schematic of a CO_2 cycle including an internal heat exchanger, an expansion valve, or an expansion work recovery device.

thermodynamic steady-state model based upon a mass specific basis energy balance of each component can be developed. The following assumptions are introduced:

- changes in kinetic and potential energies are neglected
- energy losses associated with compressor and expander irreversibilities are assumed to be heat rejected to the ambient through their housings
- internal heat exchanger is assumed to be adiabatic with respect to the environment and characterized by an effectiveness.

The specific heat absorbed through the evaporator is given by:

$$q_{\text{evap}} = h_6 - h_5 \tag{5.1}$$

Similarly, the specific heat rejected through the gas cooler from the cycle is calculated as:

$$q_{\text{gc}} = h_2 - h_3 \tag{5.2}$$

The actual specific work of the compressor is obtained by subtracting the heat losses, $q_{\text{comp,loss}}$, to the specific enthalpy difference across the compressor. That is:

$$w_{\text{comp,in,act}} = h_2 - h_1 - q_{\text{comp,loss}} \tag{5.3}$$

In the case of cycles with an expansion recovery device, the actual specific work output is calculated in a similar way to the compressor specific work:

$$w_{\text{exp,out,act}} = h_{5t} - h_4 - q_{\text{exp,loss}} \tag{5.4}$$

where $q_{\text{exp,loss}}$ accounts for the total specific heat losses of the device. If an adiabatic expansion valve is considered, then $w_{\text{exp,out,act}} = 0$ and $h_5 = h_4$.

To estimate the specific work of the compressor and the expander, an isentropic efficiency, η_{is}, and a mechanical efficiency, η_{mech}, are introduced. For the compressor, the efficiencies are defined as:

$$\eta_{\text{is,comp}} = \frac{h_{2s} - h_1}{h_2 - h_1} \tag{5.5}$$

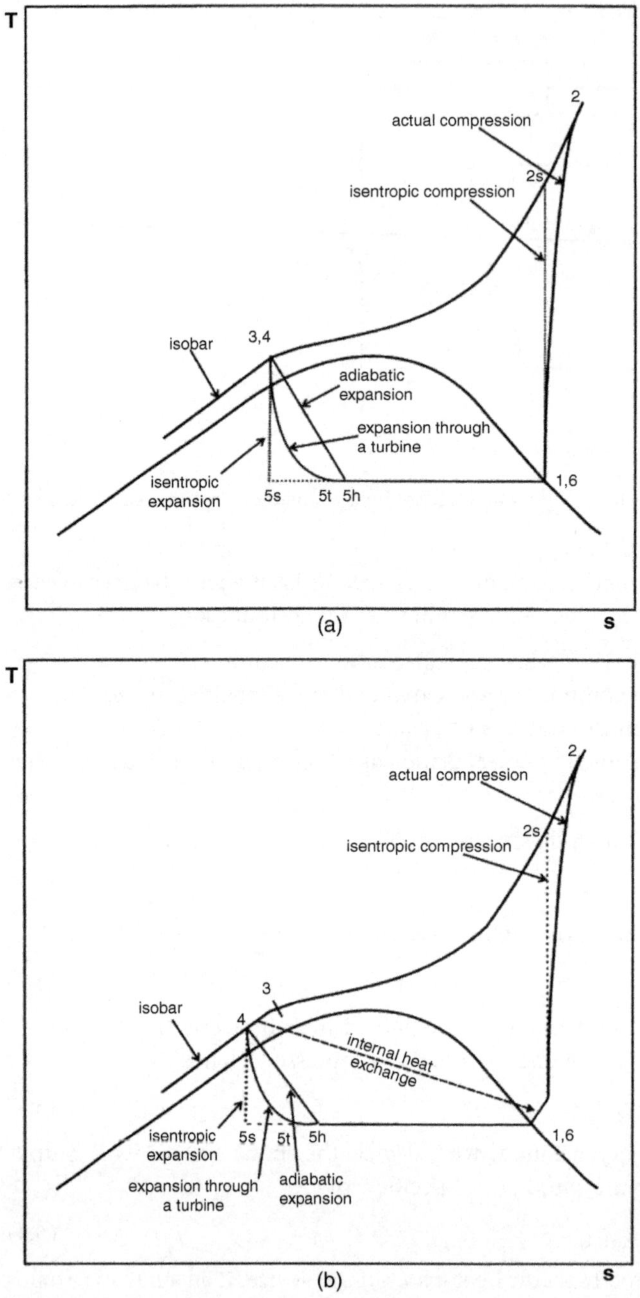

Figure 5.3 $T - s$ diagram of CO_2 cycles: (a) without internal regeneration; (b) with internal regeneration. Adiabatic, isentropic, and expansion through a turbine processes are overlaid. Source: Adapted from Robinson and Groll [6].

$$\eta_{mech,comp} = \frac{h_2 - h_1}{w_{comp,in,act}} \tag{5.6}$$

Whereas, for the expander, the efficiencies are:

$$\eta_{is,exp} = \frac{h_4 - h_4}{h_4 - h_{5s}} \tag{5.7}$$

$$\eta_{mech,exp} = \frac{w_{exp,out,act}}{h_3 - h_4} \tag{5.8}$$

It follows that the net specific work input to the cycle is:

$$w_{net,act} = w_{comp,in\ act} - w_{exp,out,act} \tag{5.9}$$

Due to the fact that the internal heat exchanger is considered to be adiabatic, the internal specific heat exchanged is calculated as:

$$q_{IHX} = h_3 - h_4 = h_1 - h_6 \tag{5.10}$$

The effectiveness of the internal heat exchanger is defined in terms of inlet and outlet temperatures of the cold and hot streams:

$$\varepsilon_{IHX} = \frac{T_{out,cold} - T_{in,cold}}{T_{in,hot} - T_{in,cold}} \tag{5.11}$$

where ε_{IHX} can range between 0 and 1.

To close the cycle model, the following energy balance equations can be established for the cycles with an expander:

$$q_{evap} + w_{net,act} = |\ q_{gc} + q_{comp,loss} + q_{exp,loss}\ | \tag{5.12}$$

and for the cycles with an expansion valve:

$$q_{evap} + w_{comp,in,act} = |q_{gc} + q_{comp,loss}| \tag{5.13}$$

In order to assess the performance of each cycle architecture, the efficiencies of the mechanical devices and the operating conditions must first be defined. The isentropic efficiency of the compressor is expressed by an empirical correlation as a function of the pressure ratio [6]:

$$\eta_{is,comp} = 0.815 + 0.022\left(\frac{p_2}{p_1}\right) - 0.0041\left(\frac{p_2}{p_1}\right)^2 + 0.0001\left(\frac{p_2}{p_1}\right)^3 \tag{5.14}$$

Whereas, the isentropic efficiency of the expansion device is assumed to be constant, $\eta_{is,exp} = 0.6$. The mechanical efficiency of both compressor and expander is set equal to $\eta_{mech,comp} = \eta_{mech,exp} = 0.9$. The heat sink is air at a constant temperature of 35°C and the outlet temperature of the gas cooler is set at 40°C. The heat source is assumed to be at constant temperature in the range from −35 to 10°C and the corresponding evaporating temperatures are in the range −40 to 5°C. The heat rejection pressure for each evaporating temperature is optimized.

For each of the cycles considered, the COP and the irreversibilities are discussed. In Figure 5.4a, the COP of the transcritical CO_2 cycle with expansion valve is compared to a similar cycle having different degrees of internal heat exchange (50% and 100%) for different evaporating temperatures. As a general comment, it can be seen that the COP increases

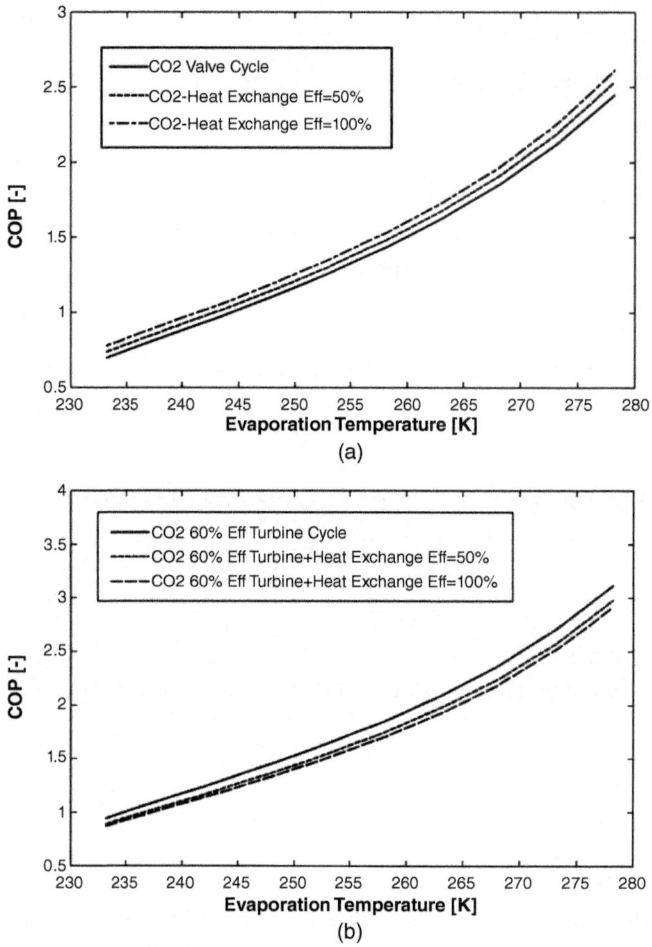

Figure 5.4 Variation of COP as a function of evaporating temperature for: (a) transcritical CO_2 cycle with internal heat exchange; (b) transcritical CO_2 cycle with internal heat exchange and expansion work recovery.

with the increase of evaporating temperature for all three cycle configurations. However, for a heat exchange effectiveness of 50%, the COP increases by approximately 4% on average with respect to the baseline cycle. Whereas, in the case of 100% heat exchange effectiveness, the percentage of improvement further increases up to 7.7%. Figure 5.4b illustrates the variations of COP at different evaporating temperatures for three cycles featuring an expansion recovery device having an isentropic efficiency of 60%, and three different degrees of internal heat exchange, i.e. 0%, 50%, and 100%. The major observation is that the use of internal heat exchange in combination with an expansion work recovery device is detrimental to the cycle performance. A heat exchange effectiveness of 50% yields a to a COP decrease by approximately 6%, and a heat exchange effectiveness of 100% decreases the COP by approximately 8% when applying an expansion work recovery device. It follows that the stream

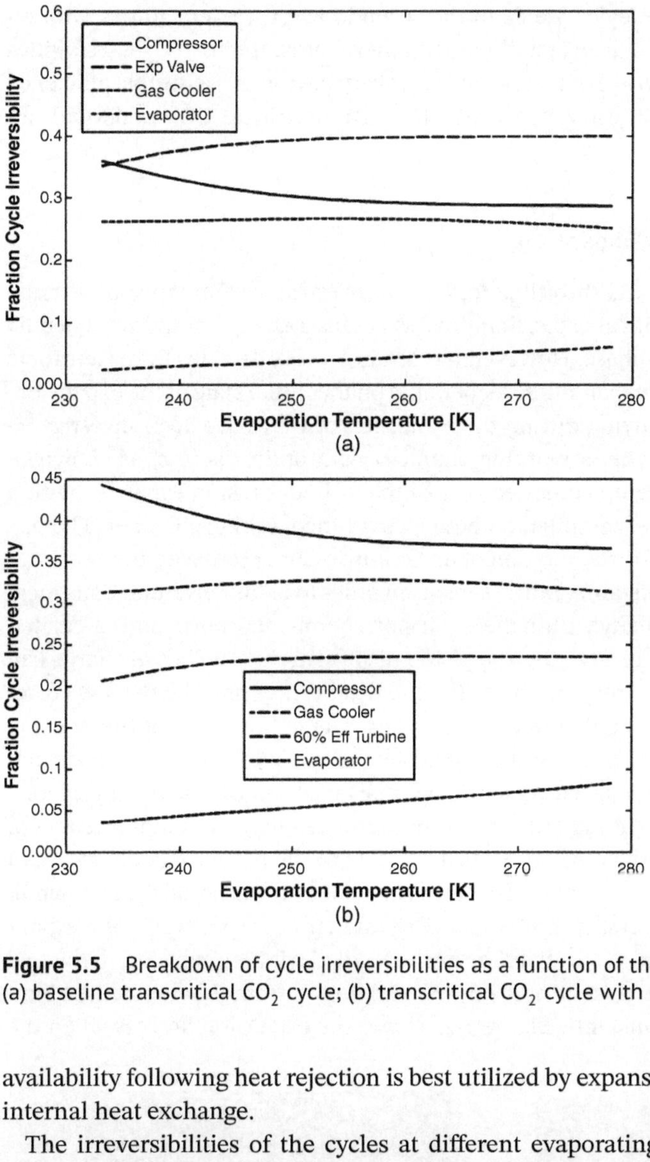

Figure 5.5 Breakdown of cycle irreversibilities as a function of the evaporating temperature: (a) baseline transcritical CO_2 cycle; (b) transcritical CO_2 cycle with expansion recovery device.

availability following heat rejection is best utilized by expansion work recovery rather than internal heat exchange.

The irreversibilities of the cycles at different evaporating temperatures are plotted in Figures 5.5a, and b. In particular, the specific irreversibilities have been normalized with respect to the specific heat absorbed at the evaporator. By considering the CO_2 cycle with expansion valve and the one with expansion work recovery device with 60% isentropic efficiency, the evaporator accounts for approximately 5% of the cycle irreversibilities throughout the range of evaporation temperatures. Whereas, the gas cooler accounts for more than 26% and 32% of the total cycle irreversibilities for the cycle with the expansion valve and the one with an expander, respectively. Although the irreversibilities associated with the heat exchangers for the two cycles are comparable, the irreversibilities of the expansion process have different magnitudes. In particular, the expansion valve leads to the highest irreversibilities, accounting for more than 39% of the total cycle irreversibilities.

The expansion work recovery device yields to the second lowest irreversibilities with an average of 23% of the total cycle irreversibilities. In other words, the total irreversibilities of the transcritical CO_2 cycle with expansion work recovery having an isentropic efficiency of 60% is only 77% of the total cycle irreversibilities of transcritical CO_2 cycle with an expansion valve.

5.2.3 Real Transcritical CO_2 Expansion

Using an expander to recover the throttling losses improves the performance of a transcritical CO_2 cycle. The transcritical expansion process entails a change in thermodynamic state from supercritical to two-phase. However, the actual process deviates from isentropic behavior due to non-equilibrium conditions depending on the initial state of the expansion. The physical phenomena occurring during the expansion of CO_2 have been investigated by Fukuta *et al.* [7] by employing a working chamber with optical access, as shown in Figure 5.6. In particular, the setup consisted of a simple cylinder-piston expander with a stroke of 10 mm and two glasses mounted on both sides of the working chamber. The piston movement was controlled by the rotation of an eccentric cam. Moreover, the expander was equipped with pressure and temperature sensors in order to obtain instantaneous measurements. The variation of density within the expansion chamber is known and calculated from an initial charge amount of CO_2 and the piston displacement. The initial charge can be easily estimated by knowing the volume at the top dead center and the density corresponding to the initial pressure and temperature values. The temperature of the housing of the expander was maintained equal to the initial state of the expansion. The expansion process was captured with a high-speed camera at a frame rate of 4000 fps with either transmitted light or reflected light. The experiments were conducted with an initial pressure of 9100 kPa and temperatures in the range of 10 to 45°C in both the presence and absence of lubricant oil (PAG-type). Three examples of expansion process visualizations are shown in Figure 5.7a–c , for initial temperatures of 30°C, 20°C, and 10°C, respectively. The expansion process is illustrated on a $p - h$ diagram to show the deviation between the actual expansion and the corresponding isentropic process and frames from the transmitted light are also reported for different instants. Figure 5.7a shows the expansion process when the

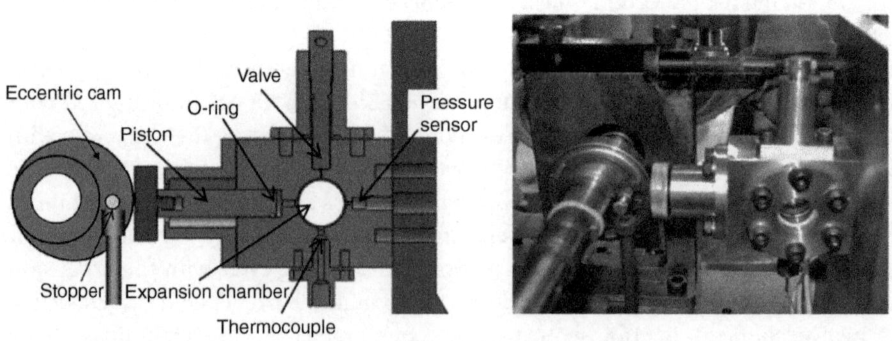

Figure 5.6 Experimental setup of the piston expander with optically accessible working chamber. Source: Obtained from Fukuta *et al.* [7].

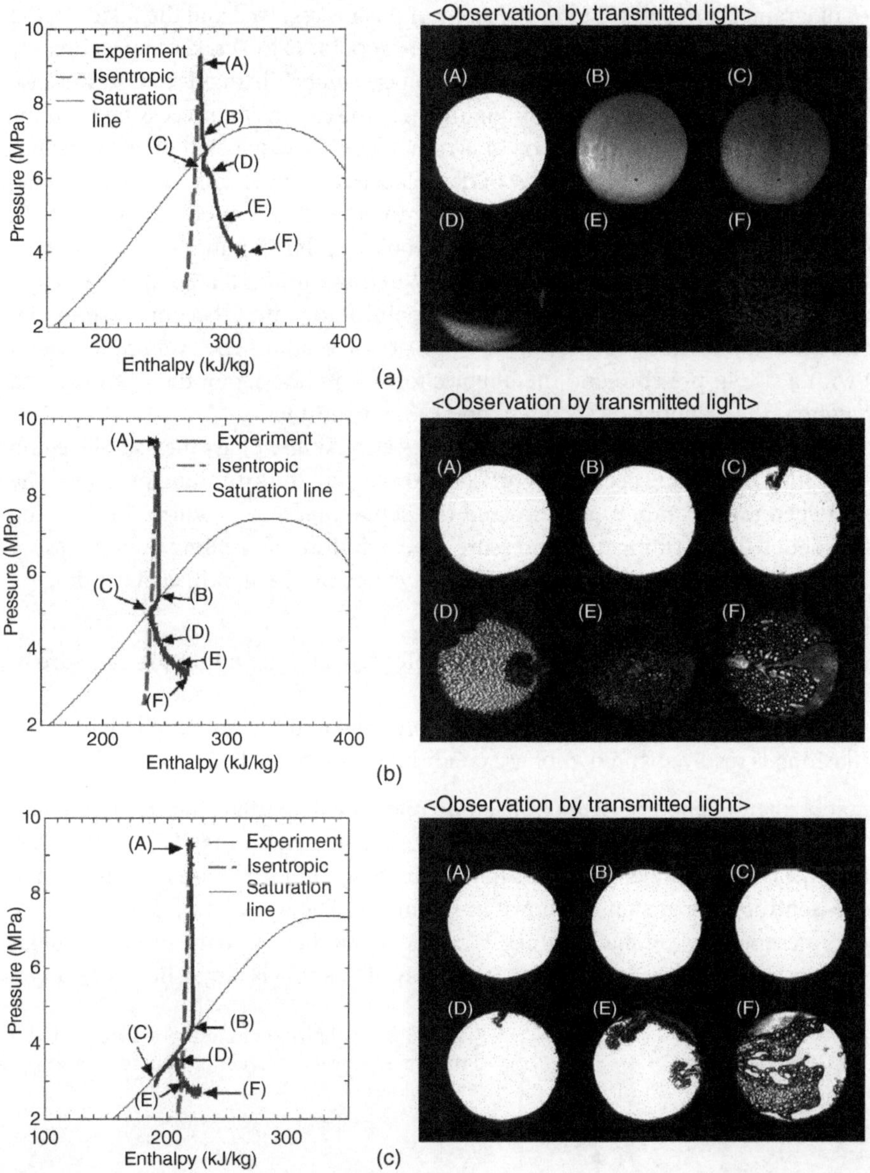

Figure 5.7 Experimental visualization of a transcritical CO_2 expansion with initial pressure of 9100 kPa and temperature of: (a) 30°C; (b) 20°C; (c) 10°C. Source: Obtained from Fukuta *et al.* [7].

initial temperature is 30°C. The expansion takes 47 ms and the corresponding rotational speed was 1276 rpm. At point B, before entering the two-phase region, the inside of the expansion chamber becomes partially fogged, followed by a slight decrease in enthalpy until point C. This phenomenon is regarded as delay of flashing. In particular, the evaporation of the liquid phase occurs with a certain time delay and CO_2 remains in the liquid condition. Consequently, the density of the two-phase mixture of CO_2 increases as represented on the $p-h$ diagram. From point C to D, the delay of flash is resolved and the inside of the expansion chamber becomes black. The transition from point C to D takes approximately 4 ms. At the end of expansion (point F), the liquid phase can be distinguished. In the case of an initial temperature of 20°C, shown in Figure 5.7b, the expansion process takes 42 ms which corresponds to a rotational speed of 1428 rpm. Once the expansion process reaches the saturate conditions at point B, the delay of flash causes the expansion to proceed along the saturated line until point C before transitioning to point D. From point C to point D, small bubbles appear in the expansion chamber. At point E, the chamber becomes darker due to scattering of the light by the tiny bubbles. When the initial temperature drops to 10°C, shown in Figure 5.7c, the delay of flash from point B to point C becomes even more extended and the chamber does not present any sign of evaporation. The evaporation starts at point D with a rise in pressure and the bubbles form from the top of the chamber and spread throughout the chamber, as shown in point E and point F.

The delay of flashing can be explained by the existence of a single-phase metastable region where the inception of flashing occurs at a pressure lower than the saturation pressure. The difference between the saturation pressure and the actual pressure at which the flashing begins can be denoted as flashing underpressure. The definition of flashing underpressure can be seen in Figure 5.8. In particular, two different values can be identified depending on the expansion process:

- UP_1: pressure difference between the saturation pressure and the minimum pressure in the single-phase metastable region.
- UP_2: pressure difference between the saturation pressure and the pressure at which the delay of flashing is resolved and two-phase conditions are established.

This unstable phenomenon of delay of flashing and associated flashing underpressure directly affect the indicated work recovered during the expansion process. In fact, the pressure inside the working chamber decreases more rapidly with the increase in volume than the reference isentropic expansion, as shown in Figure 5.9. The work losses due to delay of flashing are represented by the shaded areas. It can be noted that when the initial temperature of the expansion process is low (<10°C), the delay of flashing is larger, but its effect on

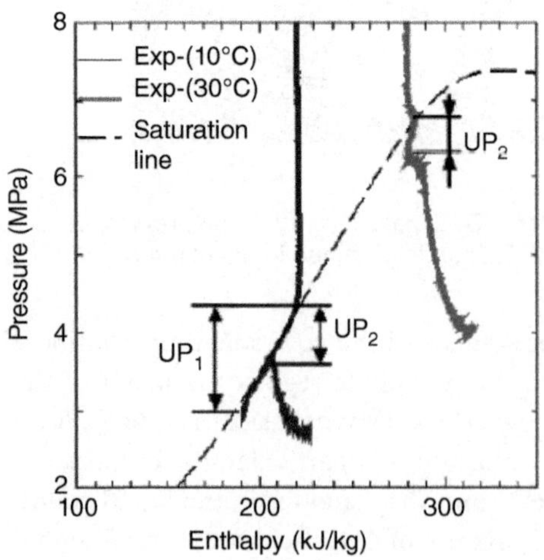

Figure 5.8 Definition of underpressure for two different expansion processes. Source: Obtained from Fukuta *et al.* [7].

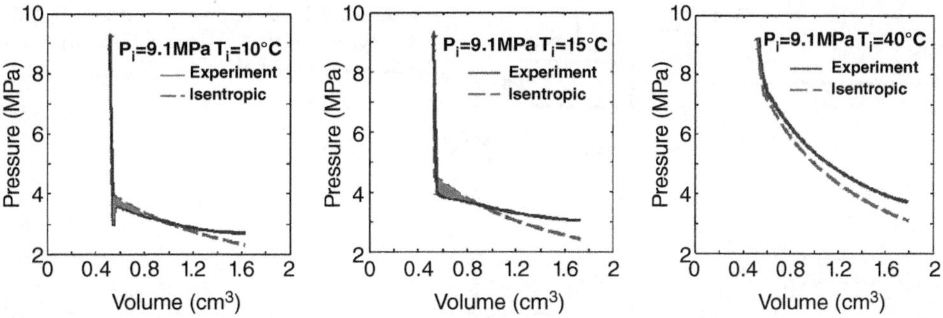

Figure 5.9 Effect of delay of flashing on the indicated expansion work. Source: Obtained from Fukuta *et al.* [7].

the work output is less detrimental because the recovery work in the two-phase region is significantly smaller than the work recovery in the supercritical region. Moreover, the heat transfer within the working chamber can also affect the work recovery. The influence of the delay of flashing on the total work recovery can be quantified in a few percentage points.

5.3 Theory of Ejector-Expansion Devices

The throttling process in the expansion valve is an intrinsic loss of a vapor compression cycle that reduced the cycle performance with respect to the associated Carnot cycle, as previously shown in Figure 5.1b. It has also been shown in Section 5.2.2 that the COP of the cycle can be improved by substituting the expansion valve with an expansion work recovery device or, in other words, the isenthalpic process with an isentropic process. The expansion device needs to handle the two-phase expansion process in an efficient way, which entails a number of challenges.

In order to recover the expansion work, Kornhauser [8] proposed the use of an ejector device to exploit the kinetic energy of the expansion process. In particular, the ejector device allows an increase in the compressor suction pressure with respect to conventional vapor compression cycles with an expansion valve or an expander. As a result, the compression work is reduced and the system COP is improved. The main advantages of the ejector device are its lower cost compared to an expander device, absence of moving parts, and the ability of handling a wide range of two-phase conditions in a robust way. The schematic of the ejector-expansion vapor compression cycle is shown in Figure 5.10 and the resulting thermodynamic cycle is illustrated in Figure 5.11 on a *p-h* diagram.

In order to better understand the working process of the ejector-expansion device, a simplified schematic of the ejector and its thermodynamic process are shown in Figure 5.12a, and b, respectively. In particular, the motive (m) stream undergoes an expansion process in the motive nozzle from the high pressure of the gas cooler, p_3, to the internal receiving chamber having pressure p_b. The specific enthalpy of the stream reduces from h_3 to h_{mb}, and the velocity increases to u_{mb}. At the same time, the suction (s) stream expands in the suction nozzle from the evaporator pressure p_7 to the chamber pressure p_b. Similarly to the motive stream, the decrease of specific enthalpy from h_7 to h_{sb} yields to increase the stream

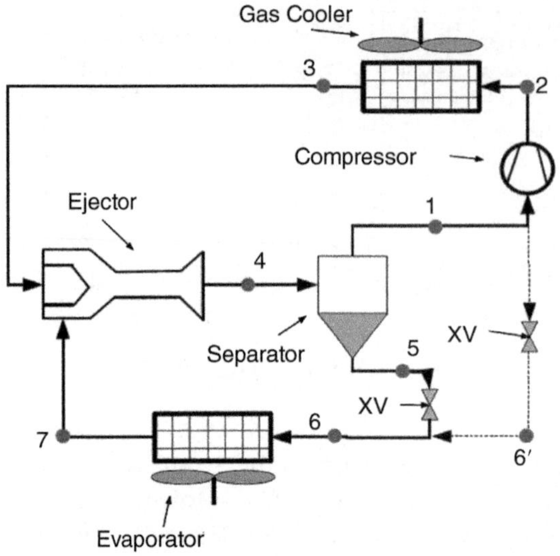

Figure 5.10 Schematic of a transcritical ejector-expansion CO_2 cycle. Source: Adapted from Li and Groll [9].

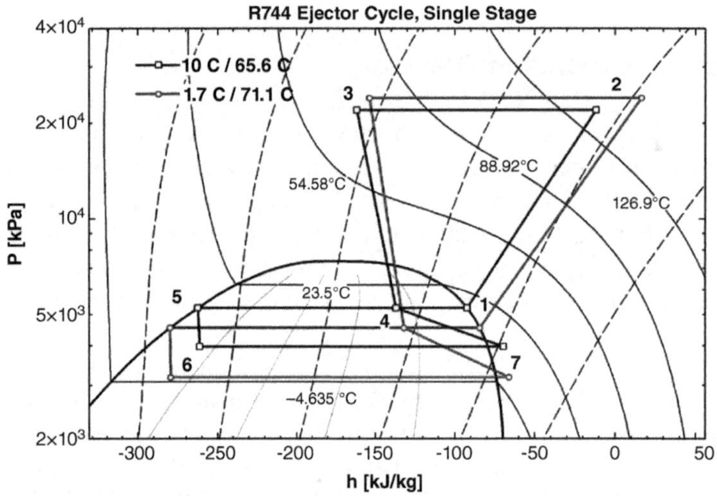

Figure 5.11 Ejector-expansion transcritical CO_2 refrigeration cycles in a p-h diagram.

velocity up to u_{sb}. The motive and suction streams mix in the mixing section of the ejector reaching an equilibrium pressure p_{mix} with a velocity u_{mix}. The mixed stream further increases its pressure to p_4 in the diffuser section of the ejector by converting the kinetic energy of the stream into internal energy. The thermodynamic process through the ejector introduces unavoidable losses that impact the overall efficiency of the device. In particular, irreversibilities are associated with the non-ideal mixing occurring in the mixing section of the ejector and deviations from the adiabatic reversible processes in the ejector nozzles and diffuser section. However, for a given ejector configuration, the entrainment ratio of the ejector is determined by the motive flow, suction flow, and the ejector outlet pressure. This causes control of the operating conditions of a real system to become difficult. To relax the

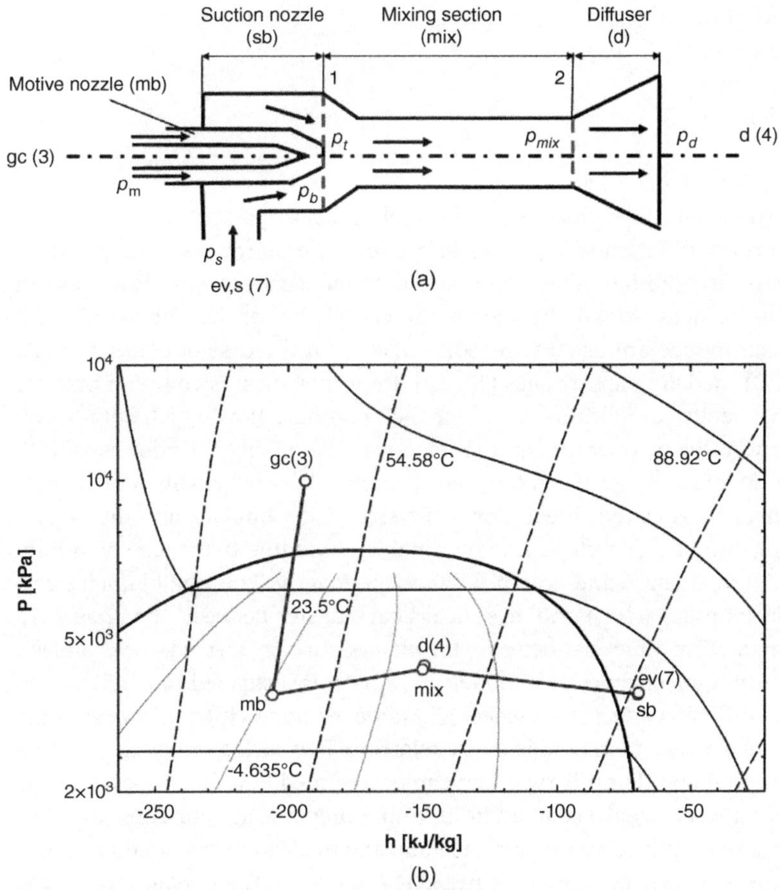

Figure 5.12 (a) Schematic of an ejector-expansion device; (b) example of working process of a CO_2 ejector-expansion device on a *p-h* diagram.

constraints between the entrainment ratio of the ejector and the quality of the ejector outlet stream, the ejector-expansion transcritical CO_2 cycle can be modified in such a way that part of the vapor from the separator is recirculated back to the evaporator inlet by means of a throttling valve that allows control of the quality [9], as shown in Figure 5.10.

5.3.1 One-Dimensional Ejector Flow Model

The theoretical aspects of the transcritical expansion occurring in the motive nozzle and the two-phase flow conditions through the ejector device can be analyzed with models having different complexity. Usually, a one-dimensional approach is employed with one of the following assumptions [8]: with mixing at constant pressure, with mixing at constant area, or with a combination of constant pressure and constant area mixing. However, in order to have a better physical-based model, a comprehensive mathematical model for a two-phase flow ejector is described in the following subsections. In particular, the ejector model consists of four sub-models [10]:

- motive nozzle flow model
- suction nozzle flow model
- mixing section flow model
- diffuser flow model.

5.3.1.1 Critical Two-Phase Flow Model

During the actual expansion through an ejector, the phases may present local regions of non-thermal equilibrium, different velocities (slip between the phases), as well as different values of pressure. In addition, depending on the system, traces of lubricant oil can also be entrained in the main stream. To reduce the complexity of the numerical models, a number of assumptions are usually introduced, even in the case of computational fluid dynamics (CFD) modeling approaches [11, 12]. In particular, it is common practice to consider a homogeneous equilibrium model for the two-phase flow which entails that the phases have same velocity, pressure, and temperature. Such simplification is valid if the Stokes number[1] is relatively small. In the ejector, high values of velocity and acceleration cause high shear stresses that break down the size of the bubbles and thus result in a low local Stokes number. Therefore, it is reasonable to assume that such small bubbles follow the continuous phase and do not flow independently. The small bubbles also present high interfacial areas which enhance local heat transfer between the phases. In other words, the temperature difference between the phases can be neglected. Nevertheless, thermal non-equilibrium effects in ejector models have been investigated in the literature [11, 14], but thermal-diffusion effects are neglected. Moreover, due to high pressures of the expanding motive stream, the small bubbles have relatively low surface tension, resulting in weak surface tension forces. For this reason, it can be assumed that the phases have the same pressure. In the present analysis, in addition to the previous assumptions, the flow through the ejector is also considered one-dimensional and in steady-state conditions, e.g. transient phenomena such as shockwaves are neglected. Based on these assumptions, the mass, momentum, and energy conservation equations for a one-dimensional homogeneous two-phase flow can be expressed as [10]:

$$\begin{aligned} &\frac{d}{dz}(A\rho_{mix}\mathrm{v}) = 0 \\ &\frac{d}{dz}(A\rho_{mix}\mathrm{v}^2) + A\frac{dp}{dz} = -\Gamma_{w}\tau_{w} + A\rho_{mix}g_z \\ &\frac{d}{dz}\left[A\rho_{mix}\mathrm{v}\left(h_{mix} + \frac{\mathrm{v}^2}{2}\right)\right] = \Gamma_h q_w + A\rho_{mix}\mathrm{v}g_z \end{aligned} \tag{5.15}$$

where ρ_{mix} and h_{mix} are obtained as a quality-weighted, x, function of the saturated properties due to the homogeneous equilibrium assumption:

$$v_{mix} = \frac{1}{\rho_{mix}} = xv_g + (1-x)v_l \tag{5.16}$$

$$h_{mix} = xh_g + (1-x)h_l \tag{5.17}$$

1 The Stokes number is defined as the ratio of the characteristic time of a droplet or particle to a characteristic time of the flow [13].

The other quantities appearing in the system of ordinary differential equations, Eq. (5.15), are the cross-sectional area A, the mean flow velocity v, the static pressure p, the wetted and heated perimeters Γ_w and Γ_h, the wall shear stress τ_w, and the wall heat transfer density q_w. In order to solve Eq. (5.15), p, v, and x can be chosen as the dependent variables of the flow, since the flow itself is considered adiabatic. It follows that Eq. (5.15) can be rearranged as:

$$\begin{bmatrix} \left(x\left(\frac{\partial v_g}{\partial p}\right)_{sat} + (1-x)\left(\frac{\partial v_1}{\partial p}\right)_{sat}\right) & -\frac{v_{mix}}{v}(v_g - v_1) \\ 1 & \frac{v}{v_{mix}} & 0 \\ \left(x\left(\frac{\partial h_g}{\partial p}\right)_{sat} + (1-x)\left(\frac{\partial h_1}{\partial p}\right)_{sat}\right) & \mathrm{v} & (h_g - h_1) \end{bmatrix} \begin{bmatrix} \frac{dp}{dz} \\ \frac{dv}{dz} \\ \frac{dx}{dz} \end{bmatrix} = \begin{bmatrix} \frac{v_{mix}}{A}\frac{dA}{dz} \\ -\frac{\Gamma_w}{A}\tau_w + \frac{g_z}{v_{mix}} \\ g_z \end{bmatrix} \tag{5.18}$$

In a transcritical ejector-expansion CO_2 cycle (see Figure 5.10), in the motive nozzle of the ejector, the CO_2 in a supercritical state (either supercritical vapor or supercritical liquid) expands into the subcritical two-phase region. Under typical operating conditions, the flow becomes critical at the nozzle throat. According to Katto's principle for two-phase critical flow [15, 16], the critical flow condition occurs when the determinant of the coefficient of the matrix on the left side of Eq. (5.15) is imposed equal to zero. Mathematically, that is:

$$\begin{vmatrix} v'_{mix} & -\frac{v_{mix}}{v} & (v_g - v_1) \\ 1 & \frac{v}{v_{mix}} & 0 \\ h'_{mix} & \mathrm{v} & (h_g - h_1) \end{vmatrix} = 0 \tag{5.19}$$

where:

$$v'_{mix} = xv'_g + (1-x)v'_1 = x\left(\frac{\partial v_g}{\partial p}\right)_{sat} + (1-x)\left(\frac{\partial v_1}{\partial p}\right)_{sat} \tag{5.20}$$

$$h'_{mix} = xh'_g + (1-x)h'_1 = x\left(\frac{\partial h_g}{\partial p}\right)_{sat} + (1-x)\left(\frac{\partial h_1}{\partial p}\right)_{sat} \tag{5.21}$$

By calculating the determinant of Eq. (5.19), the expression for the speed of sound can be obtained:

$$\mathrm{v}_c = \sqrt{\frac{\mathrm{v}^2_{mix}(h_g - h_1)}{(v_g - v_1)(\mathrm{h}'_{mix} - \mathrm{v}_{mix}) - \mathrm{v}'_{mix}(h_g - h_1)}} \tag{5.22}$$

It should be noted that the speed of sound is only a function of the flow stream quality and pressure.

5.3.1.2 Motive Nozzle Flow Model

As previously mentioned, the flow inside the motive nozzle is considered to be one-dimensional and steady. Additional simplifications are introduced to close the motive nozzle flow model:

- the motive nozzle is a converging nozzle where the throat is located at its exit
- the flow reaches critical flow conditions at the nozzle throat
- the inlet flow velocity is neglected
- the heat transfer between the flow stream and the nozzle walls is neglected
- the effect of the gravitational force on the flow is neglected
- the isentropic efficiency of the motive nozzle, η_m, is known.

Since the motive nozzle isentropic efficiency, η_m, inlet pressure, p_m, temperature, T_m, are known, the motive nozzle outlet pressure, p_t, and velocity, v_t can be calculated with an iterative scheme. In particular, by guessing a value of p_t $(0 < p_t < p_c)$, the outlet specific enthalpy, h_t, can be obtained from the definition of isentropic efficiency:

$$h_t = h_m - \eta_m (h_m - h_{t,is}) \tag{5.23}$$

The outlet velocity of the motive nozzle is calculated by imposing the energy conservation equation between the inlet and outlet of the motive nozzle:

$$h_m = h_t + \frac{v_t^2}{2} \tag{5.24}$$

Thus, the outlet quality, x_t, is known since $x_t = x_t(p_t, h_t)$. Since both x_t and p_t are determined, the speed of sound can be calculated from Eq. (5.22). The speed of sound, v_c, should be equal to the outlet velocity v_t. The outlet pressure p_t is updated until convergence. Once v_t is determined, the mass flow rate through the motive nozzle can be estimated by:

$$\dot{m}_m = \rho_t A_t v_t \tag{5.25}$$

where the throat area A_t of the motive nozzle is known from design assumptions, and the density of the homogeneous two-phase flow is given by:

$$\rho_t = \frac{1}{\frac{x_t}{\rho_{g,t}} + \frac{1-x_t}{\rho_{l,t}}} \tag{5.26}$$

The model assumes that the exit condition of the motive nozzle is in the two-phase region and that critical flow conditions occur at the nozzle exit. For each simulation, these two assumptions need to be checked for validity and the model should be updated accordingly.

5.3.1.3 Suction Nozzle Flow Model

The suction flow process through the suction nozzle usually takes place in a suction chamber with a complex geometry. In order to simplify the problem, the expansion process through the suction nozzle can be modeled as a flow through a converging nozzle by introducing similar assumptions of the motive nozzle flow model:

- the flow is one-dimensional and steady
- the inlet flow velocity is neglected
- the heat transfer between the fluid and the nozzle wall is neglected
- the effect of the gravitational force on the flow is neglected.

Since the mass flow rate through the motive nozzle, $\dot{m}_m$, is known, the mass flow rate through the suction nozzle can be obtained by introducing an injection ratio, ϕ_{inj}:

$$\dot{m}_s = \phi_{inj}\dot{m}_m \tag{5.27}$$

Similarly to the motive nozzle model, by using the suction nozzle inlet pressure, p_s and specific enthalpy, h_s, the suction nozzle isentropic efficiency, η_s, and the suction nozzle outlet throat area, A_b, the outlet pressure, p_b, and velocity, v_b can be determined with the following iterative procedure. By assuming the outlet pressure, p_b, the outlet specific enthalpy, h_b can be calculated from the definition of isentropic efficiency:

$$h_b = h_s - \eta_s(h_s - h_{b,is}) \tag{5.28}$$

The outlet flow velocity, v_b, is obtained by imposing the energy conservation equation between the inlet and outlet of the suction nozzle:

$$h_s = h_b + \frac{v_b^2}{2}$$

$$h_b = h_s - \eta_s(h_s - h_{b,is}) \tag{5.29}$$

At this point, the above calculated outlet velocity is compared with the value obtained from the continuity equation:

$$v_t = \frac{\dot{m}_s}{\rho_b A_b} \tag{5.30}$$

where $\rho_b = \rho_b\ (p_b, h_b)$. If the two values do not match, the outlet pressure p_b is adjusted until convergence is achieved. To be noted is that, under typical operating conditions, critical flow conditions do not occur at the outlet of the suction nozzle due to relatively small pressure differences between inlet and outlet sections.

5.3.1.4 Mixing Section Flow Model

By referring to Figure 5.12a, the mixing section of the ejector is located between the outlets of the motive and suction nozzles and the inlet of the diffuser. A number of assumptions are made to develop the model:

- at inlet of the mixing section, the motive stream has a velocity v_t, a pressure p_t, and occupies an area A_t
- at inlet of the mixing section, the suction stream has a velocity v_b, a pressure p_b, and occupies an area A_b
- at the outlet of the mixing section, the flow stream is assumed to be uniform with a velocity v_{mix} and pressure p_{mix}
- the heat transfer between the flow stream and the ejector wall in the mixing section is neglected
- the effect of the gravitational force is neglected.

The model of the mixing section is employed to predict the mixed flow velocity v_{mix}, pressure p_{mix}, and specific enthalpy h_{mix}. In particular, the mass, momentum, and energy conservation equations between the inlet and outlet of the mixing sections are solved simultaneously:

$$\rho_t A_t v_t + \rho_b A_b v_b = \rho_{mix} A_{mix} v_{mix} \tag{5.31}$$

$$p_t A_t + \eta_{mix} \rho_t A_t v_t^2 + p_b(A_{mix} - A_t) + \eta_{mix} \rho_b (A_{mix} - A_b) v_b^2 = p_{mix} A_{mix} + \rho_{mix} A_{mix} v_{mix}^2 \tag{5.32}$$

$$\dot{m}_m \left(h_t + \frac{v_t^2}{2} \right) + \dot{m}_s \left(h_b + \frac{v_b^2}{2} \right) = (\dot{m}_m + \dot{m}_s) \left(h_{mix} + \frac{v_{mix}^2}{2} \right) \tag{5.33}$$

where, the mixing section efficiency ηmix accounts for the friction flow losses in the mixing chamber and it is assumed the density ρ_{mix} is a function of the mixing pressure, p_{mix}, and specific enthalpy, h_{mix}. Hence, the three governing equations can be solved since the only unknowns are p_{mix}, h_{mix}, and v_{mix}. By obtaining the pressure and specific enthalpy at the end of the mixing section, the quality, x_{mix}, can also be estimated. At this point, the speed of sound of the two-phase stream can be calculated using Eq. (5.22) and compared to v_{mix} to determine whether the flow reaches critical conditions or not.

5.3.1.5 Diffuser Flow Model

In the diffuser, the static pressure of the two-phase mixed stream increases due to the fact that the kinetic energy is converted to static pressure. By making the assumption that the mixed stream at the outlet of the mixing section is in homogeneous equilibrium, the pressure recovery coefficient is defined as:

$$C_t = \frac{p_d - p_{mix}}{\frac{1}{2}\rho_{mix} v_{mix}^2} \tag{5.34}$$

where p_d is the pressure downstream the ejector. In the literature, correlations have been developed to estimate the pressure recovery coefficient. For example, Owen *et al.* [17] proposed the following correlation:

$$C_t = 0.85\rho_{mix} \left[1 - \left(\frac{A_{mix}}{A_d} \right)^2 \right] \left[\frac{x_{mix}^2}{\rho_{g,mix}} + \frac{(1 - x_{mix})^2}{\rho_{1,mix}} \right] \tag{5.35}$$

The diffuser outlet specific enthalpy can be calculated by imposing an energy balance between the inlet and outlet of the ejector:

$$\dot{m}_m h_m + \dot{m}_s h_s = (\dot{m}_m + \dot{m}_s) h_d \tag{5.36}$$

where the heat losses to the environment are neglected. The outlet quality x_d is automatically defined by knowing p_d and h_d.

By combining the models of motive nozzle flow, suction nozzle flow, mixing section flow, and diffuser flow, a simulation model of a two-phase flow ejector can be developed. Additional details regarding the solution schemes of each sub-model and the overall ejector model can be found in the work of Liu and Groll [18]. The specified parameters for the ejector model are motive nozzle throat area and efficiency, suction nozzle throat area and efficiency, cross sectional area of the mixing section and mixing section efficiency, and the diffuser outlet area.

5.3.2 Ejector Efficiencies

The COP improvements of a transcritical ejector-expansion CO_2 cycle over a baseline transcritical CO_2 cycle are heavily influenced by the efficiency of the primary ejector components. In most of the theoretical analyses proposed in the literature, values in the range

Table 5.1 Overview of common modeling assumptions found in literature for ejector component efficiencies.

Authors	η_m	η_s	η_{mix}	η_d
Elbel and Hrnjak [19]	0.9	0.9		0.9
Li and Groll [9]	0.9	0.9		0.8
Ksayer and Clodic [20]	0.85	0.85		0.75
Deng *et al.* [21]	0.7	0.7		0.8
Sarkar [22]	0.8	0.8		0.75
Elbel and Hrnjak [23]	0.8	0.8		0.8
Sun and Ma [24]	0.9	0.9		0.8
Eskandari Manjili and Yavari [25]	0.7	0.7	0.95	0.8

0.75–0.95 have been assumed for each of the individual ejector component efficiencies. A summary of commonly used values of motive, suction, mixing, and diffuser efficiencies for CO_2 applications found in the literature is reported in Table 5.1. It can be seen that ejector component efficiencies are typically assumed to be constant values, but such assumptions do not hold true under real operating conditions. The efficiencies are affected by both operating conditions and specific geometric parameters of the ejectors. To this end, empirical correlations based on experimental data can be developed to determine efficiencies of the ejector motive nozzle, suction nozzle, and mixing section. As an example, Liu and Groll [18] developed three polynomial correlations as a function of the pressure ratio (p_m/p_s), diameter ratio (D_t/D_{mix}), and injection ratio ϕ_{inj} (see Eq. (5.27)). In particular, the functional forms of the efficiencies are:

$$\begin{aligned}
\eta_m &= \eta_m\left(\frac{p_m}{p_s}, \frac{D_t}{D_{mix}}\right) \\
\eta_s &= \eta_s\left(\frac{p_m}{p_s}, \phi_{inj}, \phi_{inj}\left(\frac{p_m}{p_s}\right)^{0.02}\right) \\
\eta_{mix} &= \eta_{mix}\left(\left(\frac{D_t}{D_{mix}}\right)^{0.1}(1+\phi_{inj})^{0.35}\right)
\end{aligned} \tag{5.37}$$

5.4 Expansion Work Recovery Devices for Transcritical CO_2 Systems

In the previous sections, the benefits of using an expansion work recovery device to overcome the higher irreversibilities associated with the throttling process in the transcritical CO_2 cycle have been outlined. However, handling the two-phase expansion in an efficient way entails a number of technical challenges from a design standpoint that ultimately impact the economic viability of such devices. Extensive research has been conducted over the years to develop expanders for transcritical CO_2 cycles [26]. This section aims to provide a general understanding of the current state-of-the-art transcritical CO_2 expanders and practical challenges.

5.4.1 Positive Displacement Expanders

Due to the nature of the expansion process of CO_2 from a transcritical state into a two-phase condition, a number of positive displacement machines could be potentially suitable to recover the expansion work [27]. Various two-phase expanders for transcritical CO_2 applications have been investigated both theoretically and experimentally. A general overview of the different research efforts is provided in the following sub-sections.

5.4.1.1 Reciprocating Expanders

Early studies on expansion work recovery devices for transcritical CO_2 cycles focused on piston-type machines including conventional crank-based reciprocating expanders as well as single- and double-acting free-piston expanders [26]. Baek *et al.* [28] developed a prototype piston-cylinder expansion device based on a modified four-cycle, two-piston engine. The device was named ED-WOW (Expansion Device With Output Work). An out-of-phase firing order of the cylinders was chosen to reduce the need for mechanical inertia. The expansion process was controlled by using fast-acting solenoid valves as intake and exhaust valves. The concept design of the ED-WOW device and the prototype can be seen in Figure 5.13a, and b, respectively. Depending on the operating conditions, as shown in Figure 5.14, the maximum power output of the device was approximately 34.7 W. The expansion device led to an increase of evaporator capacity up to 5.4% and increase in COP of approximately 10%. The maximum isentropic efficiency of the device was in the order of 10%.

More recently, Kurtulus *et al.* [29] introduced an oil-free compressor for CO_2 applications that employed the Sanderson Rocker Arm Mechanism (S-RAM), as shown in Figure 5.15.

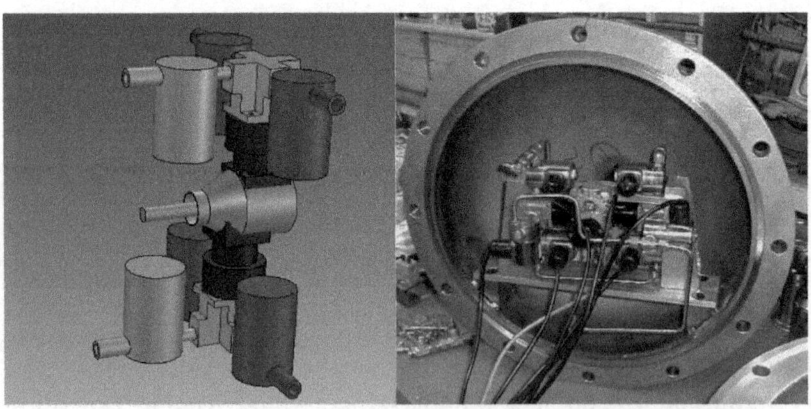

(a) (b)

Figure 5.13 ED-WOW: (a) assembly of piston-cylinder work extraction expansion device; (b) view of the prototype. Source: Obtained from Baek *et al.* [28].

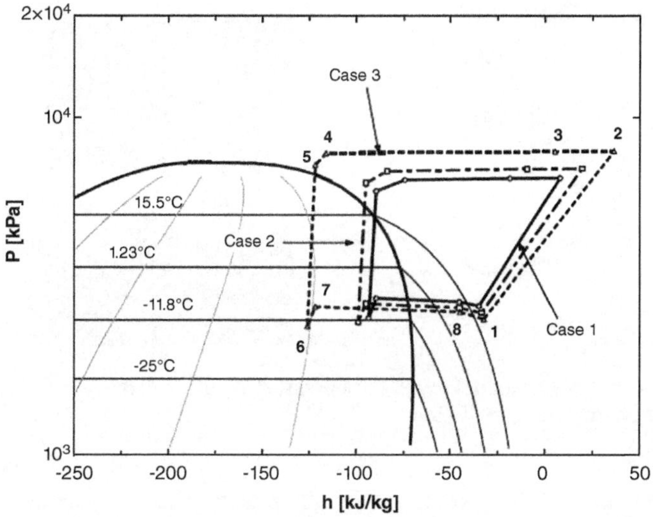

Figure 5.14 Experimental p-h diagrams for three different experimental conditions. Source: Obtained from Baek *et al.* [28].

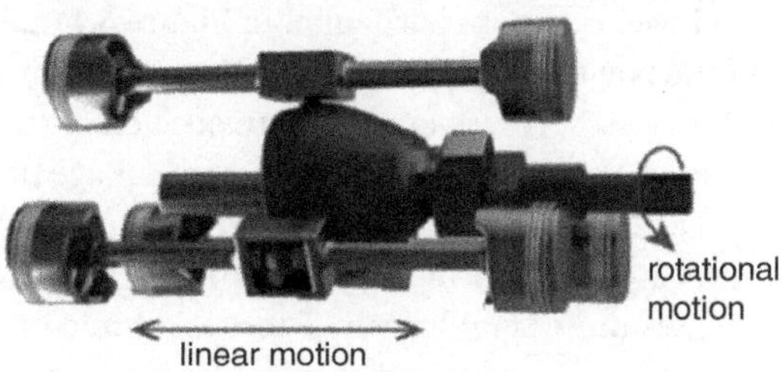

Figure 5.15 Oil-free S-RAM compressor with opposed configuration. Source: Obtained from Kurtulus *et al.* [29].

The mechanism converts the reciprocating motion into rotary motion, producing high efficiency in both directions without the energy-robbing side forces on the pistons or crossheads common to crankshaft, swash-plate or wobble-plate drive mechanisms. Moreover, the drive mechanism is able to vary the piston stroke while maintaining a fixed head clearance. Both single- and opposed-acting configurations are feasible with this mechanism. Barta *et al.* [30] investigated the feasibility of including a single-stage expander within a two-stage S-RAM compressor by employing two of the cylinders for a multi-temperature refrigerated container system.

5.4.1.2 Rolling Piston and Rotary Vane Expanders

Rolling piston machines are characterized by numerous leakage paths and friction losses. As highlighted by Zhang *et al.* [26], a single-stage rolling piston machine requires a suction control device which causes large pressure drops and noise. For these reasons, a number of studies have been conducted on two-stage rolling piston machines to recover the expansion work more efficiently. For instance, Yang *et al.* [31] developed and tested a two-cylinder rolling piston expander. As shown in Figure 5.16, the two expansion units are installed on a common crankshaft with two eccentricities to avoid run-through phenomenon during operation. Furthermore, the two expansion units are mounted such that the first discharge

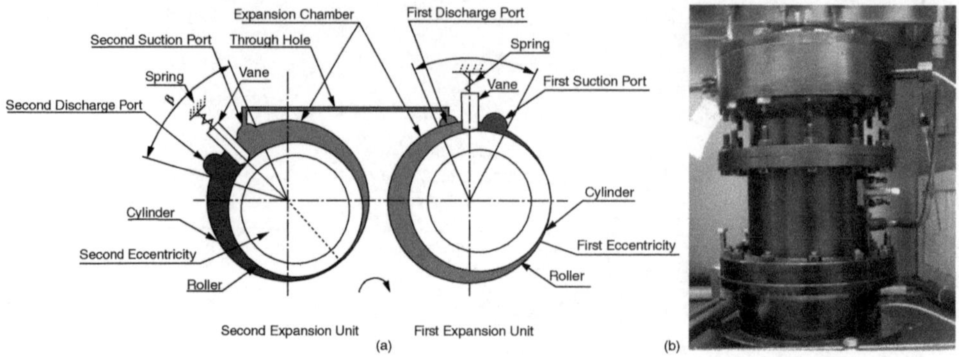

Figure 5.16 (a) Schematic view of the scroll expander with back pressure regulation; (b) Experimental cycle. Source: Obtained from Yang *et al.* [31].

port matches the second suction port by a hole in the intermediate plate. The performance of the prototype was evaluated on a transcritical CO_2 system. Nominal expander inlet conditions of pressure and temperature were set to 7200 kPa and 37°C, respectively. In order to optimize the expander, the effects of the rotational speed and the pressure ratio have been investigated. In particular, in Figure 5.17a, it can be seen that the isentropic efficiency and the power output decrease with the increase of rotational speed due to the increase of friction losses. However, the volumetric efficiency and the mass flow rate increase with the rotational speed due to the decrease in leakage flow losses. By fixing the expander rotational speed at 2000 rpm, the effect of the pressure ratio across the expander can be analyzed. The results are reported in Figure 5.17b. As the pressure ratio is increased, the power output increases quite significantly. There is a trade-off between increase in isentropic efficiency and decrease in volumetric efficiency due to higher leakage flow losses.

5.4.1.3 Scroll Expanders

Scroll machines have been demonstrated to perform well both as compressors and expanders. Leakages represent one of the major losses in a scroll machine, and the transcritical operation of the CO_2 cycle highly impacts its performance. A number of studies have been conducted on scroll expanders to recover the throttling losses. Fukuta *et al.* [32] developed and tested a scroll expander for a transcritical CO_2 cycle. In order to study the influence of transcritical leakage flows, a thermodynamic model of the expansion process based on an adjusted adiabatic exponent was developed. The expander had a swept volume of 1.53 $cm^3\ rev^{-1}$, a built-in volume ratio of 2.18, and the clearances in both radial and axial directions were assumed to be 10 μm during the calculations. The results showed that the pressure losses during the suction process increased with the increase of the rotational speed. By considering the expansion process from the supercritical region to the two-phase region, the pressure decreased rapidly with the increase of chamber volume in the supercritical region. Conversely, in the two-phase region the rate at which the pressure drops decreased with an increase in chamber volume. This is explained by the leakage flows toward the outer expansion chambers. The simulated p–V diagrams and efficiencies can be seen in Figure 5.18.

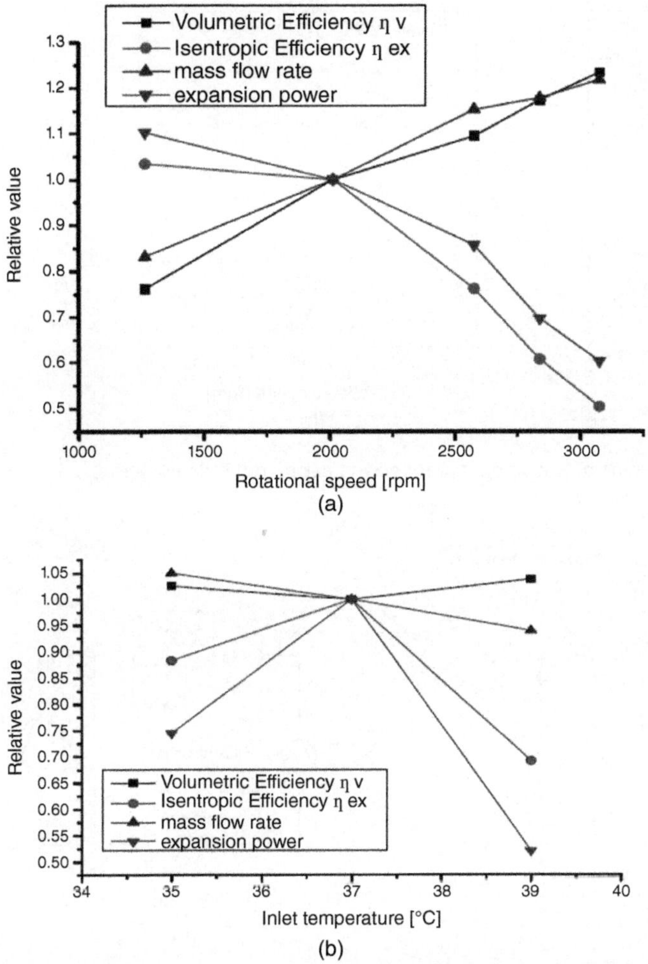

Figure 5.17 Effect of (a) expander rotational speed and (b) pressure ratio on volumetric and isentropic efficiency, mass flow rate, and expansion power. Source: Obtained from Yang *et al.* [31].

The schematic of the scroll expander prototype is shown in Figure 5.19a. Besides the scroll wraps and other typical elements of scroll designs, balancers are placed on the shaft and a pressure port is positioned on the housing to apply the same inlet pressure value to the back of the orbiting scroll to cancel the thrust force. The balancers on the shaft compensate the imbalanced radial forces and moments. The lubrication of the moving parts is ensured by having lubricant oil mixed with CO_2. The experimental CO_2 refrigeration cycle with back-pressure line is illustrated in Figure 5.19b. The experimental results in terms of volumetric efficiency and total efficiency are reported in Figure 5.20. It can be seen that the volumetric efficiency was approximately 80% for all the tested rotational speeds, which can be explained by recalling that a scroll does not have a direct flow path from inlet to outlet as in rolling piston or vane types of machines. The tight machining tolerances ensured good

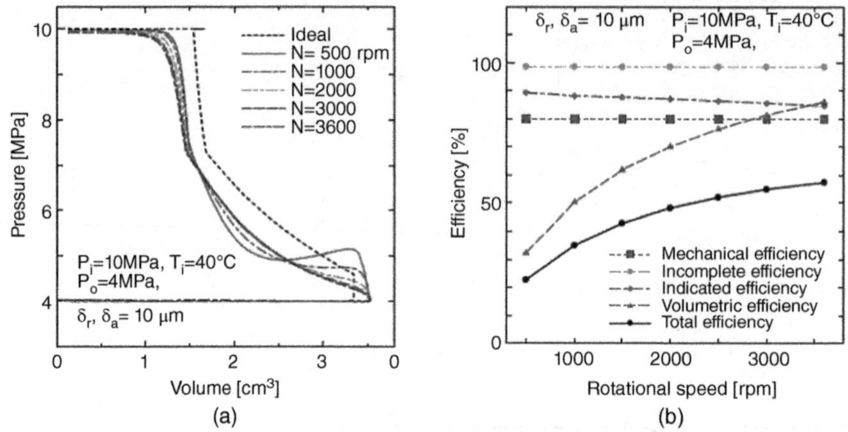

Figure 5.18 Simulated *p–V* diagrams of the scroll expander and estimated efficiencies at different rotational speeds [32].

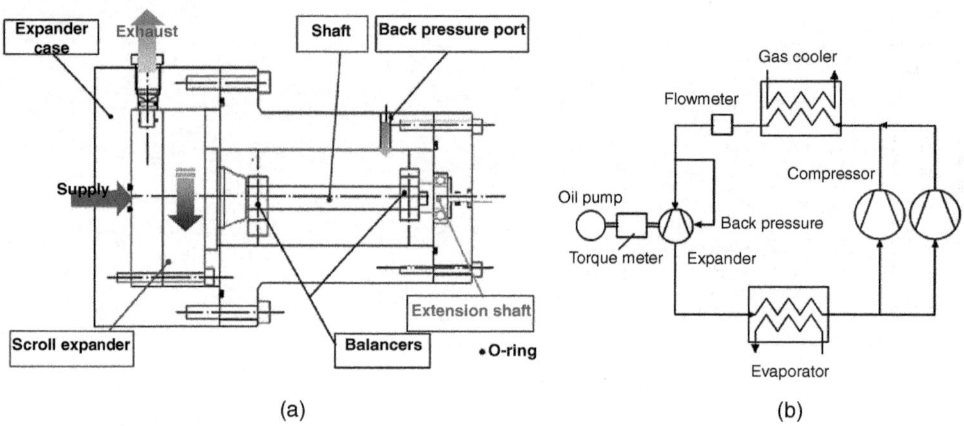

Figure 5.19 (a) Schematic view of the scroll expander with back pressure regulation; (b) Experimental cycle. Source: Obtained from Fukuta *et al.* [32].

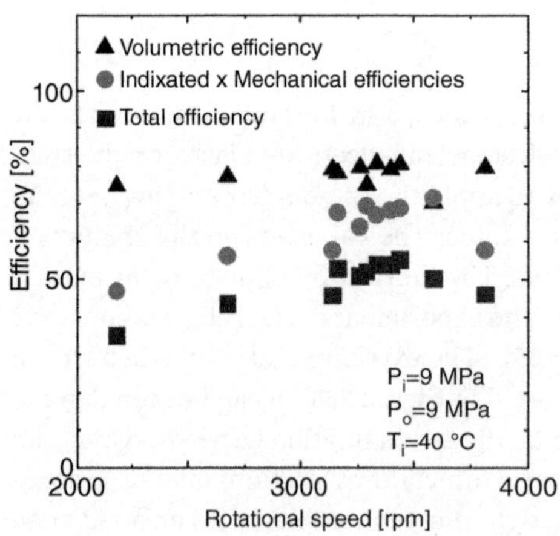

Figure 5.20 Experimental results of the CO_2 scroll expander at different rotational speeds [32].

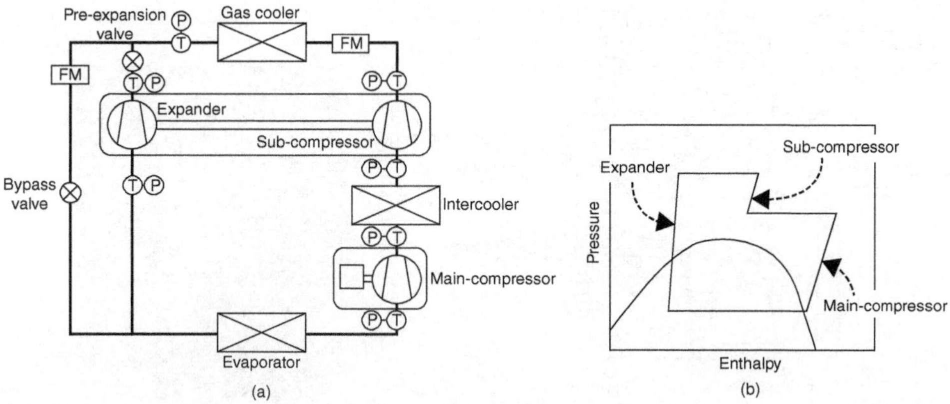

Figure 5.21 (a) Schematic of the cycle architecture; (b) p-h diagram of the cycle. Source: Obtained from Nagata *et al.* [33].

performance over a range of rotational speeds. The maximum total efficiency achieved was 55% at 3500 rpm.

Nagata *et al.* [33] proposed a combined scroll-type expander with a sub-compressor to be employed in a CO_2 refrigeration cycle with intercooler, originally proposed by Baek *et al.* [34]. The schematic of the two-stage cycle with an intercooler and an expander as well as the thermodynamic process on a $p–h$ diagram are shown in Figure 5.21a, and b, respectively. The mechanical work recovered from the scroll expander is directly used to drive the second-stage compressor after the intercooler. The expansion device is characterized by a dual-sided configuration of the scroll wraps with the expansion located on the bottom side and the compression process located on the top side. The compression and expansion sides are placed on the same center shaft and the expansion process discharges at low-pressure into the housing. The concept design of the combined scroll expander and compressor is shown in Figure 5.22a, and the actual prototype is shown in Figure 5.22b. Preliminary experimental results showed that by employing both an intercooler and an expander, the system COP improved by approximately 30% in the case of an isentropic expansion. The mechanical power generated by the expander reduced the power consumed by the second-stage compressor up to 25%. The axial loads acting on the dual-sided orbiting scroll were also calculated. Under the given operating conditions, it was estimated that the actual axial load was approximately one-tenth of the upward force resulting from the expansion side due to the balancing effect of the compression side.

5.4.1.4 Screw Expanders

Twin-screw compressors and expanders have been widely used in different applications due to their high efficiency, reliability, compactness, and relatively low manufacturing costs to achieve very small clearances. Furthermore, such technology has been proven to be suitable to handle flash expansion, as proposed by Smith *et al.* [35]. Despite the favorable characteristics over other positive displacement machines, when applied to CO_2, the high pressure differences across the rotors result in high bearing loads. Such heavy loads cause rotor deformations that are of the order of magnitude of the clearances between the rotors

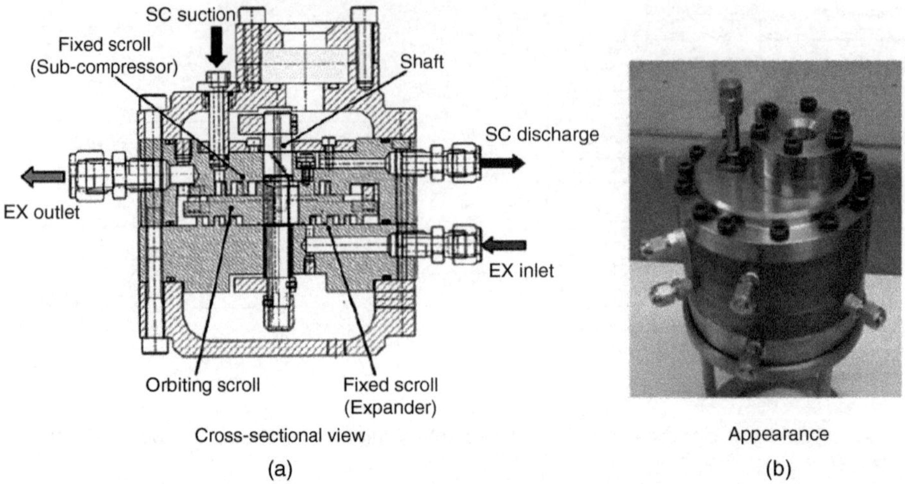

Figure 5.22 (a) Cross-sectional view of the combined scroll-expander and sub-compressor; (b) view of the scroll prototype. Source: Obtained from Nagata *et al.* [33].

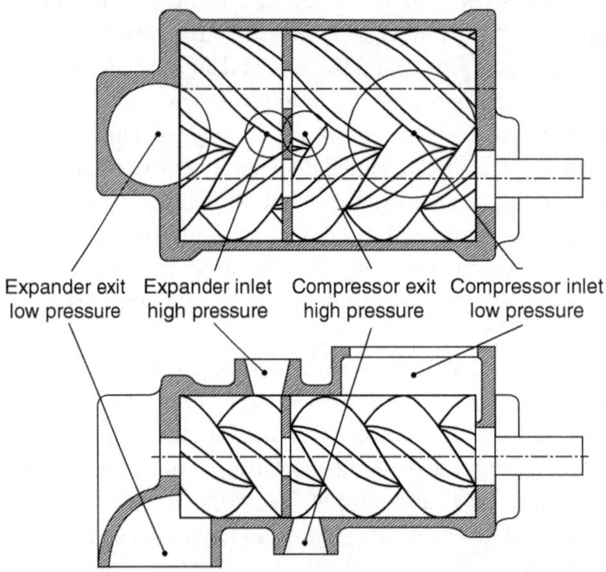

Figure 5.23 Schematic view of the combined twin-screw compressor-expander design [36].

and the casing. In order to balance the heavy loads, Stosic *et al.* [36] developed a combined compressor-expander machine, shown in Figure 5.23, to be used in a transcritical CO_2 cycle. The cycle architecture and operating conditions can be seen in Figure 5.24. The compressor and expander rotors are manufactured on the same shafts, but the compression and expansion chambers are separated within a single casing. The flow arrangements in and out of the compressor and the expander are critical to reduce the loads. As shown in Figure 5.23, the high pressure fluid enters the expander suction port located at the top of the casing, near

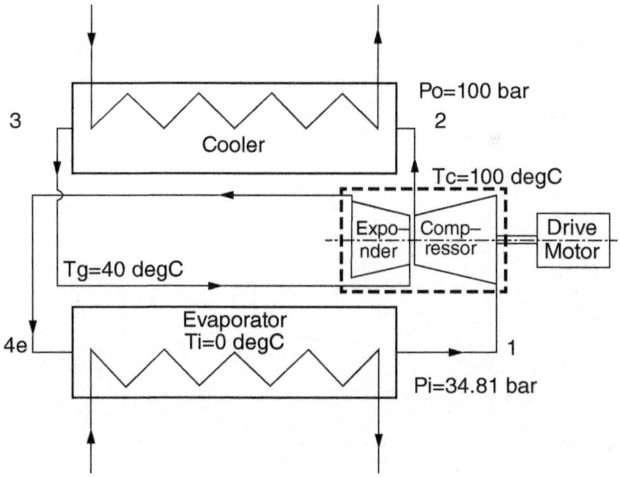

Figure 5.24 Schematic of a transcritical CO_2 heat pump with a balanced rotor compressor-expander [36].

the center, and it exits as a two-phase mixture from the low pressure port at the bottom of the casing at one end. The CO_2 from the evaporator enters the compressor part through a low-pressure port at the top of the opposite end of the casing. After being compressed, it is discharged though the high-pressure port at the bottom of the casing, near the center. Due to the fact that the high-pressure ports are positioned in the center of the machine and on the opposite sides of the casing, the high-pressure forces due to compression and expansion are opposed to each other, and, more significantly, only displaced axially from each other by a relatively short distance. Therefore, the radial forces on the bearings are also significantly reduced. In addition, since both ends of the rotors are at more or less equal pressure, the axial forces are balanced out. To ensure part-load control capabilities, the expansion section can contain a capacity control such as a slide or a lifting valve to alter the volume passing through the machine. Simulation results showed that, by introducing the compressor-expander design, the cooling capacity increased by 11% due to the different exit states of the isenthalpic and adiabatic expansion processes. At the same time, the work recovery reduced the total compressor power input by 34.6%. In the ideal case, the COP can be improved up to 72% from 2.79 to a more acceptable 4.8. Although the calculations have been carried out with idealized work input and output, the overall gain in COP is significant.

5.4.2 Turbine-Type Expanders

The expansion process of a transcritical CO_2 system starts in the supercritical region and ends in the two-phase region with a quality that can exceed 0.5, i.e. over 50% liquid by mass. Hence, utilizing radial inflow machines in the wet region leads to low performance and erosion issues from the liquid centrifuging outward [37]. Advancements in two-phase nozzle technology and two-phase impulse turbines allowed the development of commercial turbines that are cost-competitive, as stated by Hays and Brasz [37]. Usually, the shaft

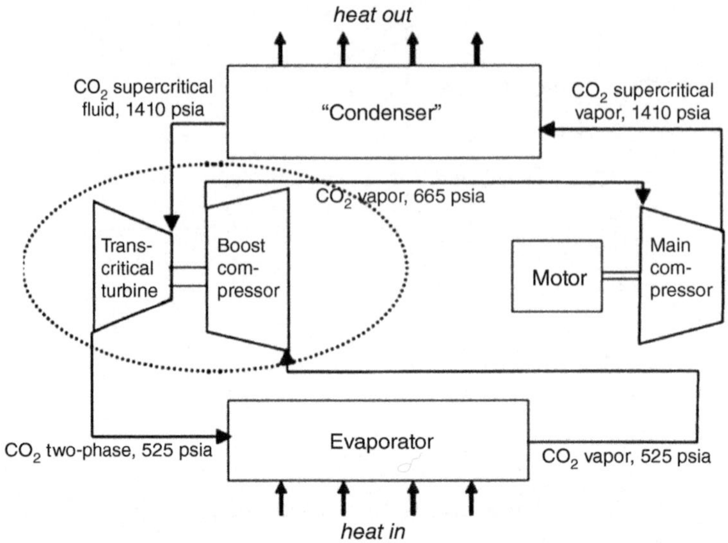

Figure 5.25 Schematic of a transcritical CO_2 heat pump with a transcritical turbine powering the boost compressor [37].

power generated by the turbines can be effectively exploited by employing several technical solutions:

- the turbine can be installed on the outboard shaft of the compressor motor, decreasing the power required by the compressor
- utilize a hermetic turbine-compressor, where the turbine shaft power directly drives a centrifugal boost compressor, reducing the power required by the main compressor, as shown in Figure 5.25
- the electrical power generated by the turbines with high-efficiency high-speed generators can be used to reduce the net power of the main compressors.

In order to deal with the two-phase conditions, Hays and Brasz [37] developed a two-phase axial-inflow turbine consisting of a two-phase nozzle to convert potential and pressure energy into kinetic energy, and an axial-flow turbine with blades designed to maximize the kinetic energy transfer from the high velocity two-phase mixture. The jet from the two-phase nozzle impinges upon the axial-flow turbine blade, as shown in Figure 5.26. In particular, if the blade has a sufficiently long axial dimension, the liquid phase tends to separate from the vapor phase, creating a liquid film on the blade that exits at an angle that is different from both the actual blade leaving angle and direction of thrust. In fact, the axial path of the two-phase flow is a key feature of an impulse turbine design to ensure that the liquid leaves the rotor (in radial inflow turbines, the centrifugal motion forces the liquid in the opposite direction to the flow). The design of an axial two-phase turbine with a single nozzle is shown in Figure 5.27a and Figure 5.27b. The turbine was initially tested with R-134a and was operated at a maximum rotational speed of 12,800 rpm. The measured efficiency was 56% at a power output of 310 W against the 61% predicted by

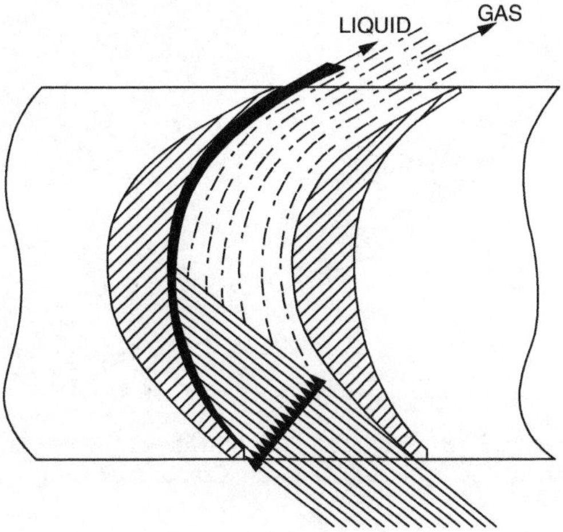

Figure 5.26 Schematic of two-phase jet interaction with axial blading.

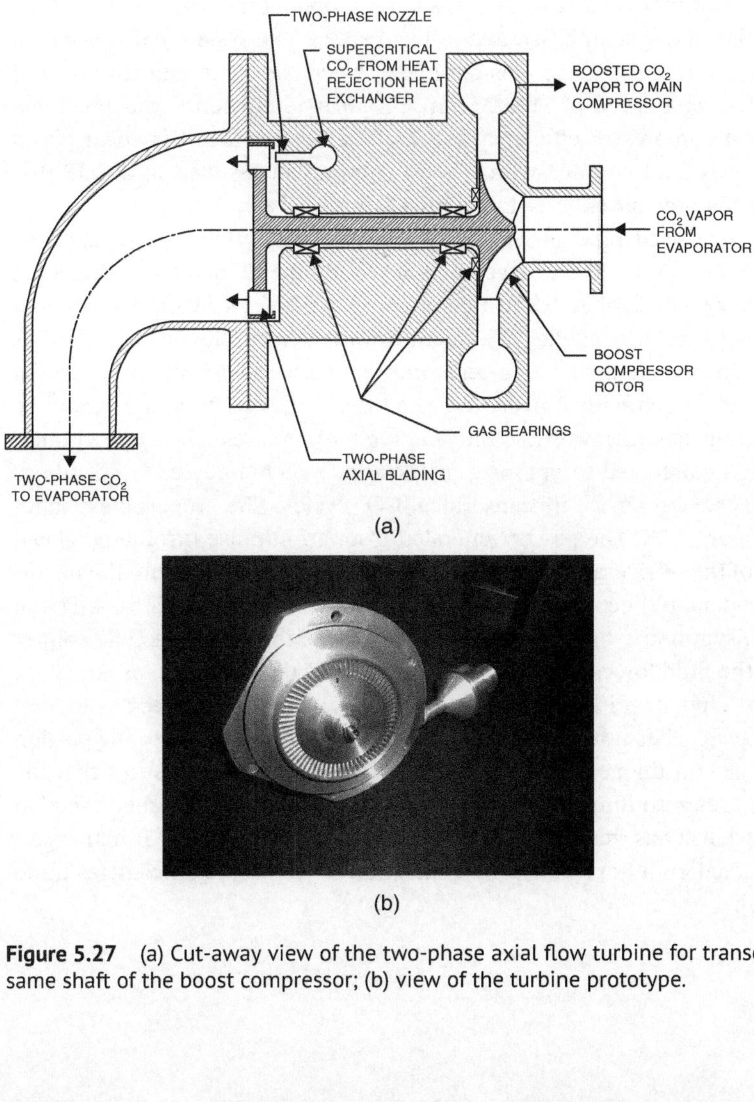

Figure 5.27 (a) Cut-away view of the two-phase axial flow turbine for transcritical CO_2 on the same shaft of the boost compressor; (b) view of the turbine prototype.

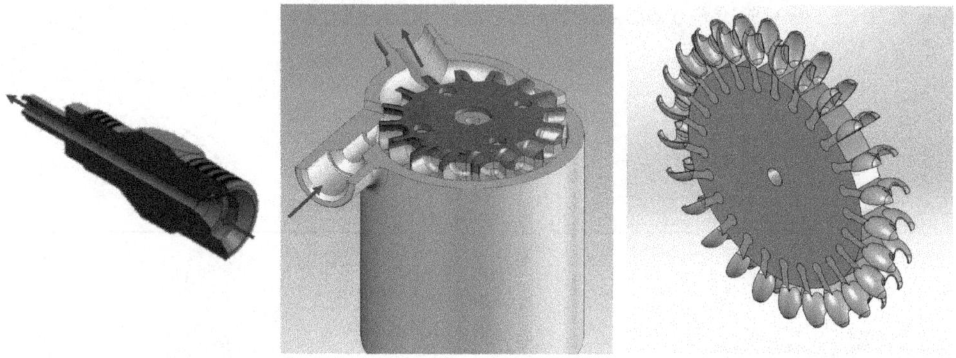

Figure 5.28 From left to right: view of the nozzle, housing, and turbine of the Viper expander device. Source: Obtained from Czapla *et al.* [38].

the design conditions. A numerical assessment was done to predict the performance of the turbine employed in the CO_2 system illustrated in Figure 5.25. The system had a nominal capacity of six tons and rotational speed was optimized by considering both turbine and compressor efficiencies. At a speed of 110,000 rpm, the turbine efficiency was predicted to be 69% with a boost compressor efficiency of 80%. The transcritical CO_2 heat pump featuring the turbine expander coupled with a boost compressor resulted in a COP that was 1.39 times higher than the baseline system with a throttling valve.

The concept of an impulse turbine, and later impulse-reaction turbine, have also been investigated by Czapla *et al.* [38] with the development of the Viper Expander device. Such a device consists of a nozzle that converts the high pressure of the working fluid into a high velocity jet that is directed to the impeller of a micro-turbine wheel. The turbine impeller is mounted on a shaft directly coupled with a generator to produce electrical energy. Initial experimental results conducted with R-410A indicated that better performance could be achieved with refrigerants having lower viscosity and higher pressures. Therefore, numerical analyses have been conducted to optimize the design of such a device to operate as an energy recovery expansion device in transcritical CO_2 cycles. The proposed expander design is shown in Figure 5.28. The power generated from an impulse turbine is related to the kinetic energy of the refrigerant impinging on the impeller and therefore the majority of the pressure drop should occur across the nozzle. The turbine impeller is a Pelton wheel which has two symmetric buckets with a splitter blade in the center. The splitter blades act to balance the fluid forces acting on the impeller and the buckets cause the fluid flow direction to turn. This magnitude of the change in flow direction relates to the rotational speed of the impeller. The nozzle is characterized by a very short converging portion followed by a long constant diameter cross-sectional area. Such design ensures that the two-phase expanding stream continues to accelerate through the constant diameter section due to the decreasing density as the fluid expands into the two-phase dome. Initial experimental results conducted on a hot-gas bypass stand showed isentropic efficiencies up to approximately 7% [39].

Nomenclature

A	area, m^2
C_t	pressure recovery coefficient
CFC	chlorofluorocarbon
CFD	computational fluid dynamics
CO_2	Carbon Dioxide
COP	coefficient of performance
D	diameter, m
δ	clearance, μm
ECU	environmental control unit
ED-WOW	expansion device with output work
ev	evaporator
EXV	electronic expansion valve
g	gravitational constant, $m\,s^{-2}$
Γ	friction perimeter, m
h	specific enthalpy, $kJ\,kg^{-1}$
HCFC	hydrochlorofluorocarbon
HCFO	hydrochlorofluoro-olefins
HFC	hydrofluorocarbons
HFO	hydrofluoro-olefins
IHX	internal heat exchanger
L	stroke, mm
$\dot{m}$	mass flow rate, $kg\,s^{-1}$
n	speed, $1\,min^{-1}$
n_{frame}	frame rate, $ft\,s^{-1}$
NH_3	Ammonia
P	pressure, kPa, MPa, bar, psia
PAG	Polyalkylene Glycol
q	specific heat transfer, $kJ\,kg^{-1}$
$\dot{Q}$	capacity, tons of refrigerant, kW
S	entropy, $kJ\,K^{-1}$
S-RAM	Sanderson rocker arm mechanism
T	temperature, °C, K
t	time, ms
τ	frictional shear stress, kPa
TXV	thermostatic expansion valve
u	velocity, $m\,s^{-1}$
V	volume, cm^3
v	velocity, $m\,s^{-1}$
$\dot{v}$	swept volume, $cm^3\,rev^{-1}$
w	specific work, $kJ\,kg^{-1}$

$\dot{W}$	power, W, kW
x	quality
XV	expansion valve
z	distance, m

Greek Symbols

η	efficiency
ε	effectiveness
ρ	density, kg m^{-3}
ν	specific volume, m^3 kg^{-1}

Subscripts

1 … i	state point
act	actual
b	receiving chamber
d	diffuser
evap	evaporator
ex	expansion
exp	expansion
comp	compressor
cond	condenser
g	vapor
gc	gas cooler
h	isenthalpic expansion outlet
H	heat source
in	inlet
inj	injection
is	isentropic
l	liquid
L	heat sink
m	motive
mb	motive – receiving section interface
mech	mechanical
mix	mixing section
out	outlet
ϕ	injection ratio
s	isentropic, suction
sb	suction – receiving section interface
sat	saturated
t	turbine expansion outlet, throat
v	volumetric
w	wall
z	gravitational direction

References

1 Lorentzen, G. and Pettersen, J. (1992). Development of a transcritical carbon dioxide environmental control unit. *Proceedings of the IIR International Symposium on Refrigeration, Energy, and Environment*, Trondheim, Norway, pp. 147–163

2 Kim, M.H., Pettersen, J., and Bullard, C.W. (2004). Fundamental process and system design issues in CO_2 systems. *Progress in Energy and Combustion Science* 30: 119–174.

3 Lorentzen, G. (1994). Revival of carbon dioxide as refrigerant. *International Journal of Refrigeration* 17: 292–301.

4 Robinson, D.M. and Groll, E.A. (2000). Theoretical performance comparison of CO_2 transcritical cycle technology versus HCHC-22 technology for a military packaged air conditioner application. *HVAC&R Research* 6: 325–348.

5 Groll, E.A. and Kim, J.H. (2007). Review of recent advancements toward transcritical CO_2 cycle technology. *HVAC&R Research* 13: 499–520.

6 Robinson, D.M. and Groll, E.A. (1997). Efficiencies of transcritical CO_2 cycles with and without an expansion turbine. *International Journal of Refrigeration* 21: 577–589.

7 Fukuta, M., Nakamura, Y., and Yanagisawa, T. (2012). Characteristics of CO_2 transcritical expansion process. In Proceedings of International Refrigeration and Air-Conditioning Conference at Purdue, Paper 1265. https://docs.lib.purdue.edu/iracc/1265/.

8 Kornhauser, A. A. (1990). The use of an ejector a refrigerant expander. Internationa Refrigeration and Air Conditioning Conference at Purdue. Paper 82.

9 Li, D. and Groll, E.A. (2005). Transcritical CO_2 refrigeration cycle with ejector-expansion device. *International Journal of Refrigeration* 28: 766–773.

10 Liu, F., Groll, E.A., and Li, D. (2012). Investigation on performance of variable geometry ejectors for CO_2 cycles. *Energy* 45: 829–839.

11 Lucas, C., Rusche, H., Schroeder, A., and Koehler, J. (2014). Numerical investigation of a two-phase CO_2 ejector. *International Journal of Refrigeration* 43: 154–166.

12 Banasiack, K., Palacz, M., Hafner, A. et al. (2014). A CFD-based investigation of the energy performance of two-phase R774 ejectors to recover the expansion work in refrigeration systems: an irreversibility analysis. *International Journal of Refrigeration* 40: 328–337.

13 Crowne, C.T. (2006). *Multiphase Flow Handbook*. CRC Press/Taylor & Francis Group.

14 Colarossi, M., Trask, N., Schmidt, D., and Bergander, M. (2012). Multidimensional modeling of condensing two-phase ejector flow. *International Journal of Refrigeration* 35: 290–299.

15 Katto, Y. (1968). Dynamics of compressible saturated two-phase flow (critical flow). *Bulletin of JSME* 11: 1135.

16 Katto, Y. (1969). Dynamics of compressible saturated two-phase flow (critical flow-sequel, and flow in a pipe). *Bulletin of JSME* 12: 1417.

17 Owen, I., Abdul-Ghani, A., and Amini, A.M. (1992). Diffusing a homogenized two-phase flow. *International Journal of Multiphase Flow* 18: 531–540.

18 Liu, F. and Groll, E.A. (2013). Study of ejector efficiencies in refrigeration cycles. *Applied Thermal Engineering* 52: 360–370.

19 Elbel, S. W. and Hrnjak, P. S. (2004). Effects of internal heat exchanger on performance of transcritical CO_2 systems with ejector. *Proceedings of 10th International Refrigeration and Air Conditioning Conference at Purdue*, Purdue University, USA. Paper R166.

20 Ksayer, E. B. and Clodic, D. (2006). Enhancement of CO_2 refrigeration cycle using an ejector: 1D analysis. *Proceeding of International Refrigeration and Air Conditioning Conference at Purdue*, Purdue University, USA. Paper R058.

21 Deng, J., Jiang, P., Lu, T., and Lu, W. (2007). Particular characteristics of transcritical CO_2 refrigeration cycle with an ejector. *Applied Thermal Engineering* 27: 381–388.

22 Sarkar, J. (2008). Optimization of ejector-expansion transcritical CO_2 heat pump cycle. *Energy* 33: 1399–1406.

23 Elbel, S. and Hrnjak, P. (2008). Experimental validation of a prototype ejector designed to reduce throttling losses encountered in transcritical R744 system operation. *International Journal of Refrigeration* 31: 411–422.

24 Sun, F. and Ma, Y. (2011). Thermodynamic analysis of transcritical CO_2 refrigeration cycle with an ejector. *Applied Thermal Engineering* 31: 1184–1189.

25 Eskandari Manjili, F. and Yavari, M.A. (2012). Performance of a new stage multi-intercooling transcritical CO_2 ejector refrigeration cycle. *Applied Thermal Engineering* 40: 202–209.

26 Zhang, Z., Ma, Y., Li, M., and Zhao, L. (2013). Recent advances of energy recovery expanders in the transcritical CO_2 refrigeration cycle. *HVAC&R Research* 19: 376–384.

27 Kruse, H., Russmann, H., Martin, E., and Jakobs, R. (2006). Positive displacement carbon dioxide expansion machines – changes and limitations. In Proceedings of International Compressor Engineering Conference at Purdue. Paper 1766.

28 Baek, S. J., Groll, E. A. and Lawless, P. B. (2002). Development of a piston-cylinder expansion device for the transcritical carbon dioxide cycle. In Proceedings of International Refrigeration and Air Conditioning Conference at Purdue. Paper 584

29 Kurtulus, O., Lumpkin, D., Yang, B. (2014). Performance and operating characteristics of a novel positive-displacement oil-free carbon dioxide compressor. *Proceedings of International Compressor Engineering Conference at Purdue.* Paper 2375

30 Barta, R. B., Groll, E. A., Hugenroth, J. J. (2018). Modeling of S-RAM energy recovery compressor integration in a transcritical carbon dioxide cycle for application in multi-temperature refrigerated container system. In Proceedings of 13th IIR Gustav Lorentzen Conference, Valencia. http://dx.doi.org/10.18462/iir.gl.2018.1117

31 Yang, J., Zhang, L., Zhang, L., and Li, H. Y. (2010). Development of a two-cylinder rolling piston CO_2 expander. *Proceedings of International Compressor Engineering Conference at Purdue.* Paper 2022.

32 Fukuta, M., Yanagisawa, T., Kosuda, O., and Ogi, Y. (2006). Performance of Scroll expander for CO_2 Refrigeration Cycle. *Proceedings of International Compressor Engineering Conference at Purdue.* Paper 1768.

33 Nagata, H., Kakuda, M., Sekiya, S. (2010). Development of a scroll expander for the CO_2 refrigeration cycle. *Proceedings of International Compressor Engineering Conference at Purdue.* Paper 1952.

34 Baek, J. S., Groll, E. A., and Lawless, P. B. (2002). Development of a piston-cylinder expansion device for the transcritical carbon dioxide cycle. *Proceedings of International Refrigeration and Air Conditioning Conference at Purdue.*

35 Smith, I.K., Stosic, N., and Aldis, C.A. (1996). Development of the trilateral flash cycle system: part 3: the design of high-efficiency two-phase screw expanders. *Proceedings of the Institution of Mechanical Engineers, Part A: Journal of Power and Energy* 210: 75–93.

36 Stosic, N., Smith, I. K., and Kovacevic, A. (2002). A Twin Screw Combined Compressor and Expander for CO_2 Refrigeration Systems. *Proceedings of International Compressor Engineering Conference at Purdue.* Paper 1591.

37 Hays, L. and Brasz, J. J. (2004). A transcritical CO_2 turbine-compressor. *Proceedings of International Compressor Engineering Conference at Purdue.* Paper 1628

38 Czapla, N., Inamdar, H., Barta, R., and Groll, E. A. (2016). Theoretical analysis of the impact of an energy recovery expansion device in a CO_2 refrigeration system. *Proceedings of International Compressor Engineering Conference at Purdue.* Paper 2504.

39 Barta, R. B. and Groll, E. A. (2017). Application of viper energy recovery expansion device in transcritical carbon dioxide refrigeration cycle. *Proceedings of 12th IEA Heat Pump Conference*, Rotterdam.

6

Transcritical Carbon Dioxide Compressors

Xin-Rong Zhang

Department of Energy and Resources Engineering, College of Engineering, Peking University, Beijing, China

6.1 Introduction

The extensive and widespread use of refrigerants chlorofluorocarbons (CFCs) and hydrochlorofluorocarbons (HFCs) has brought increasingly serious problems to the environment. Because natural refrigerants cause minimal damage to the environment, the selection and use of a range of natural refrigerants has become a trend. Among them, CO_2 has become one of the best alternative refrigerants for CFCs and HFCs due to its good thermo-physical properties. At present, a large number of researchers have begun to conduct research into the use of CO_2 as the most alternative refrigerant. Numerous studies have shown that CO_2 can be used as a replacement for conventional working fluids in heat pumps [1]. In addition, under the Montreal Protocol, Fluorocarbons will be phased out [2]. Ammonia (R717, ASHRAE safety level B2) sometimes has better performance, but if you want to meet both technical and safety requirements, there are actually not many natural refrigerants. For another example, water (R718, ASHRAE safety level A1) is not too non-toxic. However, due to the low working pressure and low density of water, it cannot be used as the best choice for the vapor compression refrigeration cycle [3]. Another disadvantage of water is that its performance heating coefficient (COP) is very low and it is not cost-effective [4]. Table 6.1 compares the characteristics of some of the most commonly used refrigerants and carbon dioxide. Table 6.1 compares the thermal properties of CO_2 and common refrigerants, such as critical pressure, critical temperature, heat production, and safety, such as toxicity and flammability. Meanwhile, the environmental protection of ozone depletion potential (ODP)/global warming potential (GWP) is also compared in detail.

Among the alternative natural refrigerants, CO_2 (ASHRAE safety level A1) is one of the few refrigerants that is non-toxic and non-combustible. As a refrigerant in the thermal cycle, CO_2 does not need to consider its exhaust gas and exhaust pollution problems too much. It can be directly released into the air without worrying about excessive emissions of harmful gases. CO_2 has a very low GWP compared to other commercial refrigerants. In addition,

Transcritical CO_2 Heat Pump: Fundamentals and Applications,
First Edition. Xin-Rong Zhang and Hiroshi Yamaguchi.

Table 6.1 Basic properties of CO_2 (R744) compared with other refrigerants [5].

Properties	R744	R22	R134A	R410A
ODP/GWP	0/1	0.06/1700	0/1300	0/1900
Flammability/toxicity	N/N	N/N	N/N	N/N
Molecular mass (kg kmol^{-1})	44.0	86.5	102.0	72.6
Critical pressure (MPa)	7.38	4.97	4.07	4.79
Critical temperature (°C)	31.1	96.0	101.1	70.2
Refrigeration capacity (kJ m^{-3})	22 545	4356	2868	6763

using CO_2 as a refrigerant does not have the responsibility of being over-regulated by the relevant agencies, because it has zero ODP. A large amount of carbon dioxide in the environment (0.04% of the atmosphere) makes it cost-effective. Chemically, carbon dioxide is an inert gas. According to ASHRAE 15 and 34 and ISO 5149 safety standards, carbon dioxide is a safe refrigerant. Therefore, there are almost no leakage issues.

In addition, because of its high fluid density and working pressure, CO_2 can provide great help with manufacturing light heat pump systems for a given specific energy and pumping power [6]. The higher exothermic temperature sliding in the gas cooler further enhances the heating performance of the heat pump system [5]. Similarly, compared with fluorocarbons, CO_2 leads to higher isentropic efficiency in heat pump systems due to its lower compression ratio [7]. These favorable factors make CO_2 an excellent candidate for effectively replacing traditional working fluids such as CFC and HFC. However, the main problems due to the use of CO_2 are its lower critical point (31.1°C and 7.38 MPa) and higher working pressure (8.0–11.0 MPa) [8]. Therefore, special attention needs to be paid to these two issues when designing system components. The research into supercritical carbon dioxide (SCO_2) compressors is very important because of the large power consumption of the compressor and its safety and reliability. This chapter focuses on the research status and practical application challenges of the SCO_2 compressor. Several types of SCO_2 compressors are introduced and described as following.

6.2 Sliding Vane CO_2 Compressor

A sliding vane compressor is a type of positive displacement compressor. The leakage path of gas in the compression process of the sliding vane compressor includes peripheral sealing gap, rotor surface gap, sliding vane end gap and re-expansion. Because the leakage loss is an important factor that affects the volumetric efficiency in a sliding vane CO_2 compressor, and the pressure drop loss has a small effect on the volumetric efficiency, the volumetric efficiency will increase with the increase of the rotational speed. For sliding vane CO_2 compressors, the suction pressure drop is small even at high speeds, so it is more suitable for working at high speeds compared to R134a compressors [9]. Since the leakage loss decreases with the increase of the rotation speed and the pressure loss increases with the increase of the rotation speed, the indicated efficiency will be slightly convex as the rotation speed increases. Due to the pressure loss during inhalation, the amount of refrigerant sucked in is less than the ideal amount, so the benchmark for indicating efficiency will exceed 1.0. At the same time, the higher the speed, the greater the inertial force on the sliding plate, which will increase the friction at the top of the sliding plate, causing the mechanical efficiency to slowly decrease with the increase in the rotating speed.

Due to the high CO_2 pressure, the leakage is relatively large and the volumetric efficiency is relatively low. Compared with the R134a sliding vane compressor, the total clearance must be reduced to two-thirds of the R134a to achieve the same volumetric efficiency. Effective sealing, that is, reducing peripheral clearance and rotating surface clearance, that is, reducing the width of the stator and increasing the thickness of the sliding plate, can increase the volumetric efficiency. However, an increase in the thickness of the sliding plate will also lead to a decrease in mechanical efficiency. On the other hand, the flow velocity of the suction and exhaust gas in the CO_2 compressor is small, and the flow resistance is small, so that the indication efficiency of the CO_2 sliding vane compressor is high and hardly changes with the change of the clearance.

The perimeter seal length is a very important parameter to measure the efficiency of the compressor. Studies have shown that increasing the length of the peripheral seal is an effective way to reduce leakage, and the volumetric efficiency will increase significantly with the increase of the length of the peripheral seal.

Sliding vane compressors can also be designed as two-stage compression. Leakage around the seals around the first and second compression rooms can be ignored, so the compressor efficiency can be improved. The pressure difference acting on the sliding plate is also small, which is beneficial for improving the reliability of the valve strength and the mechanical efficiency of the compressor. Using the second flow channel of the sliding vane compressor as an expander, it becomes an expansion compressor.

Table 6.2 shows the structural parameters of the single-stage sliding vane, double-stage sliding vane and sliding vane expansion compressors developed by Shizuoka University in Japan. Their performances are compared in Table 6.3 [10].

Table 6.2 Specification parameters of the sliding vane CO_2 compressor developed by Shizuoka University, Japan [10].

Compressor type		Distance from stator center to edge (mm)	The width of the stator (mm)	Slider thickness (mm)
Single stage compressor		25	10	2
Two-stage compressor	First stage	15	15	2
	Second stage	25	15	2
Expansion compressor	Compressor	25	15	2
	Expander	25	15	2

Table 6.3 Comparison of working performance of various sliding vane CO_2 compressors developed by Shizuoka University in Japan [10].

Compressor	Clearance (μm)	η_v	η_i	η_m	Torque (N)
First-level compression	15	0.76	0.93	0.81	11.4
Secondary compression	15	0.83	0.87	0.86	10.8
	15	0.71	0.88		
Expansion compressor	15	0.91	0.90	0.74	9.8
	15	0.87	0.89		

η_v: volume efficiency; η_i: indicated efficiency; η_m: machine efficiency.

6.3 Screw CO_2 Compressor

The twin-screw expansion compressor was developed by Stosic, N. and others at the Positive Compressor Technology Center of City University of London [11, 12]. The test results show that the pressure range at which single-stage screw compressors can run stably is below 6.5 MPa, and under special conditions, it can reach 8.5 MPa. Under high pressure (HP), the axial load of the rotor is reduced due to the inclusion of a balance piston; however, a high radial load is still unavoidable. The developed twin-screw expansion compressor not only balances the axial load but also reduces the radial bearing load, and solves the design problems caused by the high bearing load of the screw compressor used in the CO_2 system. Aiming at the characteristics of high CO_2 trans-critical cycle pressure, the twin-screw compressor of the traditional working fluid system was improved, and the expansion and compression were performed in two independent chambers to avoid leakage in theory. The expansion compressor has been proved to reduce axial loads by 20%.

The CO_2 single-stage screw compressor introduced by Japan's MAYCOM is mainly used in refrigeration and air-conditioning systems [13]. The twin-screw expansion compressor is shown in Figure 6.1. The refrigerant runs from right to left of the Figure. First, the low-pressure refrigerant is compressed from the inlet to the compressor to high-pressure gas. Secondly, the high-pressure gas from the outlet of the compressor enters the expander, and the high-pressure gas is transformed into low-pressure gas through the expander, and is discharged with the outlet of the expander. The CO_2 screw compressor uses an oil injection method. The compressed CO_2 gas must be separated from oil and CO_2, for which a self-differential oil separation system is used. The oil and gas mixture then enters the separation tank, the separated oil is heat exchanged with the cooling water from the cooling tower and then returned to the compressor after the temperature is lowered. Finally, the CO_2 enters the gas cooler for heat exchange. The design of the whole unit is to use both cold and heat. The exhaust of the compressor is used to heat the hot water. The unit is equipped with a water storage tank and low-temperature CO_2 is used for refrigeration.

Wang, and others from Xi'an Jiaotong University [10], developed a CO_2 screw compressor for an NH_3/CO_2 cascade refrigeration system. The profile of the rotor was designed using the envelope theory. The number of teeth of the male rotor was five and the number of teeth of the female rotor was eight. Based on this, the exhaust port of the compressor was

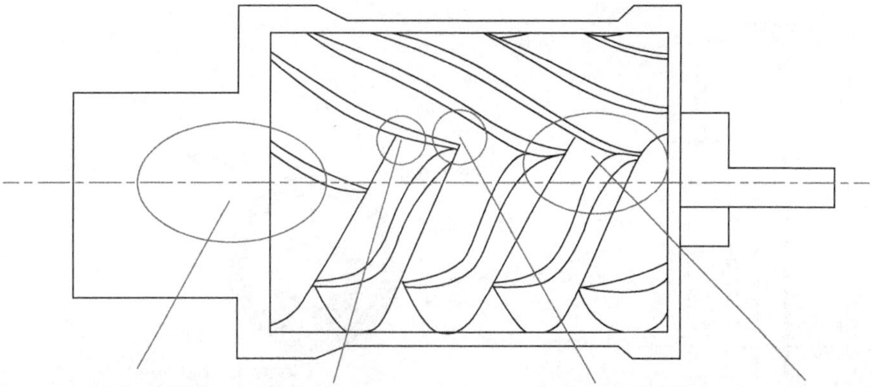

Figure 6.1 Structure diagram of CO_2 twin-screw expansion compressor.

designed. As a result, it is seen that the rotor deformation and the bearing can meet the requirements. The test proves that the designed CO_2 twin-screw compressor can be used in the NH_3/CO_2 cascade refrigeration system.

6.4 CO_2 Rolling Rotor Compressor

6.4.1 CO_2 Compressors Developed by the Company

Daikin Industry Co., Ltd. developed a high-performance and high-reliability rolling-rotor compressor for CO_2 heat pump water heaters and automotive air conditioners in February 2002 [14]. The compressor schematic diagram and mechanism schematic diagram are shown in Figure 6.2. The advantages and disadvantages of the rolling-rotor compressor and the scroll compressor are compared. In terms of pressure bearing, leakage and reducing friction, the rolling-rotor compressor has great advantages. The heat pump water heater rolling rotor compressor has a size of $\Phi 126$ mm × 265 mm, a working volume of 3.7 ml, and a built-in permanent magnet synchronous direct current (DC) motor. Research shows that reducing the ratio of the height to the diameter of the cylinder will increase the efficiency of the compressor because the leakage through the gap between the rolling rotor and the cylinder is reduced [14]. The eccentricity of the CO_2 compressor is small, which results in its maximum stress not being greater than that of the R-134a compressor; under severe conditions such as acceleration, the liquid film on the bearing is maintained well and has good reliability.

Japan's SANYO has designed rolling-rotor compressors for domestic water heaters [15]. In the design, double-stage compression is used to make the shaft's resistance torque change smoothly; the shaft shape and slides within the eccentric with concentrated stress are improved and the first-stage exhaust is divided into two ways to ensure the pressure is intermediate pressure and facilitates lubrication.

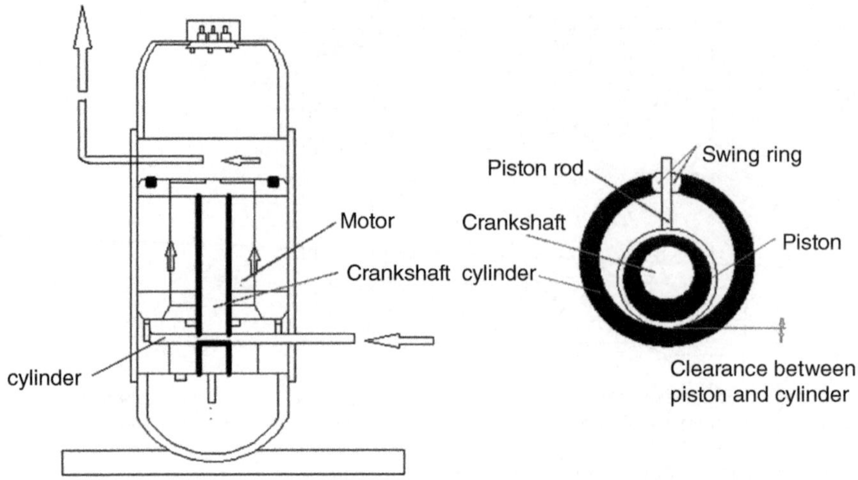

Figure 6.2 Schematic diagram of rolling-rotor compressor and mechanism.

6.4.2 Two-Stage Rolling Piston CO_2 Compressor

Tadano *et al.* [16] developed a two-stage rolling-piston CO_2 compressor for small refrigeration equipment such as refrigerators. Figure 6.3 is a layout diagram of the compression equipment and a flow chart of the air flow. In the first stage of compression, low-pressure gas is sucked in from the lower part of the compression equipment, compressed to an intermediate pressure, and sent to the casing and its external pipes respectively. The refrigerant passing through the two channels is exchanged in the shell, and then enters the second stage compression from the upper part of the compression device. Two rolling pistons are used to achieve two-stage compression. Two rotary compression units use a single drive shaft to maintain a 180° phase difference. This type of compressor has the advantages of high efficiency, low vibration, and low noise; due to the high-pressure working conditions, it is appropriate to control the pressure difference in each stage within 2–4 MPa.

The University of Maryland developed a two-stage rolling-piston CO_2 compressor [17]. A sectional view of the compressor is shown in Figure 6.4. The compressor lubrication system design, suction, and exhaust flow pattern design, exhaust valve design, radial, and thrust bearing design were analyzed in detail. The compressor cooling capacity and COP increase as the suction temperature increases. Compressor mechanical efficiency is 40–60% and volumetric efficiency is 40–80%. As the ambient temperature increases, the cooling capacity decreases significantly, and the COP also shows a downward trend. It has also been found that the compressor volume efficiency is less affected.

Research by Yang *et al.* [18] showed that the increase in ambient temperature significantly reduced the cooling capacity, and the COP also showed a downward trend.

Tecumseh's Yap and Haller developed a two-stage rolling piston compressor for a commercial heat pump [19]. The plan view of the mechanism is shown in Figure 6.5. The outer diameter is 207.8 mm, the height is 445.8 mm, and the working volume is 21.8 ml. The first stage exhaust pressure is 7–8 MPa, and the second stage exhaust pressure is 11–12 MPa.

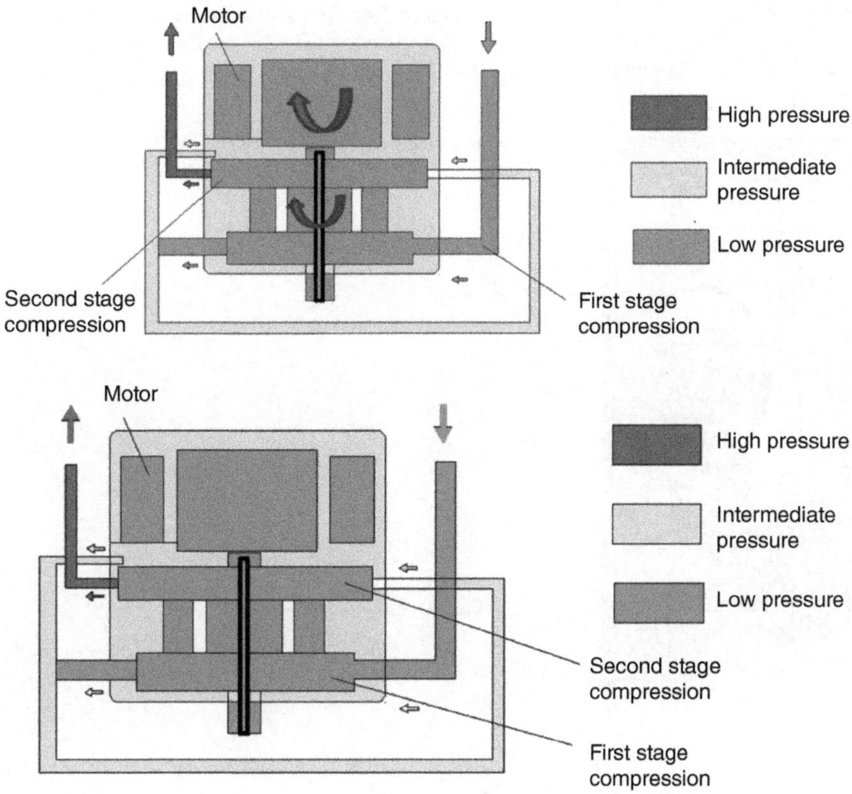

Figure 6.3 Schematic diagram of a two-stage rolling piston compressor mechanism.

Compressor performance tests showed that a COP of 2.25 was obtained at an intake pressure of 4.48 MPa and an exhaust pressure of 10.17 MPa. From the compressor performance parameters, the feasibility of the prototype for commercial heat pump water heaters is very high. Further work is focused on improving efficiency.

6.5 SCO_2 Scroll Compressor

Fagioli B.E. of the Norwegian University of Science and Technology has established axial and radial leakage models. This model simulates the working performance of a CO_2 scroll compressor and compares the performance of different working fluids. If the volumetric efficiency or isentropic efficiency is equal, when compared with other refrigerant compressors, the gap value of the CO_2 compressor should be smaller [20].

Xi'an Jiaotong University developed a fully enclosed CO_2 scroll compressor prototype [21]. It designed a new type of axial flexible mechanism and opened a circular hole for installing a radial slider in the eccentric part of the main shaft so that the radial gap value between the moving disk and the static disk was kept in a small range. In addition, an annular groove is provided on the side of the endplate of the moving disk near the main shaft,

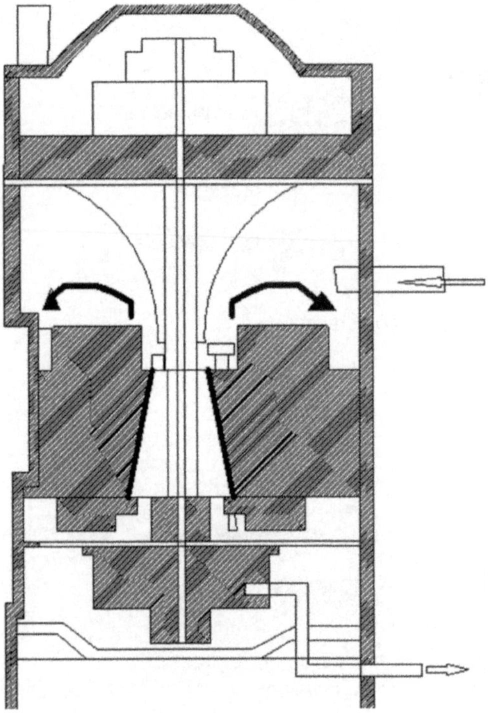

Figure 6.4 The section view of a two-stage rolling piston CO_2 compressor.

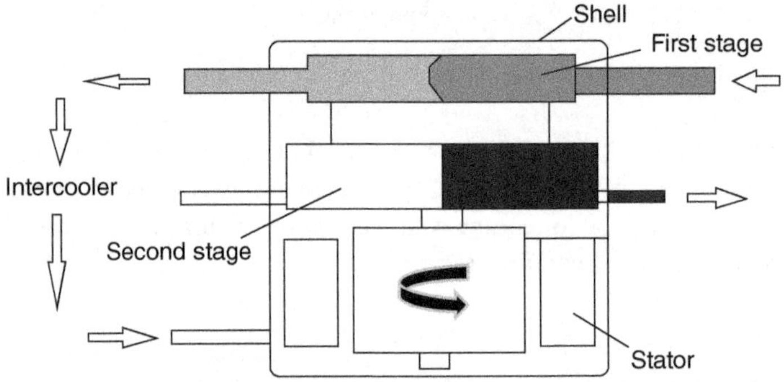

Figure 6.5 Plan view of a rolling piston compressor mechanism.

and the eccentric rotation of the annular groove causes the lubricating oil to be intermittently supplied into the suction chamber and the cross-slip ring groove, thereby enhancing lubrication.

Iwata, *et al.* [22] of Matsushita Electric Industrial Co., Ltd. studied the relationship between the injection rate and performance efficiency of the compression chamber in detail through experiments, and considered the difference between different oil film thicknesses and reducing circumferential and radial leakage. It was determined that the optimal injection rate range is 2–15%. It shows that the correlation between the optimal fuel injection rate and the refrigerant flow rate is strong.

Japan's Denso designed and developed a scroll compressor for CO_2 water heaters [13]. The compressor has a volume of $3.3 cm^3$ and uses a DC motor and an inverter. In order to reduce friction losses, rolling bearings are used. Precision machining and assembly can reduce leakage losses and enable the compressor to run efficiently.

Based on the R-134A scroll compressor, Matsushita Co., Ltd. of Japan has developed a CO_2 scroll compressor [13] with a system cooling capacity of 215–510 kW. The research results show that improving its thrust bearings is a very effective way to improve compressor efficiency.

Japan's Mitsubishi Heavy Industries (Mitsubishi) has developed a CO_2 scroll compressor for a CO_2 water heater. The moving scroll and static scroll are specially designed to reduce leakage; the oil pressure pipe is used instead of the thrust bearing to generate an axial back force to play a thrust role [13]. Japan's Mitsubishi Heavy Industries conducted a mechanical friction loss analysis of a SCO_2 scroll compressor for automotive air conditioners, and the loss caused by leakage and heat transfer accounted for a large proportion [23]. Taking the top contact mechanism and selecting thrust rolling bearings and other measures will greatly reduce this part of the loss.

Hasegawa, *et al.* carried out experimental verification on the developed prototype and compared it with theoretical simulation results [24]. The results show that the volumetric efficiency and compressor efficiency of a fully enclosed CO_2 scroll compressor increase with the increase of the rotational speed and the pressure difference between the suction and discharge. In addition, the volumetric efficiency of the CO_2 scroll compressor is not much different from that of the R-134A, but the compressor efficiency of the CO_2 scroll compressor is lower than that of the R-134A compressor.

6.6 SCO_2 Turbo-Compressor

6.6.1 SCO_2 Turbo-Compressor Applications and Challenges

Huge pressure differences in CO_2 during compression and expansion leads to high throttling losses. Cycle efficiency when using CO_2 as a refrigerant, compared with the current HCFs refrigerant, may be low. However, if the trans-critical CO_2 refrigeration cycle with an expander is compared with the HCFs refrigeration cycle with an expansion valve it was found that the former has better performance. In addition, using an expander with an efficiency of only 60% compared with a CO_2 refrigeration cycle with a throttle valve, the former has a 33% improvement in refrigeration cycle efficiency. Similarly, compared with the CO_2 refrigeration cycle with the largest internal heat exchange throttle valve, the CO_2 refrigeration cycle efficiency with a 60% efficiency expander is increased by 25%.

Another application related to gaseous CO_2 is liquefied CO_2, for transport or storage. If you make extensive use of the CO_2 refrigeration cycle, it is clear that an expander is needed to be used to recover energy. However, the expansion, which starts in the supercritical region, enters the two-phase region, producing over 50% liquid by mass. This part of the energy loss problem needs to be solved urgently. Until recently, two-phase expanders have not been developed for commercial applications. Attempts to use radial inflow machines in the wet region have not been successful due to poor performance and erosion from the liquid centrifuging outward. Attempts to use volumetric machines are also costly. However, the latest results of two-phase nozzle technology and two-phase pulse turbine design have enabled commercial turbines to recover energy from two-phase refrigerant expansion [25–27]. The units designed and produced by Hays has a precise structure, high efficiency, and low cost [28]. They have been used as part of more than 125 large commercial chiller OEM installations [28]. Another requirement of the expander for improving efficiency of the cooling system is that the cost-effective method of generating shaft power is used. This has been accomplished by installing a two-phase turbine on the outboard shaft of the compressor motor, unloading the power required for refrigerant compression [26].

Another method uses a closed two-phase turbo compressor, in which the centrifugal booster compressor is directly driven by the power of the two-phase turbine shaft so that the power consumed by the main compressor is reduced [29]. For small systems, this method has a low manufacturing cost and highest reliability.

The third option is to use the expansion energy to generate electricity. Expansion work can greatly reduce the compressor's net power flow. Recent advances in high-speed generators may make this option very effective for large systems.

6.6.2 The Two-Phase Axial-Flow Turbine

The two-phase axial-flow turbine is comprised of two basic elements:

(1) A two-phase nozzle to convert thermal and pressure energy into directed kinetic energy.
(2) Axial-flow turbine blades designed to maximize kinetic energy transfer from the high-velocity two-phase mixture.

Two-phase nozzles are nozzles that expand to high speed and low pressure under high pressure. The nozzle is filled with this gas-liquid mixture, which is composed of a flashing liquid or condensed gas. If the gas phase is the water vapor of an accelerated liquid, this flow is called a single component. The other can also be a chemical different from a liquid, and in this case, the flow is called two-component.

The gas in the two-phase nozzle is accelerated by the pressure difference between the inlet and outlet. Because the gas shears the liquid phase into relatively small droplets, which results in a relatively large surface, a good coupling between the phases is achieved. The gas transfers momentum to the droplets so that the gas forms a homogeneous mixture at high speed. By programming the most basic conservation equations for droplet formation and rupture, and heat transfer and momentum exchange between phases, nozzle performance can be better predicted and nozzle design improved.

Some laboratories and sites have performed multiple verifications for multiple working fluids and successfully designed a suitable two-phase nozzle design in real life [26]. Many designs include not only the design of turbine nozzles for geothermal and waste heat recovery using water as a working fluid, axial flow turbines that use refrigerants in chiller applications are also included.

A typical two-phase nozzle profile generated by the code is shown below in Figure 6.6 [27]. Note that the nozzle is of a converging-diverging geometry with a throat because the flow exiting the two-phase nozzle is usually supersonic. An additional

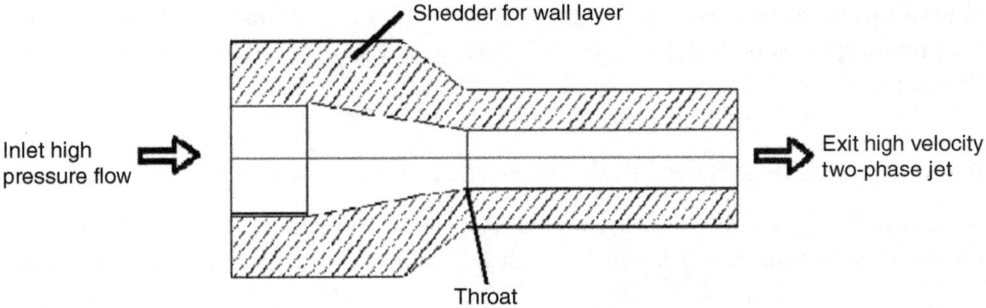

Figure 6.6 Typical two-phase nozzle profile generated from computer code.

feature shown is a shedder to strip liquid from the wall into the bulk stream where it can be efficiently atomized.

Jets from two-phase nozzles impinge on axial-flow turbine blades, as shown in Figure 6.7 [28]. In this position, the liquid in the gas can be separated, so that a thin layer can be formed on the blade. If the axial dimension of the blade is longer, the liquid will separate and flow on the blade (there is a tendency to move in a straight line) will exit the blading at an angle that is different from the blade-leaving angle and direction of thrust. Therefore, it should be noted that the design should keep the chord length of the blade as small as possible.

The most important part of the pulse turbine design is to provide an axial path for two-phase flow. As previously mentioned, a radial inflow machine will centrifuge the liquid in the opposite direction of the flow. Serious corrosion sometimes occurs between the nozzle and the rotor blade, as most of the liquid is concentrated between the nozzle and the rotor blade. In the axial design, most of the liquid leaves the rotor in the form of a vortex, so that it can be collected on the housing wall. A small portion after centrifugation is collected on the shroud, which also directs the airflow to the casing wall. An axial

Figure 6.7 Schematic of two-phase jet interaction with axial blading.

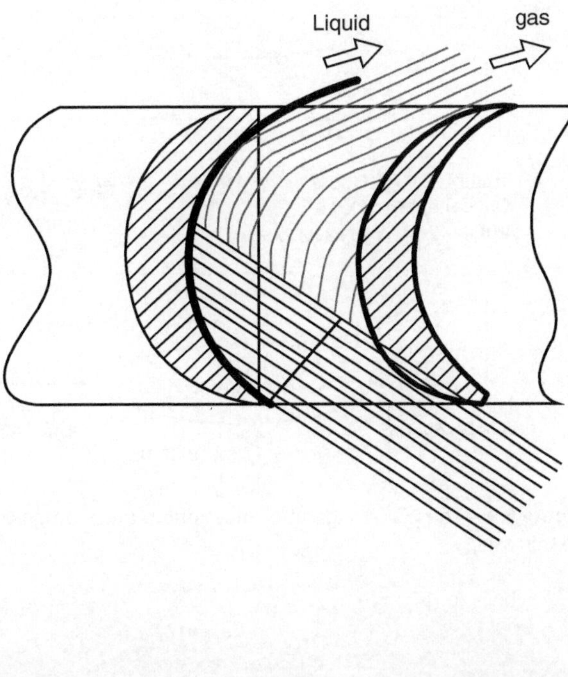

flow two-phase turbine having a single nozzle was also designed and tested for a high lift heat pump application using R-134a [29]. A schematic drawing of this unit is shown in Figure 6.8.

6.6.3 Application of Transcritical Turbine to CO_2 Refrigeration Systems

As shown in Figure 6.9, this cycle is a CO_2 refrigeration-heat pump cycle using a trans-critical turbine and a booster compressor [30]. First, the supercritical CO_2 from

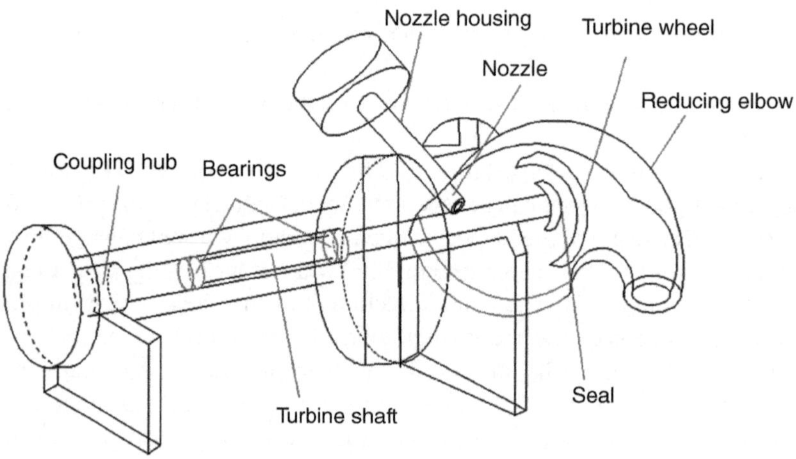

Figure 6.8 Cut-away drawing of two-phase axial flow turbine for heat pump.

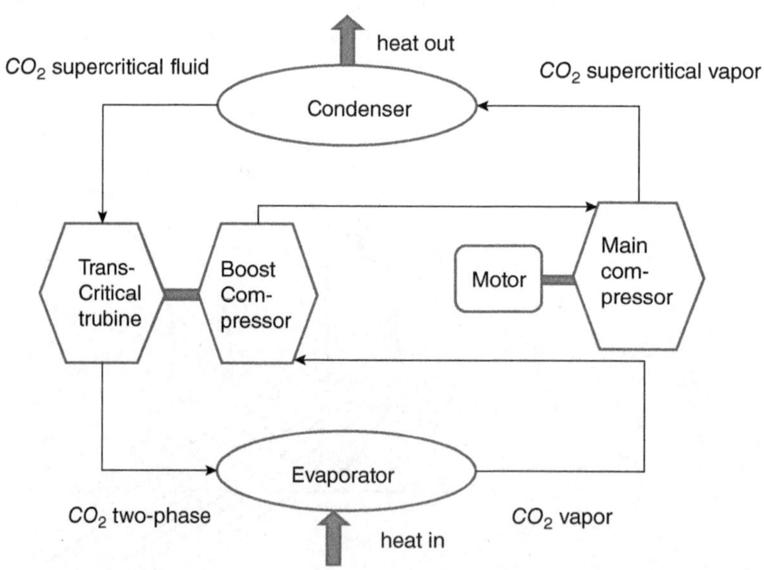

Figure 6.9 The CO_2 refrigeration-heat pump cycle with a trans-critical turbine powering the boost compressor.

the main compressor is cooled in a heat exchanger. Next, the cooled supercritical CO_2 is expanded into a two-phase region in a trans-critical turbine. Immediately afterward, the liquid CO_2 in the two-phase stream is evaporated in an evaporator. CO_2 gas then enters the booster compressor driven by the turbine drive shaft. Finally, the pressure in the booster compressor flows from CO_2 to the main compressor, so that the pressure in the main compressor is increased to the maximum circulating pressure.

6.7 SCO_2 Piston Compressor

Carbon dioxide (CO_2) has nowadays become the standard refrigerant choice for many applications, often granting better COP levels than the previously adopted technology based on HFCs: this trend is progressively leading to both a lower usage of high GWP refrigerants and sensible energy cost reductions for several industrial sectors. However, specific components and appropriate system design are mandatory when approaching CO_2 technology, due to the unique thermodynamic characteristics of this refrigerant.

In particular, the compressor has been subject to a deep engineering investigation and modeling, which resulted in the introduction of many technical solutions arising from the automotive sector: far more advanced than the typical state-of-the-art HFCs compression technology.

Given the very large differential pressures induced in a CO_2 heat pump (see later section), piston compression technology seems to be the most effective, reliable and durable. In fact, in a typical rotative compression process, the sealing between the high-pressure and low-pressure side vanes is assured only by the lubrication oil; during operation, especially in case of low ambient temperatures, a CO_2 heat pump brings differential pressures as large as 80 bar, which in case of rotative compression technology leads to severe high-pressure leak-back, leading to poor heating capacities and often calling for more complicated double stage solutions and/or use of external electric heaters to assure enough capacity is delivered during very cold ambient operation.

On the other hand, reciprocating (piston) compression technology offers much more effective performance, even with the very large differential pressures induced by the heat pump operation when very low ambient temperatures occur: in fact, the typical piston assembly with one or more compression rings fitted on its external skirt offers a formidable way to significantly decrease the leak-back from the compressor high-pressure side to its low-pressure side. For this reason, reciprocating (piston) design represents the best technology to cope with CO_2 heat pump applications, this being the reason why this chapter will focus only on the reciprocating kind of technology.

6.7.1 CO_2 Challenges from a Compressor Perspective

Taking into proper consideration the thermo-physical properties and the fluid dynamic features of carbon dioxide, it can clearly be seen how the compressor is to be engineered with specific features and advanced design solutions. Particular mention should be made about the following parameters, which are the drivers for the proper understanding of the challenges to be overcome for an appropriate and durable compressor design.

6.7.1.1 High Polytropic Exponent and Discharge Temperatures

When a gas undergoes a reversible process, this process frequently takes place in such a manner that a plot of log P vs log V is a straight line (1), thus leading to the following equation:

$$PV^n = \text{constant}$$

where n is defined as the polytropic exponent.

If we assume the process is fully reversible and if we assume constancy for specific heats, n is equal to the ratio of the same specific heats:

$$n = Cp/Cv.$$

The polytropic exponent is typically used to calculate the isentropic end of compression discharge temperatures (IECDT). The table below lists various refrigerants' data, based on a specific modeling (2) and given a fixed return gas temperature (RGT) of 4.44°C and specific suction pressure (SP) and discharge pressure (DP) (Table 6.4).

It is therefore clear that CO_2 brings severe challenges connected with its high polytropic exponent which, in turn, leads to high discharge temperature in consideration of the mechanical and electrical compression inefficiencies. In fact, the real end of compression discharge temperature (RECDT) will be higher than the isentropic one, by a factor which is a function of how the compressor performs in real life. Therefore, it is worth comparing real-life operation with the following table, which shows RECDT values for various refrigerants. Data are taken by using DORIN software, release 18-03, rating compressors in typical medium temperature (MT) applications, with a given evaporating temperature (ET), a given ambient temperature (AT) and a given superheat (SH) (Table 6.5).

Besides real-life operation, compressor manufacturers typically design their equipment with a given safety margin, testing and qualifying the machine in operating conditions which are supposed not to occur, but which assure safe operation throughout its lifetime. In this particular case, it is likely to test the compressor with a very high-pressure ratio

Table 6.4 IECDT values for various refrigerants.

Refrigerant	SP (kPa)	DP (kPa)	n (–)	IECDT (°C)
R404A	484, 80	2250, 95	1,004830	65,08
R290 (propane)	242, 48	1671, 83	1,004783	64,81
R744 (CO2)	4195, 21	11 376, 75	1,289373	98,14

Table 6.5 RECDT values for various refrigerants.

Refrigerant	ET (°C)	AT (°C)	SH (K)	RECDT (°C)
R404A	−20	35	10	62
R290 (propane)	−20	35	10	69
R744 (CO2)	−20	35	10	145

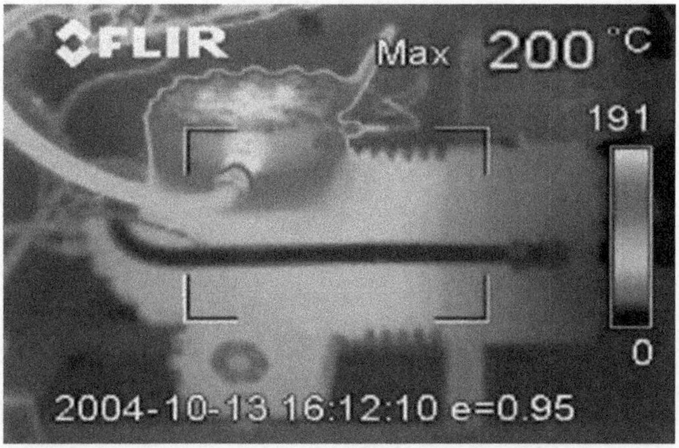

Figure 6.10 Thermal picture of a CO_2 compressor during qualification tests.

(PR = 8) with a suction pressure (SP = 13 bar) and a discharge pressure (DP = 104 bar) with 10 K at the suction superheat (SH). The picture below shows the thermal behavior of the compressor running in such conditions (Figure 6.10).

It is clearly visible that the RECDT is approaching 200°C, more than double the discharge temperature which typically occurs in the discharge plenum of a standard HFC compressor. Obviously, this significantly affects the engineering solutions which have to be adopted in order to cope with this problem. Design adaptations cannot be achieved with standard HFCs' compressors' platforms: this extreme thermal load is a challenge which needs to be faced properly; no synergy between HFCs' compressors' platforms and CO_2 compressors' platforms can be in place. The following solutions may be considered, especially if the compressor is used in a heat pump circuit.

6.7.1.2 Lubricant

A new generation of lubricants is used to cope with the various challenges carbon dioxide imposes. These lubricants, especially in heat pump applications, need to withstand the very high thermal loads induced in the compressor operation, without cracking and without compromising their lubrication properties at such high temperatures. In this sense, Polyalchalineglyocol (PAG) oils have proven to be a reliable solution, with a flash point above 200°C, assuring appropriate lubrication during compressor operation in a heat pump circuit.

6.7.1.3 Discharge Plenum

Specific design for the compressor discharge plenum is necessary in order to cope with the very high compression temperatures occurring with carbon dioxide. In this sense, efforts must be concentrated on achieving the best thermal insulation between low pressure and high-pressure compressor sides; refrigeration compressors, shells, and crankcase are typically made of special cast iron, a material which features excellent thermal conduction properties. Thus, if the high discharge temperatures are not properly isolated, these will immediately propagate to the compressor body, leading to higher discharge

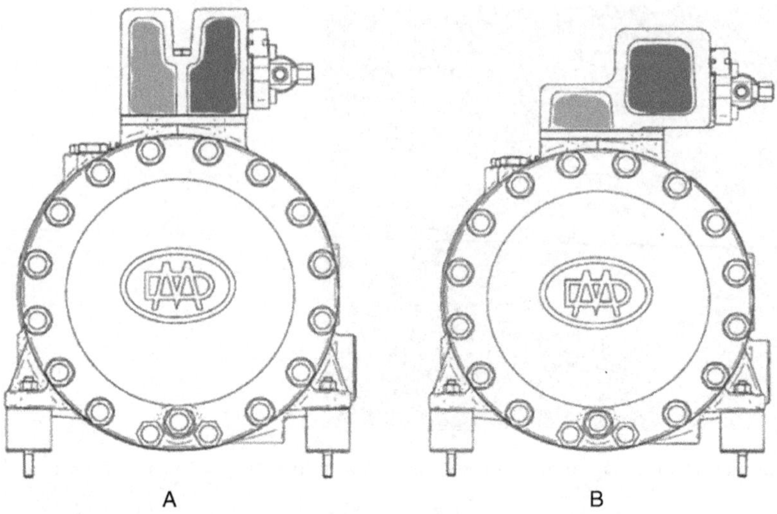

Figure 6.11 Different discharge plenum designs.

temperatures and higher lubricant temperatures, thus reducing the compressor lifetime. Below, Figure 6.11 shows two possible discharge plenum designs (A and B), which have been investigated by compressor manufacturer, Dorin.

Configuration "A" brings too much unsatisfactory thermal insulation between the compressor high-pressure and low-pressure sides, while configuration "B" features an enlarged discharge plenum which offers an additional heat exchange area, leading to good heat dissipation toward the surrounding ambient. Below Figures show measurement data relating to the end of compression discharge temperature (T_D) and oil temperature (Oil T), both expressed in °C as functions of the pressure ratio (PR) (Figure 6.12).

Test results clearly confirm what was previously predicted; configuration "B" allows for much better thermal insulation between the compressor high-pressure side and low-pressure side, assuring appropriate heat dissipation with the surrounding ambient, leading to lower discharge temperatures and lower oil temperatures, thus assuring a more reliable and durable compressor design.

6.7.1.4 Pistons and Compression Rings

The very high discharge temperature occurring at the end of the compression stroke makes it necessary to find specific solutions for a safe and durable piston operation. In this case, the very high thermal level induces several challenges, including:

① Severe material thermal dilatation

The compressor crankcase is often built with a different material than the piston; this implies different dilatation processes, which may result in premature wear or even seizure. To prevent this, either similar core materials are used for both crankcase and piston, or, if different core materials are selected, larger tolerances and a specific coating are to be applied.

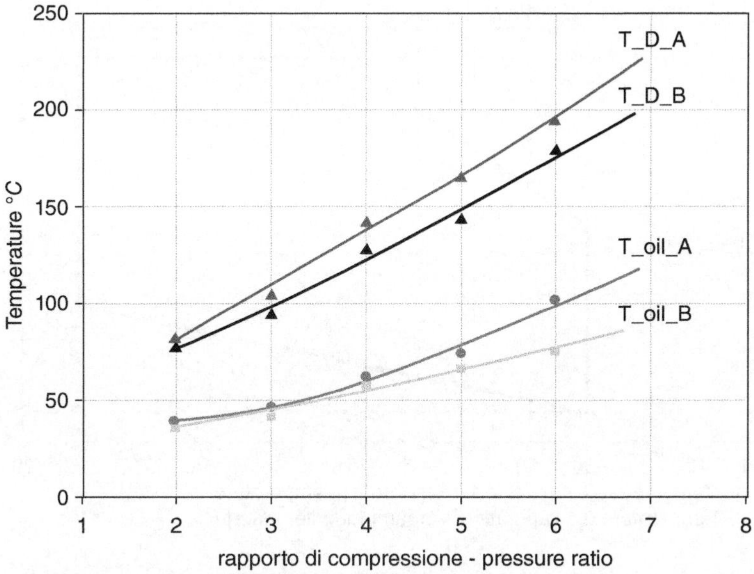

Figure 6.12 Oil temperatures and discharge temperature profiles.

② Poor lubrication

With the increasing end of compression temperatures, lubricant properties will severely deteriorate, ending up in small residual centistokes available to perform proper lubrication. Because of this, specific low friction coatings should be used both on the external piston skirt and for the compression rings, thus enabling a durable and reliable operation.

6.7.2 Design Pressures

Carbon dioxide thermo-physical properties make it necessary to cope with very large design pressures. The below table shows typical suction pressures (SP) and discharges pressures (DP) occurring inside a compressor working with R404A and with carbon dioxide in a simple air-cooled cycle with 35°C as the ambient temperature (AT) (Table 6.6).

In the case of a heat pump application, the discharge pressure is likely to rise further, up to values approaching 120 bar. In addition, compressors (and all the other heat pump components) must be designed in order to safely maintain the refrigerant charge in case of a prolonged standstill occurring. Typically, a safe compressor design foresees that its low-pressure side shall be designed for 100 bar pressure containment.

Table 6.6 Typical R404A and CO_2 operating pressures.

Refrigerant	ET (°C)	AT (°C)	SP (bar)	DP (bar)
R404A	−10	35	4,35	16,07
R744 (CO2)	−10	35	26,5	90,0

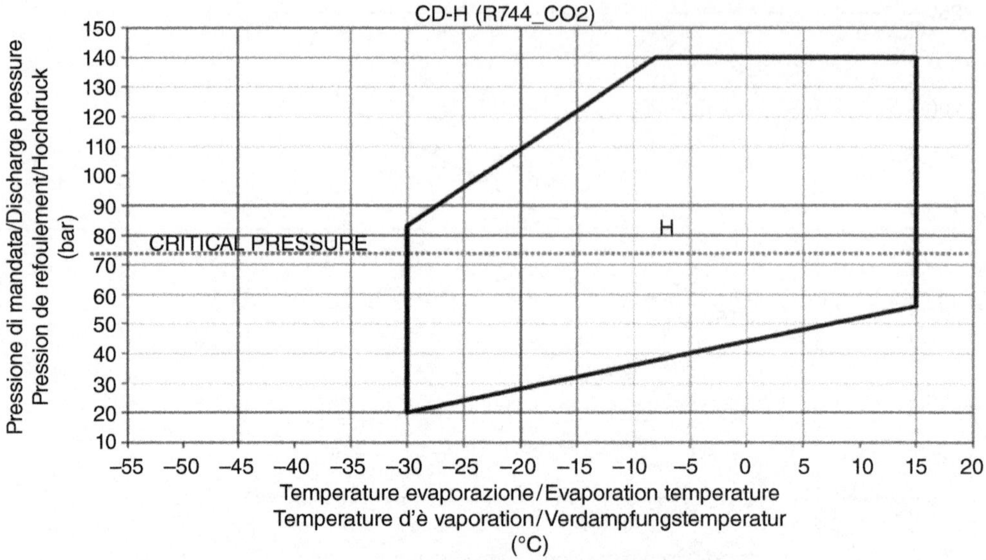

Figure 6.13 Application envelope for trans-critical CO_2 compressors.

CO_2 trans-critical compressors are also subject to severe challenges in terms of operating pressures. The following Figure 6.13 shows a typical application envelope for a compressor suitable for operating:

① Low-temperature applications (such as frozen food cold rooms and the like)
② Medium temperature applications (such as fresh food cold rooms and the like)
③ High-temperature applications (such as air-conditioning, heat pumps and the like).

The aforementioned application envelope clearly shows that compressor capabilities shall be very different depending on the application: high differential and high-pressure ratio are occurring for typical refrigeration applications while air-conditioning and heat pump applications are imposing very strong challenges in terms of oil miscibility and lubrication, since higher evaporating temperatures will induce higher refrigerant solubility into the lubricant (see later section). Nevertheless, the compressor must be able to cope with all operating conditions in order to fit the various criticality with a single platform.

Therefore, once again, it is evident how carbon dioxide induces totally different challenges into the compressor, making it impossible to adopt an HFC compressor platform to a CO_2 compressor platform. Specific solutions must be adopted to cope with such pressure requirements.

6.7.2.1 Materials

HFC compressor envelopes (crankcase, covers, heads) are typically made of gray iron, which is very fragile and not suitable for withstanding the pressure load induced by carbon dioxide. It is therefore necessary to consider alternative materials with much greater ductility.

6.7.2.2 Wall Thickness and Envelope Shapes

CO_2 Compressor envelopes (crankcase, covers, heads) must be properly designed, with wall thicknesses that are much larger than with conventional HFC compressors. In addition, rounded envelope shapes shall be the preferred solution in order to allow an even distribution of the pressure fields.

6.7.2.3 Safety Valves

Depending on the compressor design pressures, specific safety valves should be used, the most typical design being one which is spring-based: a spring is inserted in a disk and when the internal pressure exceeds the spring force, the refrigerant is vented outside. It is very important to notice that the safety valve spring material has a specific interest curve; therefore, safety valve relieving pressure will decrease with each valve opening. It is, therefore, a good practice to replace the entire safety valve whenever an opening has occurred.

Despite the specific solutions adopted, compressor manufacturers are requested to cope with local normative and law requirements in terms of safe compressor deployment; safety coefficients against burst compressors burst are country-dependent and shall be duly respected. For instance, European EN12693 imposes a safety factor of 3 against compressor envelopes burst over the nameplate data, as declared by the manufacturer, while American UL regulations impose a safety factor of 5.

6.7.3 Performances

Compressor performances are the most important data for the correct calculation of any kind of equipment. Typically, refrigeration systems are dimensioned around the required cooling capacity and the various components of power consumption. In this scenario, being the compressor main system energy consumer, appropriate equations are in place to be used.

In particular, compressor-wise, while power consumption is a typical performance that can easily be measured, refrigeration capacity is more directly understandable. Truly, the real measurable compressor performance is not the refrigeration capacity but the compressor mass flow. Refrigeration capacity can then be calculated by using refrigerant enthalpy difference at the evaporator inlet/outlet.

Below the typical performance equation is provided in the form of a polynomial expression:

$$\begin{aligned} y = {} & C1 + C2 \times to + C3 \times pc + C4 \times to^2 + C5 \times to \times pc + C6 \times pc^2 + C7 \times to^3 \\ & + C8 \times pc \times to^2 + C9 \times to \times pc^2 + C10 \times pc^3 \end{aligned}$$

where "y" is the given specific performance (mass flow or power consumption); "C1, …, C10" are the polynomials coefficients which are compressor dependent; "pc" is the discharge pressure; "to" is the evaporating temperature.

6.8 Future Trends

Over time, compressor technology for supercritical CO_2 systems has reached a relatively advanced level. At this stage, researchers are focusing on two-stage compression and expansion technologies to improve compressor efficiency [31].

6.8.1 Two-Stage Compressor

The CO_2 pressure differential between the discharge pressure and the suction pressure of the refrigerant of HFC than traditional (hydrofluorocarbon) refrigerant is several times larger. This leads to increased gas leakage and mechanical losses. However, with the introduction of two-stage compression, this greatly reduces the pressure difference in each stage, thereby achieving higher system efficiency and reliability. As shown in Table 6.7, Sato *et al.* [32] compared the performance of single-stage and two-stage compressors by comparing structural differences. From Table 6.7, the two-stage compressor is superior to the single-stage compressor in many aspects. Although compared with single-stage compressors in terms of equipment cost and stability, two-stage compressors are slightly insufficient. On the one hand, in the high-power practical application of the two-stage compressor, the cost saved from power consumption can be made up for, and the economy is comprehensively considered in practical applications. On the other hand, stability is mainly considered from the perspective of circulating liquid return and equipment vibration. This is also the main challenge to overcome for the two-stage compressor in the future. The structure of the compressor has a great impact on the performance of the compressor, especially for CO2 compressors with high pressure. Figure 6.14 shows the difference between the structure of the rotary compressor and the swing compressor.

Suzai *et al.* [33], Tadano *et al.* [34], and Yamasaki *et al.* [35] developed a two-stage rotary compressor shown in Figure 6.15, to mitigate the impact of the CO_2 inherited high-pressure differential. By compressing the compression process into two separate processes, it can reduce the leakage at the seals, so as to obtain higher compression efficiency. Further, the two pistons are placed out of phase (180°). This arrangement on opposite sides greatly reduces the vibration and noise problems. CO_2 from the basic properties can achieve high efficiency and a small volume of the piston size. Since the rotational inertia of CO_2 balance and torque is small, it is possible to achieve a smooth rotation of the shaft when up and running. In addition, the internal intermediate pressure design enabled the shell wall thickness to be 35% thinner than that of the high-pressure operation. Accordingly, this helps to reduce the total weight which is found to be about the same as the conventional R-410A

Table 6.7 Comparison of single and double stage compressor performance parameters.

Compressor type	Discharge temperature	COP	Equipment cost	Reliability	Applicability at high power	Ultra-low temperature suitability
Single stage	High	Low	High	High	Low	Low
Two-stage	Low	High	Low	Low	High	High

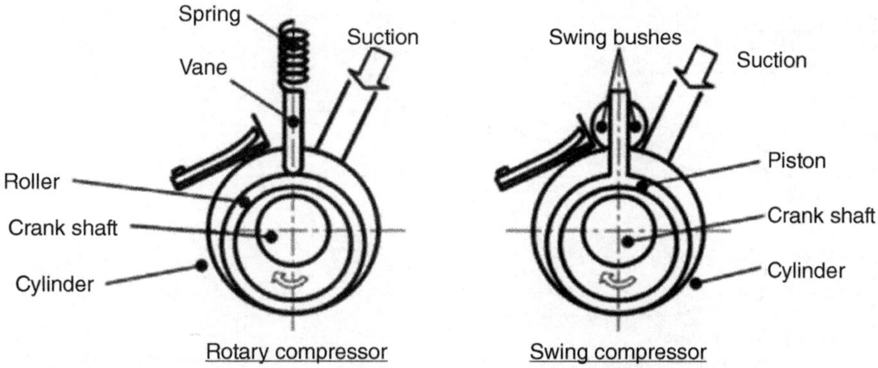

Figure 6.14 Rotary compressor vs swing compressor – a schematic.

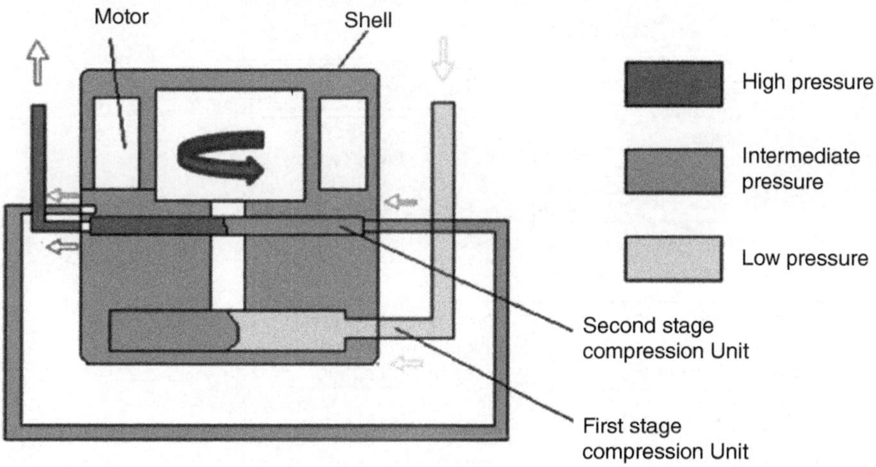

Figure 6.15 Schematic of a hermetic two-stage CO_2 compressor.

compressor. Moreover, the internal intermediate pressure design makes the pressure difference between on and off periods smaller than that of the one-stage design. This eventually raises high reliability to prevent fatigue of the shell material through a periodically high/low pressure cycle.

In addition, two-stage compressors are suitable for energy-saving cycles. As shown in Figure 6.15, Masahiro, K. [36] uses this compressor in heat pump and water heater systems in cold climates to achieve economical thermal cycles. With this design, when the outdoor air temperature is –20°C, a heat capacity of about 8 kW and a water heating performance of 4.5 kW can be obtained. At the same time, compared with systems without the economizer cycle, the design capacity and COP increased by 17%.

Yokoyama *et al.* [35] developed a two-stage rotary compressor for CO_2 heat pump systems with refrigerant injection. Figure 6.16 shows the performance at various rotational speeds in association with the single-stage type having the same specifications. It is found that when a two-stage compressor is at a low speed or high-pressure ratio, its efficiency is higher than a single-stage compressor.

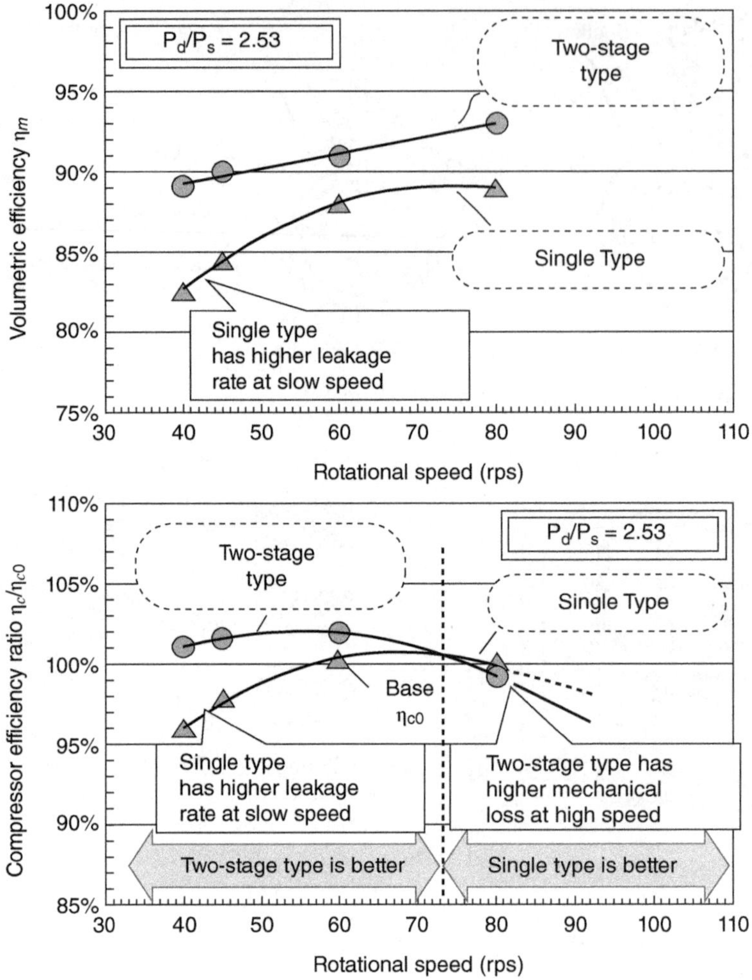

Figure 6.16 Comparison of the two-stage and single-stage CO_2 compressors at $P_d/P_s = 2.53$ [35].

Obviously, because two-stage compression is related to a lower compression ratio at each stage, it is possible to reduce gas leakage in the compression device. The two-stage type is also superior to the single-stage type in compressor efficiency and heating capacity because it has the ability to improve these performances with refrigerant injection during high pressure-ratio operation.

As shown in Figure 6.17, Sato *et al.* [32] designed a new commercial CO_2 heat pump water heater. In order to minimize leakage and mechanical losses, the design uses rotary and scroll mechanisms in the first and second stages, respectively. In addition, the new compressor uses an intermediate gas injection method to improve it in order to obtain greater heat capacity and efficiency under a wider range of conditions. Their test results show a 15% improvement under rated conditions is obtained. Under the condition of a higher pressure

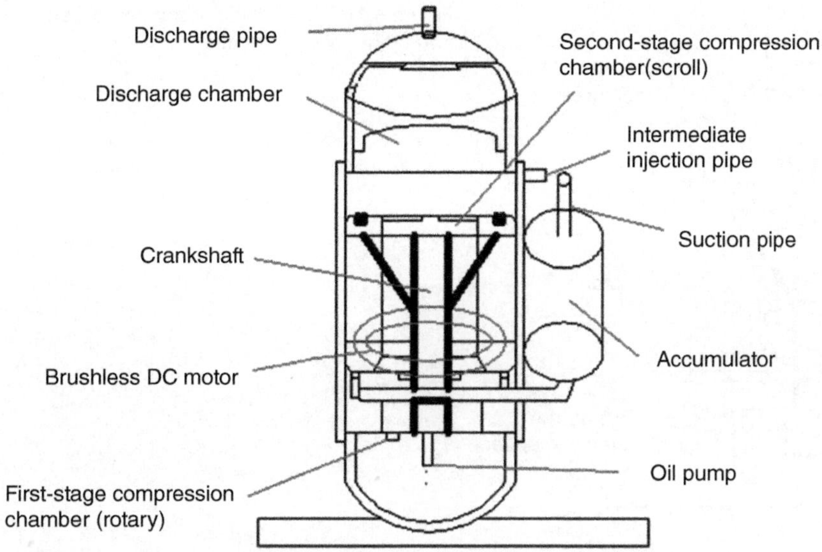

Figure 6.17 Two-stage compressor employs rotary and scroll mechanisms in the first and second stages.

ratio, it can increase by more than 30%. This result is in contrast to a conventional prototype single-stage scroll compressor.

6.8.2 Expander and Expander–Compressor

One of the major challenges of carbon dioxide in air conditioning applications is low energy system efficiency at higher operating temperatures. The main loop is the loss of CO_2 throttling associated with the expansion process. Lost during the extended availability of the device can be recovered by expanding the produced work (a so-called spreading unit [37]). Matsui *et al* [38] used CO_2 and R-134a as the working fluid in system performance calculation and comparison. Their results show that the use of low-cost replacement CO_2 expanders can improve the performance of the system. An expander is coaxially connected to a compressor, which helps to improve recycling. They are the most efficient recovery rates of up to 14.5% of the compressor input power.

As shown in Figure 6.18, a two-stage rotary expander was developed, which was designed to connect its shaft to the shaft of a commercial scroll compressor [38]. At the same time, an expansion chamber surrounded by a piston and a cylinder was added to the second stage, which was much larger than the first stage. After examining the impact of each design parameter on the performance through the simulation, an optimized design had been made and the expander efficiency was increased by as much as 60%, which is equivalent to a 6% rise of system COP. Hiwata *et al.* [39] designed the scroll profile of a scroll expander suitable for this refrigerant based on the CO_2 characteristics. Their design makes use of overexpansion to control the axial force on the thrust bearing. Through this design, an exceptionally high volumetric efficiency of 96% is demonstrated.

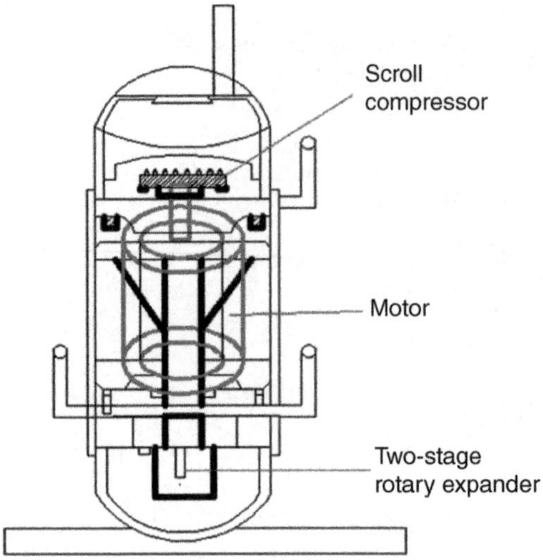

Figure 6.18 Schematic of the developed two-stage rotary expander.

6.9 Some Key Technical Problems of CO_2 Compressor

6.9.1 Mechanical Strength

Because the CO_2 trans-critical cycle operates at higher pressures, up to 10 MPa, and the lowest is about 3 MPa, the compressor parts, especially the crankcase, have higher pressure resistance performance, that is, less deformation under large pressure differences. When processing and manufacturing compressors, material strength and stiffness analysis should be performed and redesigned if necessary. For the screw compressor, the shaft and the yin-yang rotor must consider the pressure, so the yin-yang rotor tooth pairs may need to be changed.

6.9.2 Lubricant Problems

The difficulties in selecting lubricants for CO_2 trans-critical cycle systems are as follows.

(1) Supercritical CO_2 is easy to dissolve in the lubricating oil, and the viscosity of the diluted lubricating oil will be greatly reduced.
(2) The compressor bears a high-pressure load.
(3) The lubricating oil film may be damaged by the flowing supercritical CO_2.
(4) The lubricating oil film may be destroyed by the evaporation of CO_2 dissolved in the lubricating oil.
(5) Chemical reaction and corrosion may occur between supercritical CO_2 and lubricating oil.

6.9.2.1 Miscibility of Lubricant and CO_2

In supercritical conditions, CO_2 is an effective solvent for various types of hydrocarbons. No matter what kind of lubricant is selected, it will cause lubricant carryover due to the ability of CO_2 to dissolve[40]. In order not to hinder heat transfer, it is necessary to ensure that the lubricating oil can flow back into the compressor, which makes low-temperature fluidity and mixing important. CO_2 viscosity is very small. After dissolving CO_2 in the lubricating oil, the viscosity of the solution will be significantly lower than that of pure lubricating oil. Therefore, the choice of lubricating oil should be based on the diluted viscosity rather than the nominal viscosity. In PAG, PAO, and POE lubricants, the solubility of CO2 in POE is extremely low, which has a relatively small effect on the viscosity of the lubricant. POE lubricants show good miscibility.

6.9.2.2 Lubricant Stability

Aging lubricants can cause corrosion, filter clogging, and reduced system efficiency. Therefore, whether the lubricant has an aging reaction is an important basis for measuring the stability of the lubricant. PAO and AB are stable in a CO_2 environment, and the lubricant neither ages nor reacts with the catalyst. Additives added to PAG to improve the lubricating ability will cause an aging reaction, and the aging products may react with copper and steel. Because POE will produce an aging reaction, it is less stable than PAO and AB.

6.9.2.3 Choice of Lubricant

Proper lubrication is one of the most important parameters in order to assure a reliable compressor operating within its lifetime. Several investigations have been conducted in the last two decades in order to find out the most appropriate lubricants for carbon dioxide systems and compressors.

For instance, PAG, PVE, and POE oil have been investigated through a number of experimental tests, such as steel ball wear tests and the like. The figure below shows some of the outcomes. Though all the three lubricants were assuring proper and sufficient lubrication across the passages, as it is visible, the best lubrication performance was obtained by using PAG oil, which is assuring smaller wear for the amount and also a smoother sliding surface appearance (Figure 6.19).

However, wear rate and surface appearance are not the sole parameters to be taken into account when selecting the most appropriate lubricant for carbon dioxide applications.

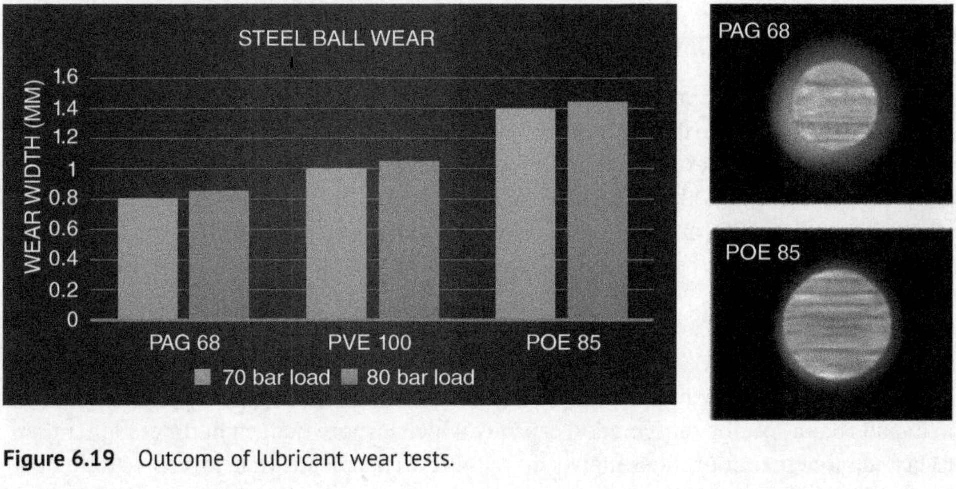

Figure 6.19 Outcome of lubricant wear tests.

In fact, good oil return performance from the evaporator to the compressor suction port must also be assured. In this sense, POE oil is definitely the best performer, which makes it possible to more easily get lubricant back to the compressor, especially for parallel compressors rack assembly. Therefore, the lubricant choice is not only compressor dependent, but must be accurately made in conjunction with the complete system designer.

6.9.3 Oil Dilution

Oil dilution into the refrigerant is also a key parameter when approaching the design of CO_2 compressors. Carbon dioxide is a strong detergent and it is, therefore, crucial to assure correct lubrication for the entire lifetime of the machine. In order to do so, the specific behavior of oil-to-refrigerant miscibility curves shall properly be taken into account. For instance, one may consider the aforementioned curves pertaining to one of the most commonly used lubricants in carbon dioxide applications, e.g. a special POE oil specifically designed and produced for carbon dioxide trans-critical applications.

The miscibility curve of CO_2 and POE oil is shown in Figure 6.20. This special lubricant is rated for 85 CST (centistokes) as the nominal viscosity at 40 °C, in pure conditions (100% lubricant – 0% refrigerant). However, these are just ideal conditions that are never reproducible in real-life system operation: depending on the specific operating conditions, a certain amount of carbon dioxide will always be diluted into the oil: the specific amounts vary with changing temperature and pressure conditions.

For instance, if 30 bar pressure and 50° C temperature are reasonably taken into account, the aforementioned graph shows a residual viscosity of only 16 CST: a very strong viscosity drop has taken place, thus potentially impacting in a very negative way on the compressor durability.

Therefore, special attention is paid to the reasons for compressor lubricants. Because of this characteristic, we need to consider not only the precise design but also the choice of materials for specific bearings, especially to deal with the decline of strong viscosity. The principle of material selection is to favor a material with a lower boundary friction coefficient.

6.9.4 Large Pressure Differences

Taking again into consideration the aforementioned Table 6.8, another important parameter can be highlighted: the differential pressure (delta_P) induced by carbon dioxide operation is more than five times larger than that occurring in an R404A application. This brings extra-severe challenges to all those components implied in the compression stroke and once again highlights how CO_2 applications unavoidably call for a dedicated engineering process: no synergy can be in place between an HFC compressor and a trans-critical CO_2 compressor.

6.9.4.1 Wrist Pin

In particular, this very large pressure difference shall not be underestimated, especially in conjunction with the very high specific refrigeration capacity featured by CO_2. In fact, carbon dioxide has a specific refrigeration capacity which is more than eight times higher than R134a and more than four times higher than R404A. This means that, given a certain duty

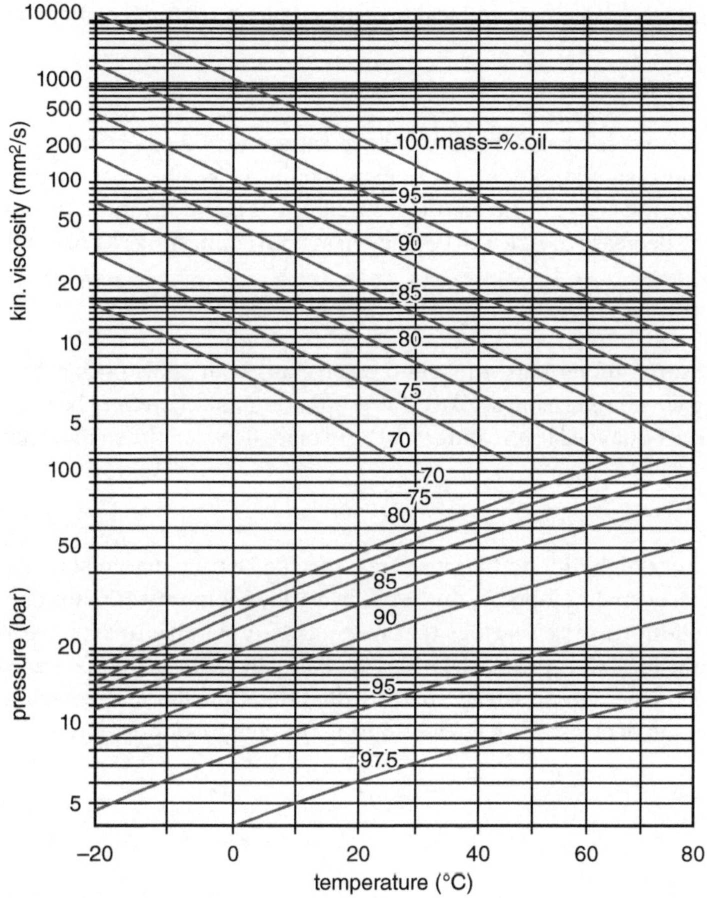

Figure 6.20 Miscibility curves for CO_2 and POE oil.

Table 6.8 Typical R404A and CO_2 operating pressures.

Refrigerant	ET (°C)	AT (°C)	SP (bar)	DP (bar)
R404A	−10	35	4,35	16,07
R744 (CO2)	−10	35	26,5	90,0

at specific and same boundary conditions, CO_2 compressors will feature the same magnitude smaller displacement: in case of piston compressors this will unavoidably lead to the same magnitude smaller cylinder diameters.

Therefore, the combination of very high pressure differentials and very small cylinder walls ends up in a tremendous increase in the specific load appearing at the wrist pin level: the wrist pin shall be considered the most challenging part in a CO_2 trans-critical compressor and specific means to decrease to local and impede high friction coefficients shall be implemented.

6.9.4.2 Connecting Rod

At the same time, very large pressure differentials induce extra-severe loads in the connecting rod during its compression stroke. A large safety coefficient shall be used when dimensioning the connecting rod body and, in addition, specific means and advanced tribology shall be implemented to decrease the friction coefficient at both connecting rod big-end and small-end (Figure 6.21).

6.9.4.3 Crankshaft

Due to the very high-pressure differentials, the compressor crankshaft is also subject to very strong challenges at each compression stroke. Main supports shall, therefore, be very generously dimensioned and enhanced lubrication shall be in place between the shaft necks and the various supports.

6.9.4.4 Bearings

In order to cope with the very high differential pressures, specific bearing design shall be in place. New concepts have been developed in this sense by various compressor manufacturers, with multi-layer self-lubricating bearings that have made it possible to cope with these very demanding challenges. Compressor bearings are basically made of a strong and robust core material, which is then progressively coated with different materials, featuring a smaller friction coefficient to decrease local stresses during the compressor operation.

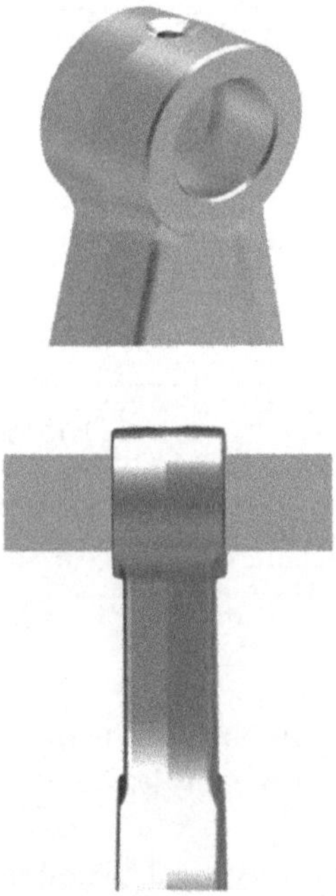

Figure 6.21 Typical connecting rod and wrist pin assembly.

6.9.4.5 Valve Plate

With the extraordinary large differential pressures in place, valve plates are subject to severe challenges, especially at their discharge side level. In fact, the discharge vale reeds shall be able to keep a proper sealing between the compressor's LP and HP sides. Without proper sealing, severe leaks back to the compression chamber will occur, thus leading to very poor performance and compressor overheating.

Thus, specific criteria shall be in place in order to assure both a durable discharge reed valve operation and a correct and durable resistance of the valve plate seat: in case of premature wear on the valve seat, back leakage to the compression chamber will occur unavoidably, with consequent severe performance drop and discharge temperature increase.

6.10 Conclusion and Perspectives

This chapter presents a detailed introduction to the development status and challenges of different types of SCO_2 compressors. At the same time, the research directions and problem-solving measures of SCO_2 compressors are summarized.

In the last decade, CO_2 has strongly gained interest as a mainstream refrigerant for a large number of applications, due to its environmentally benign characteristics as well as to its very interesting performance. However, its specific thermodynamic features lead to severe challenges, especially in the compressor. CO_2 compressors have to withstand very large pressure differences, very high discharge temperatures, and critical lubrication conditions. Therefore, the design criteria of CO_2 compressors must be specifically taken into account. The research directions of future supercritical CO_2 compressors are as follows:

(1) Development of oil-free compressor
Since CO_2 is soluble to oil, reducing the viscosity of oil and directly affecting the lubrication effect, and CO_2 circulating heat transfer is particularly sensitive to oil, Yanagisawa [41] showed that the more oil, the faster the pressure rise in the compression process, leading to the decline in indicated efficiency. Oil-free compressors can avoid oil-related problems, especially in the food refrigeration industry.
As described by Stosic *et al.* [42], the oil-free piston compressor developed by the University of Zurich in Switzerland is in good continuous operational condition and has achieved initial results. The next step of development focuses on the more compact design and lower cost. The oil-free compressor is the trend of future development.

(2) Improvement of a two-stage compressor
The two-stage compressor has a compact structure and a more flexible system arrangement. More importantly, because two-stage compression can reduce pressure difference, reduce leakage and mechanical loss, and can significantly improve the efficiency of the system and the compressor, a two-stage compressor will be the future large-scale development and production form.

(3) Development of expansion compressor
In the CO_2 trans-critical cycle, reducing the loss of the expansion part is an effective way to solve the efficiency. At the same time, the output power of the expander drives the compressor to complete the compression process. The expansion compressor will be a unique component of the CO_2 cycle.

Nomenclature

COP	coefficient of performance
HFCs	hydrochlorofluorocarbons
GWP	global warming potential
SCO_2	supercritical carbon dioxide
CFCs	chlorofluorocarbons
ASHARE	American Society of Heating, Refrigerating and Air-Conditioning Engineers
ODP	ozone depletion potential
ISO	International Organization for Standardization
η_v	volume efficiency
η_i	indicated efficiency
η_m	machine efficiency
DC	direct current
P	pressure, bar
V	volumetric
Cp	heat capacity at constant pressure, J
Cv	specific heat at constant volume, J
IECDT	isentropic end of compression discharge temperatures, °C
RECDT	real end of compression discharge temperature, °C
RGT	return gas temperature, °C
SP	suction pressure, kPa
DP	discharge pressure, kPa
MT	medium temperature, °C
ET	evaporating temperature, °C
AT	ambient temperature, °C
SH	degree of superheat, K
PR	pressure ratio
PA	polyalkylene glycol
PAO	poly alpha olefin
POE	polyol ester
PVE	polyolvinyl ether
AB	alkylbenzene
y	refers to the mass flow or power consumption
C1 ~ C10	refer to the polynomials coefficients
pc	discharge pressure
to	evaporating temperature
HP	high pressure
LP	low pressure

References

1 Bolaji, B.O. and Huan, Z. (2013). Ozone depletion and global warming: case for the use of natural refrigerant – review. *Renewable and Sustainable Energy Reviews* 18: 49–54.

2 Varotsos, C. (2008). The 20th anniversary of the Montreal Protocol and the unexplainable 60% of ozone loss. *Environmental Science and Pollution Research* 15: 448–449.
3 Lachner, B.F., Nellis, G.F., and Reindl, D.T. (2007). The commercial feasibility of the use of water vapor as a refrigerant. *International Journal of Refrigeration* 30: 699–708.
4 Yang, D. (2014). *ASHRAE Position Document on Natural Refrigerants*. Atlanta, GA: American Society of Heating, Refrigerating and Air-Conditioning Engineers, Inc.
5 Yang, D., Song, Y., Cao, F. et al. (2016). Theoretical and experimental investigation of a combined R134a and transcritical CO_2 heat pump for space heating. *International Journal of Refrigeration* 72: 156–170.
6 Kim, M.H., Pettersen, J., and Bullard, C.W. (2004). Fundamental process and system design issues in CO_2 vapor compression systems. *Progress in Energy and Combustion Science* 30: 119–174.
7 Rony, R. U. (2018). Life Cycle Climate Performance Energy Tool. Emerson Climate Technologies.www.emersonclimate.com (accessed 3 November 2018). Energies, 12, 457 30 of 35.
8 Baek, C., Heo, J., Jung, J. et al. (2013). Optimal control of the gas-cooler pressure of a CO_2 heat pump using EEV opening and outdoor fan speed in the cooling mode. *International Journal of Refrigeration* 36: 1276–1284.
9 Dexi, Y. and Bingfeng, Y. (2006). Research progress of carbon dioxide transcritical compressors. *Refrigeration and Air Conditioning* 6 (3): 1–8.
10 Wang, B.B., Li, J.-F., Wu, H. et al. (2009). Research and development of CO_2 screw compressor in NH_3/CO_2 cascade refrigeration system. *Fluid Machinery* 37 (3): 16–18, 63.
11 Stosic, N. Smith, I., Kovacevic. A, et al. (2002). A twin-screw combined compressor and expander for CO_2 refrigeration system[C]. *The Proceedings of the 16th International Compressor Engineering Conference*. Purdue: Purdue University.
12 Guan, H.Q., Ma, Y.T., Yang, J. et al. (2004). Research review of CO_2 transcritical cyclic expansion compressor. *Mechanical Design* 21 (7): 1–4.
13 Li, M.X., Ma, Y.T., Li, L.X. et al. (2004). Research progress of CO_2 transcritical cycle refrigeration compressor. *Compressor Technology* 187 (5): 38–42.
14 Ohkawa, T., Kamakura, E., Higashi, H., et al. (2002). Development of hermetic swing compressor for CO_2 refrigerants[C]. *The Proceeding of the 16th International Compressor Engineering Conference*. Purdue: Purdue University, p. 499.
15 Yang, D. and Bingfeng, Y. (2006). Research progress of carbon dioxide transcritical compressor. *Refrigeration and Air Conditioning* 6 (3): 8–14.
16 Tadano, M., Ebara, T., Oda, A., et al. (2000). Development of the CO_2 hermetic compressor[C]. *The Proceedings of the4th IIR-Gustav Lorentzen Conference on Natural Working Fluids*. Purdue: Purdue University, p. 323–330.
17 Dreamin, N., Bunch, R., Hwang, Y., et al. (2004). Two-stage rolling piston carbon dioxide compressor[C]. *The Proceedings of the 4th IIR-Gustav Lorentzen Conference on Natural Working Fluids*. Purdue: Purdue University.
18 Yang, J., Lu, P., Zhang, L. et al. (2008). Development and performance test of a new fully enclosed rotary CO_2 compressor. *Journal of Shanghai Jiao Tong University* 42 (3): 426–529.
19 Kai, Y.Z. and Haller David, K. (2004). Development of a hermetic rotary carbon dioxide compressor. *International Compressor Engineering Conference*. Purdue.

20 Fagerli, B.E. (1998). On the feasibility of compressing CO_2 as working fluid in scroll compressors[C]. *International Refrigeration and Air Conditioning Conference*. Purdue: Purdue University, pp. 165–170.

21 Hao, P.Y., Li, L.S., Zhao, Y.-y. et al. (2010). Research progress of CO_2 scroll refrigeration compressor. *General Machinery* 10: 82–85.

22 Iwata, A., Iida, N., Futagami, Y., et al. (2002). Performance investigation with oil-injection to compression chambers on CO_2 -scroll compressor[C]. *International Compressor Engineering Conference*. Purdue: Purdue University, pp. 18-04.

23 Hagita, T., Kobayashi, H., Takeuti, M., et al. (2000). The development of CO_2 scroll compressor for automotive air-conditioning systems[C]. *The International Symposium On HCFC Alternative Refrigerants and Environ-mental Technology*. pp. 48–51.

24 Hasegawa, H., Ikoma, M., Nishiwaki, F., et al. (2000). Experimental and theoretical study of hermetic CO_2 scroll compressor[C]. *The 4th IIR Gustav Lorentzen Conference on Natural Working Fluids*. Purdue: Purdue University, pp. 347–353.

25 Brasz, J.J. (1995). Improving the Refrigeration Cycle with Turbo Expanders. *Proceedings of the 19th International Conference of Refrigeration*, The Hague, Netherlands, vol. IIIa, pp 246–253.

26 Hays, L. and Brasz, J.J. (1996). Two-Phase Turbines for Compressor Energy Recovery. *Proceedings of the 1998 International Compressor Engineering Conference at Purdue*, vol. II, pp. 657–662

27 Hays, L. and Brasz, J.J. (1998). Two-Phase Turbine as Stand-Alone Throttle Replacement Units in Large 2000–5000 Ton Centrifugal Chiller Installations. *Proceedings of the 1998 International Compressor Engineering Conference at Purdue*, vol. II, pp. 797–802.

28 Hays, L., (1999). History, and Overview of Two-Phase Flow Turbines. *Proceedings of the IMechE International Conference on Compressors and their Systems*, London, England, pp. 159–170

29 Hays, L. (2003). Transcritical Turbine and Method of Operation, U.S. Patent 6,644,062 B1.

30 Hays. (2003). Transcritical turbine and method of operation: US, 29401097[p]. 2003-11-11.

31 Jarn (2013). ATW heat pump market in Japan. JARN, p. 62–4.

32 Sato, H., Kimata, Y., Hotta, Y. et al. (2012). Development of two-stage compressor for CO_2 heat-pump water heaters. *Mitsubishi Heavy Industries Technical Review* 49: 92–97.

33 Suzai, T., Sato, A., Tadano, M., et al. (1999). Development of a carbon dioxide compressor for refrigerators and air conditioners. *Paper presented in the conference of the Japan Society of Refrigerating and Air Conditioning Engineers*. Tokyo.

34 Tadano, M., Ebara, T., Oda, A., et al. (2000). Development of the CO_2 hermetic compressor. *Proceedings of the fourth IIR-Gustav Lorentzen Conference on natural working fluids at Purdue*. p. 323–30.

35 Yokoyama, T., Sasaki, K., Sekiya, S., and Maeyama, H., (2008). Developing a two-stage rotary compressor for CO_2 heat pump systems with refrigerant injection. *Proceedings of the 19th international compressor engineering conference*. Purdue, USA. Paper 1193.

36 Masahiro, K. (2008). CO_2 heat pump heating and water heater system for the cold area. *Sanyo Technical Review* 38: 10–15.

37 Huff, H.J., Radermacher, R. (2003). CO_2 compressor–Expander Analysis. Technical Report ARTI-21CR/611-10060-01.

38 Matsui, M., Wada, M., Ogata, T., and Hasegawa, H. (2008). Development of high-efficiency the technology of two-stage rotary expander. *Proceedings of the 19th inter-national compressor engineering conference*. Purdue USA. Paper 1237.

39 Hiwata, A., Ikeda, A., Morimoto, T. et al. (2009). Axial and radial force control for CO_2 scroll expander. *International Journal of HVAC&R Research* 19: 759–770.

40 Ma, Y.T., Liu, Z.Y., and Tian, H. (2013). A review of transcritical carbon dioxide heat pump and refrigeration cycles. *Energy* 55: 156–172.

41 Yanagisawa, T., Fukuta, M., et al. (2000). Basic operating characteristics of reciprocating compressor for CO_2 cycle. *The Proceedings of the 4th IIR-Gustav Lorentzen Conference on Natural Working Fluids*. Purdue, pp. 331–338.

42 Stosic, N., Smith, I.K., Kovacevic, A., et al. (2002). A twin-screw combined compressor and expander for CO_2 Refrigeration Systems. *The Sixteenth International Compressor Engineering Conference*. Purdue.

7

CO_2 Subcooling

Rodrigo Llopis, Daniel Sánchez, Laura Nebot-Andrés, Jesús Catalán-Gil and Ramón Cabello

Mechanical Engineering and Construction Department, Jaume I University, Castellón de la Plana, Spain

7.1 Introduction

Initial expansion of CO_2 refrigeration and heat pump systems was originated in the coldest regions of the world, especially in the Northern countries of Europe, where favorable environmental conditions allowed CO_2 to obtain an even higher energy efficiency level than systems relying on artificial and traditional refrigerants. However, its extension to warm and hot territories was limited due to its reduced energy efficiency, particularly when the cycles operated in transcritical conditions. In this context, the irreversibilities during the expansion processes and in the gas-cooler limited the performance of the cycles. During recent years, scientists around the world have worked to surpass these limitations. Great effort has been made in the improvement of individual components [1] or on the definition of alternative refrigeration schemes [2], which quickly outlined that the useful CO_2 layouts are very different from the traditional ones, since CO_2 cycles require a devoted optimization system of the heat rejection pressure, as analyzed by [3–6] or experimentally evaluated by [7–9]. Later, scientists attempted to reduce the irreversibilities during the expansion processes by using "*energy recovery*" systems. On the one side, development of expanders (Figure 7.1a), which perform a more reversible expansion process than the isenthalpic devices, was done [9–12]. However, their current state of development has not yet made possible its implementation in the commercial field. On the other hand, ejector technology (Figure 7.1b) in most cases is developed with a fixed geometry. This element, used instead of a throttling valve, recovers some kinetic energy during the expansion process, which is generally used to increase a compressor's suction pressure, thus resulting in reductions of the compression work and improvements in the system efficiency [13–15]. This device, working as "multi-ejector" is used worldwide to improve the performance of CO_2 systems [16]. Finally, the last strategy to reduce irreversibilities in the expansion process, which has received great attention during recent years, is CO_2 subcooling (Figure 7.1c). CO_2 subcooling or "after-cooling" is based on cooling the CO_2 at the exit of the condenser/gas cooler by means of an additional heat exchanger, commonly referred to as "subcooler" or "after-cooler" [17]. Subcooling allows, on the one hand, reduction of the irreversibilities

Transcritical CO_2 Heat Pump: Fundamentals and Applications,
First Edition. Xin-Rong Zhang and Hiroshi Yamaguchi.

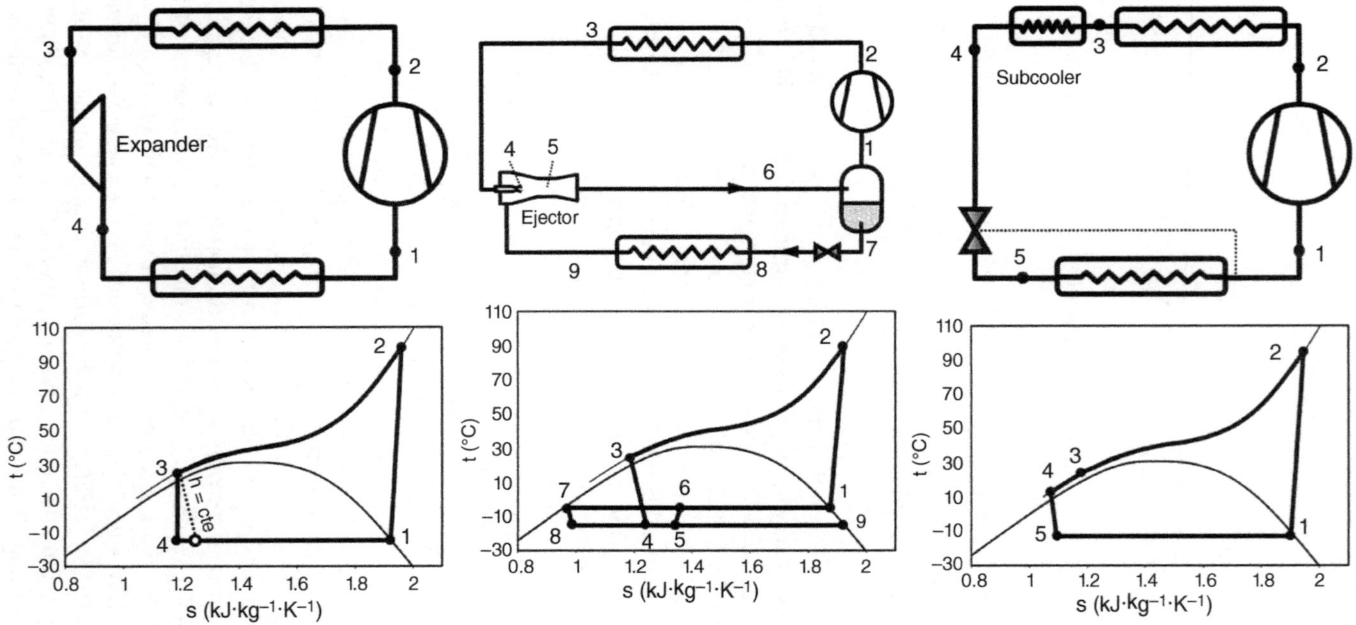

Figure 7.1 Schematic representation of CO_2 cycle with ideal expander (a), with ejector (b) and with subcooling (c).

during the expansion process, and on the other hand, enhancement of the energy efficiency of the thermodynamic cycles, through improvement of the compressor's performance, since the optimum heat rejection pressure is reduced, as analyzed in [18]. Additionally, subcooling always brings about an increment in the cooling capacity of the systems, as well as an increment in the specific refrigerating effect, thus reducing the refrigerant mass flow rate in the cycle.

Subcooling can be achieved by natural heat transfer with environment as long as a temperature difference between condensation and environment exists. Pottker and Hrnjak [19] analyzed theoretically the effect of condenser subcooling in single-stage compression cycles with water-cooled-condensers, and they concluded that liquid subcooling reduces the throttling losses in the expansion device, resulting in increments in the refrigerating effect and coefficient of performance (COP). Their simulations for air conditioning devices at the optimum subcooling degrees predicted COP improvements of 8.4% with R-1234yf, 7.0% with R-410A, 5.9% with R-134a and 2.7% with R-717, at condenser and evaporator inlet temperatures of 14 and 0°C, respectively.

Referring to CO_2 cycles, the subcooling method mentioned above would only be possible when operating in subcritical conditions, thus it is limited to operation at environment temperatures below 25°C, in practice.

For the highest surrounding temperatures, where the greatest improvement of subcooling is located [20], the possibilities are: to perform subcooling by using the cycle itself, such as using an internal heat exchanger (IHX) or liquid-to-suction heat exchanger [21, 22], an economizer [23, 24], or even the integrated mechanical subcooling (IMS) cycle [24–26]; or execute subcooling by using an auxiliary cycle, such as a dedicated mechanical subcooling (DMS) cycle [27–29], a thermoelectric subcooling system [30–32] or other hybrid systems [33, 34]. Energy improvements of some representative subcooling methods are collected in Table 7.1. It has been measured that the IHX brings about up to 12% COP increment, and predicted that the enhancement reaches 22% with economizers, 25.6% using thermoelectric systems, 21.3% with an IMS system or 30.3% using a DMS system. However, these values must be considered only as representative data, since in most cases they are not completely optimized. Furthermore, as analyzed in this chapter, the improvement depends on the COP of the auxiliary device used for subcooling purposes, thus wider studies are required.

The purpose of this chapter is to set the thermodynamic basis of the CO_2 subcooling to develop both wider and more specific future studies. Therefore, first, a thermodynamic approach including properties, ways, optimization and costs is addressed. Second, the chapter focuses on the subcooling device which has been most investigated, the IHX, the function of which clearly produces positive results for CO_2 cycles. Third, the chapter focuses on the application and developments made with the DMS cycles. And finally, on the application of the IMS system. As it can be seen in the chapter, special attention is paid to subcooling methods which could have a direct application in industry now. Other subcooling systems, which are being researched now or which implementation in medium to large plants facing difficulties are not detailed in this text.

Table 7.1 Predicted or measured improvements of CO_2 refrigeration systems with subcooling methods.

Subcooling system	Reference system	*COP of reference system*	t_O (°C)	$t_{gc,out}$ (°C)	Capacity increment in relation to reference system (%)	COP increment in relation to reference system (%)	Type	References
Internal heat exchanger	Basic cycle	1.16 ($t_O = -15.0°C$, $t_{gc,out} = 33.9°C$) to 1.91 ($t_O = -5.1°C$, $t_{gc,out} = 31.0°C$)	−15 to −5°C	31 and 34°C	12% max	12% max.	E, O	[21]
Economizer	Double-stage cycle with intercooling	2.62 ($t_O = 2.7°C$, $t_{gc,out} = 33.0°C$) and 2.87 ($t_O = 2.7°C$, $t_{gc,out} = 22.0°C$)	2.7°C	22, 33°C	—	22.1%, 21.0%	T, O	[23]
Thermoelectric	Basic cycle	2.412 ($t_O = 5.0°C$, $t_{gc,out} = 40.0°C$)	−15 to 5°C	30–50°C	—	7.0–25.6%	T, O	[32]
Integrated mechanical subcooler	Basic cycle	Not provided	−10°C	30–42°C	—	20.5–21.3%	T, O	[25]
Dedicated mechanical subcooler	Basic cycle	1.32, 1.93, 2.57 ($t_O = 0.0°C$, $t_{w,in} = 24, 30.2, 40, °C$) and 0.98, 1.44, 1.91 ($t_O = -10.0°C$, $t_{w,in} = 24, 30.2, 40, °C$)	0, −10°C	24, 30, 40°C	23.1–39.4% ($t_O = 0.0°C$) and 24.2–55.7% ($t_O = -10.0°C$)	10.9–26.1% ($t_O = 0.0°C$) and 6.9–30.3% ($t_O = -10.0°C$)	E	[29]

T = Theoretical, E = Experimental, O = optimized cycle, Basic cycle: single-stage cycle without IHX

Table 7.2 Results under optimal operating conditions.

$T_{O,CO2}$ (°C)	$T_{W,in}$ (°C)	$P_{GC\text{-}K.opt}$ (bar)	$\dot{Q}_{O.CO2.opt}$ (W)	COP_{opt} (–)	$\dot{W}_{elec\ opt}$ (W)	$\Delta\dot{Q}_{O.CO2.opt}$ (%)	ΔCOP_{opt} (%)	$\Delta P_{GC\text{-}K.opt}$ (bar)	$\Delta\dot{W}_{elec\ opt}$ (%)
Base Cycle									
0.2	34.9	86.2	834.2	1.92	433.8	—	—	—	—
0.2	30.1	79.6	930.2	2.33	399.5	—	—	—	—
−9.9	34.7	87.0	571.9	1.39	411.5	—	—	—	—
−9.8	30.0	77.0	646.5	1.69	383.5	—	—	—	—
IHX Cycle									
0.1	35.0	85.6	852.7	2.03	421.0	+2.2	+5.3	−0.6	−2.9
0.0	29.9	77.8	970.9	2.43	399.2	+4.4	+4.4	−1.8	−0.1
−9.7	34.6	86.6	604.4	1.48	409.6	+5.7	+6.2	−0.4	−0.5
−9.8	29.9	75.0	651.6	1.75	381.8	+3.3	+3.7	−2.0	−0.4

7.2 CO_2 Thermodynamic Properties and Approach

CO_2 subcooling generally occurs in the proximities of the critical and pseudocritical regions, therefore this section summarizes the CO_2 most important properties and their variation in these regions and establishes the thermodynamic approach of CO_2 subcooled cycles.

7.2.1 Thermodynamic Properties of CO_2

The thermodynamic state of the CO_2 when it is subcooled after gas-cooler/condenser exit, is inside the shaded region shown in Figure 7.2. The pressure limits are established for compressor safety operation (upper limit) and proper expansion valve operation (bottom limit). Meanwhile, temperature limits are indicative. The maximum is considered in relation to the maximum environment temperature (40°C) and the minimum in relation to the minimum temperature to subcool CO_2 (0°C). Finally, if the vapor compression cycle is equipped with a backpressure before a liquid receive, the CO_2 enthalpy at the gas-cooler outlet must be lower than the critical enthalpy for a proper operation of the facility.

CO_2 could be in different matter states in this region (subcooled liquid, compressed fluid or supercritical fluid), and its thermodynamic and transport properties related to heat transfer suffer abrupt variations when it goes through those matter states around the critical pressure (see Figure 7.3).

The maximum variation in properties is not only in the critical point but around the so-called pseudocritical points. These are defined as points at a pressure ($p_{pc} > p_{ccrit}$) and at a temperature ($t_{pc} > t_{crit}$) corresponding to the maximum value of the specific heat at this particular pressure. All of those singular behaviors of properties have a significant influence on heat transfer next to the critical conditions, that is why this region is called the near-critical region (region around the critical point, where all thermophysical properties of a pure fluid

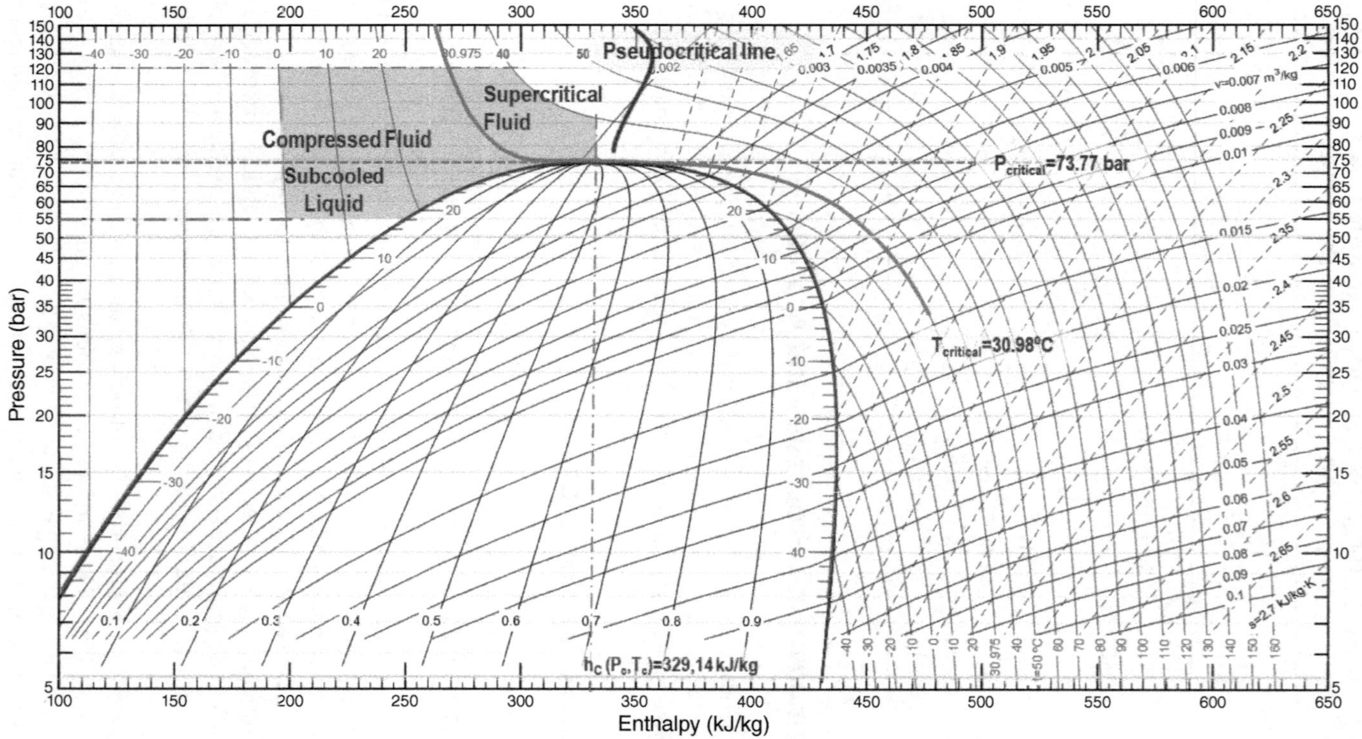

Reference: International Institue of Refrigeration h = 200 (kJ/kg), s = 1 (kJ/kg·K) saturated liquid at T = 0°C.
Lemmon E.W. McLinden M.O. and Huber M.L. 2002 REFPROP NIST Standard Reference Database 23, v7.0. National Institue of Standards and Technology. Gaithersburg, MD.
©Grupo de Ingenieria Terrnica (G.I.T.) (www.git.uji.es) Universidad Jaurne I de Castellon

Figure 7.2 Subcooled CO_2 working region.

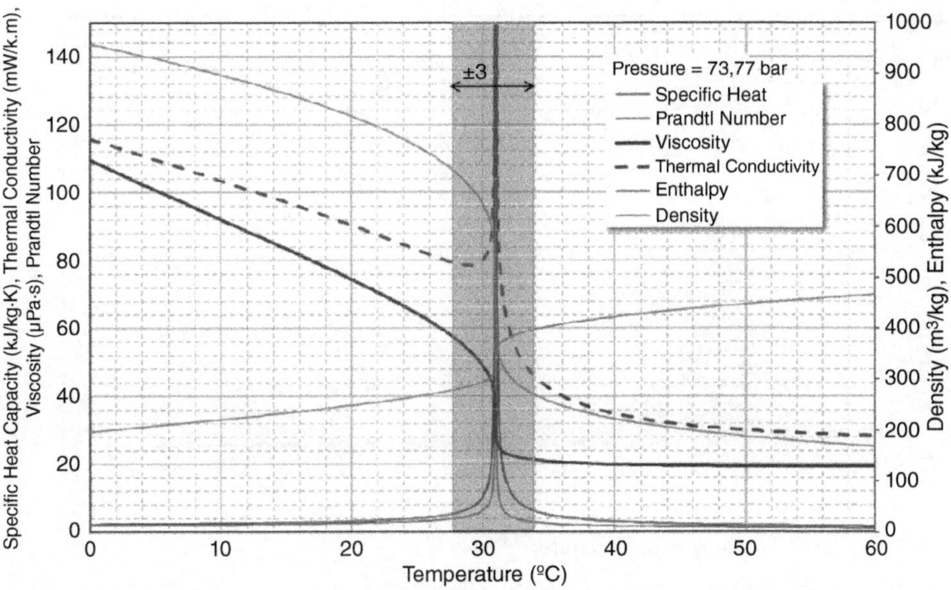

Figure 7.3 Thermodynamic and transport properties vs temperature and at critical pressure. Maximum variation zone.

exhibit rapid variations) and it is where CO_2 operates at the exit of the gas-cooler/condenser and is subcooled before its expansion.

Liao *et al.* [35], proposed a correlation of the pseudocritical temperature as a function of pressure for carbon dioxide with data obtained from REFPROP database [36]. The correlation is shown in Eq. (7.1)

$$t_{pc} = -122.6 + 6.124 \cdot p_{gc} - 0.1657 \cdot p_{gc}^2 + 0.01773 \cdot p_{gc}^{2.5} - 0.0005608 \cdot p_{gc}^3 \tag{7.1}$$

CO_2 thermodynamic properties, in a wide range of pressures and temperatures covering the near-critical region and the adjacent regions for comparison purposes, are shown in Figure 7.4, specific heat at constant pressure, and in Figure 7.5 density, while transport properties are detailed in Figure 7.6, Prandtl number in Figure 7.7, thermal conductivity, and in Figure 7.8, dynamic viscosity. In all those figures, the region where CO_2 operates when it is subcooled after exit from the gas-cooler/condenser has been shaded in green.

It is observed from the above figures that density, viscosity, and thermal conductivity present higher values at greater pressures and lower temperatures. Besides, density and dynamic viscosity undergo a significant drop that it is almost vertical near the critical point within a very narrow temperature range, while specific heat and the Prandtl number have peaks near the critical and pseudocritical points. These peaks decrease in magnitude very quickly with an increase in pressure. Also, the "peaks" transform into soft maximum profiles at pressures beyond the critical pressure. The maximums of thermal conductivity have similar characteristics with those of specific heat, but they do not coincide with the pseudocritical temperature. In addition, it should be noted that the dynamic viscosity and thermal conductivity present a minimum value right after the critical and pseudocritical points.

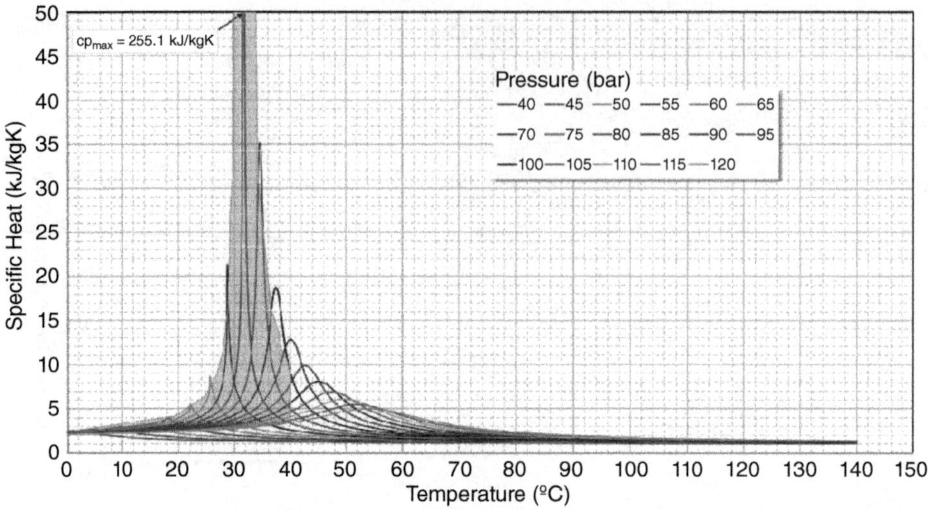

Figure 7.4 Specific enthalpy vs temperature.

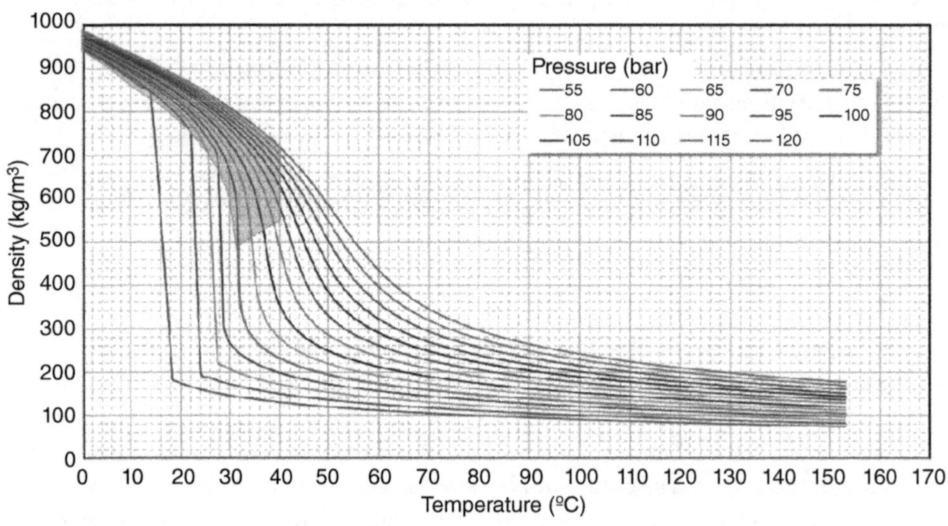

Figure 7.5 Density vs temperature.

It could be said that passing through the pseudocritical line at constant pressure is like crossing the saturation line from liquid into vapor. The major difference in crossing these two lines is that all changes (even drastic variations) in thermophysical properties at supercritical pressures are gradual and continuous, and take place within a certain temperature range. On the contrary, at subcritical pressures we have discontinuity of properties through the saturation line: one value for liquid and another for vapor. Therefore, supercritical fluids are considered as single-phase substances.

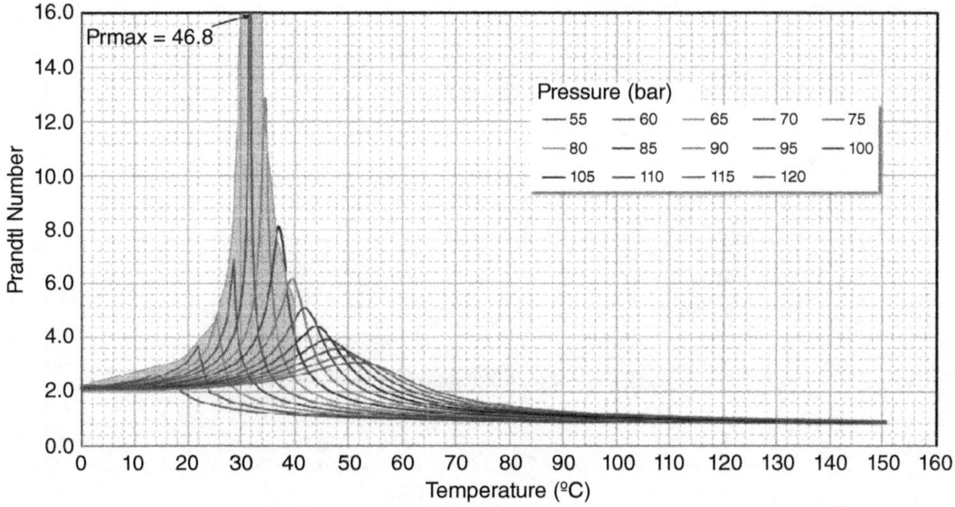

Figure 7.6 Prandtl number vs temperature.

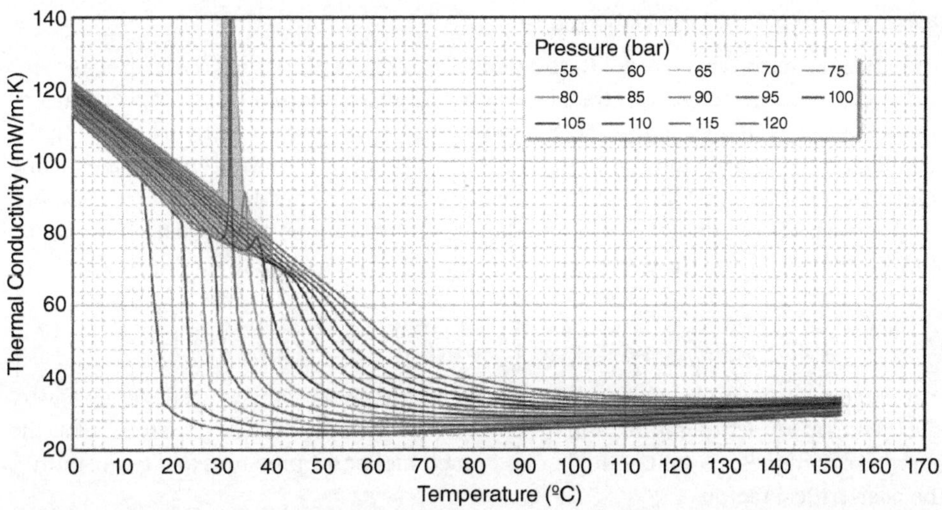

Figure 7.7 Thermal conductivity vs temperature.

As already noted, the heat transfer coefficient is affected by the variations of these thermophysical properties. Some characteristics of CO_2 heat transfer in the near-critical region, according to Hendricks *et al.* [37] are:

- Nonlinearities in heat flux against temperature difference
- Wall temperature excursions due to peaks
- Similarities to the two-phase regime
- Oscillations
- Large momentum pressure drops

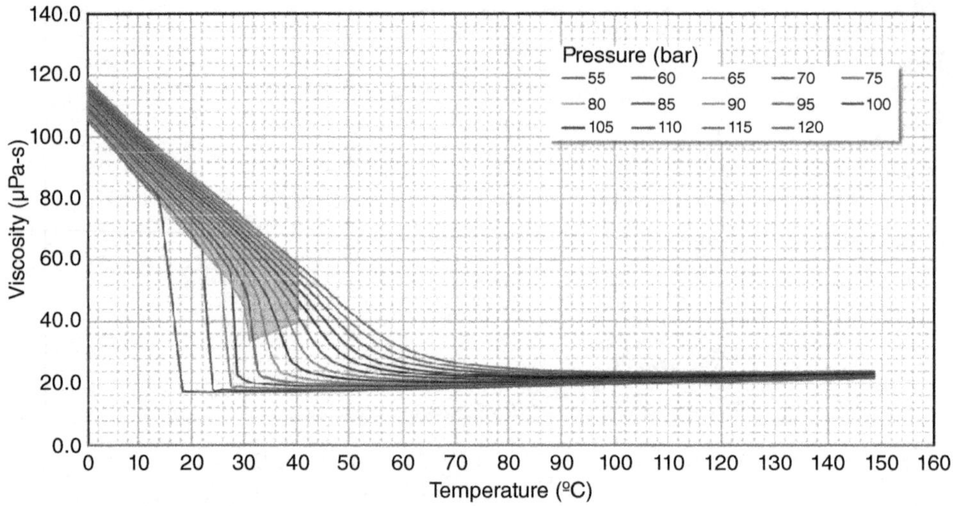

Figure 7.8 Dynamic viscosity vs temperature.

- System-dependent results
- Failure of conventional correlations

For indicative purposes, in order to show the influence of properties' variations as mentioned above, the convective heat transfer coefficient in an internal forced convection scenario has been depicted in Figure 7.9, in a range of pressure and temperature covering the subcooling region (that shadowed in Figure 7.2). The coefficient has been evaluated using the Gnielinsky correlation [38], Eq. (7.2), considering a 2 mm inner diameter tube transporting 1 kg s^{-1} mass flow rate. Transport and thermodynamic properties have been evaluated using REFPROP [36] and the wall effect has been neglected to simplify in Eq. (7.2).

$$\mathrm{Nu} = \frac{\xi/8 \cdot \mathrm{Re} \cdot \mathrm{Pr}}{1 + 12.7 \cdot \sqrt{\xi/8} \cdot (\mathrm{Pr}^{2/3} - 1)} \cdot \left[1 + \left(\frac{D_H}{L}\right)^{2/3}\right] \cdot \left(\frac{t_B}{t_W}\right)^{0.15} \cdot \left(\frac{Pr_B}{Pr_W}\right)^{0.11} \quad (7.2)$$

From Figure 7.9, it can be observed that the convection heat transfer coefficient has very similar values at low temperatures and they are not much dependent on pressure, but as the temperature increases, its value is higher and much higher for pressures and temperatures in the near-critical region.

7.2.2 CO_2 Subcooling Approach

The subcooled CO_2 cycle is analyzed using the basic layout used to control the heat rejection pressure and the degree of superheat at the evaporator exit simultaneously, as presented in Figure 7.10, whose main elements are:

- A compression system that consumes mechanical energy, P_c.
- A gas-cooler/condenser that rejects heat ($\dot{Q}_{gc}$) to the hot sink at a temperature t_H.
- A heat exchanger to cool or "subcool" CO_2 leaving the gas-cooler or condenser. In the subcooling device heat is absorbed from CO_2 at an intermediate temperature t_I, which is lower than the hot sink value (t_H).

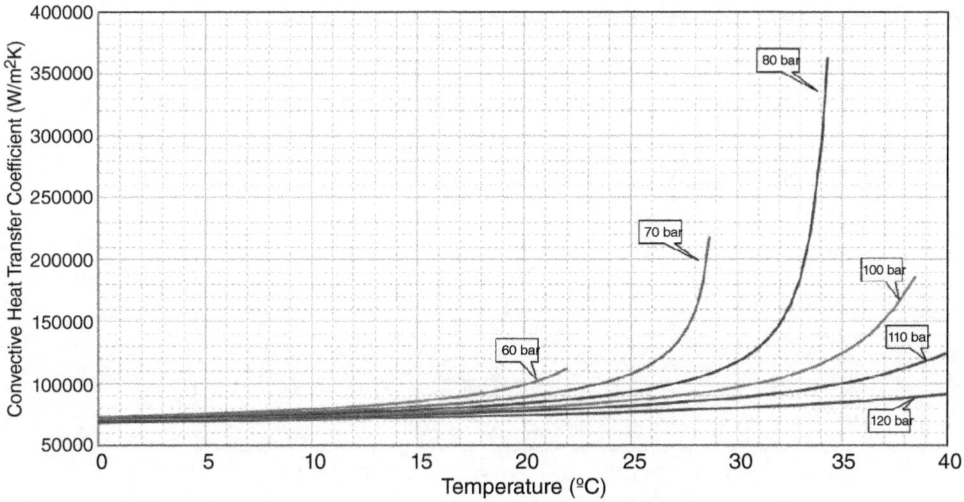

Figure 7.9 CO_2 internal forced convective heat transfer coefficient through a smooth tube.

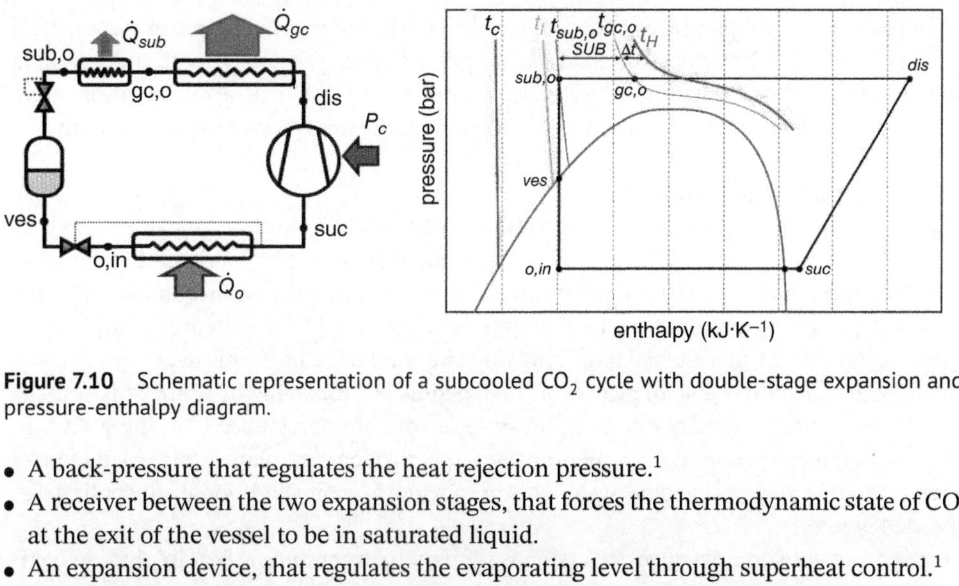

Figure 7.10 Schematic representation of a subcooled CO_2 cycle with double-stage expansion and pressure-enthalpy diagram.

- A back-pressure that regulates the heat rejection pressure.[1]
- A receiver between the two expansion stages, that forces the thermodynamic state of CO_2 at the exit of the vessel to be in saturated liquid.
- An expansion device, that regulates the evaporating level through superheat control.[1]
- An evaporator that absorbs heat ($\dot{Q}_o$) from the cold source at a temperature t_C.

The low critical temperature of CO_2 ($t_{crit} = 30.978°C$) causes the CO_2 subcooled cycle to work in two modes of operation: at low heat rejection levels (low t_H), the cycle works in subcritical conditions with condensation at constant temperature; whereas at high heat rejection levels (theoretically at a rejection temperature higher than the critical value,

1 CO_2 cycle presents two degrees of freedom, heat rejection pressure and degree of superheat at evaporator exit. If a double-stage expansion system is used (back-pressure + vessel + expansion valve), cycle is able to regulate high rejection pressure and evaporation, however, if a single-stage expansion system is used only one of those two variables could be controlled, therefore the reasoning could differ.

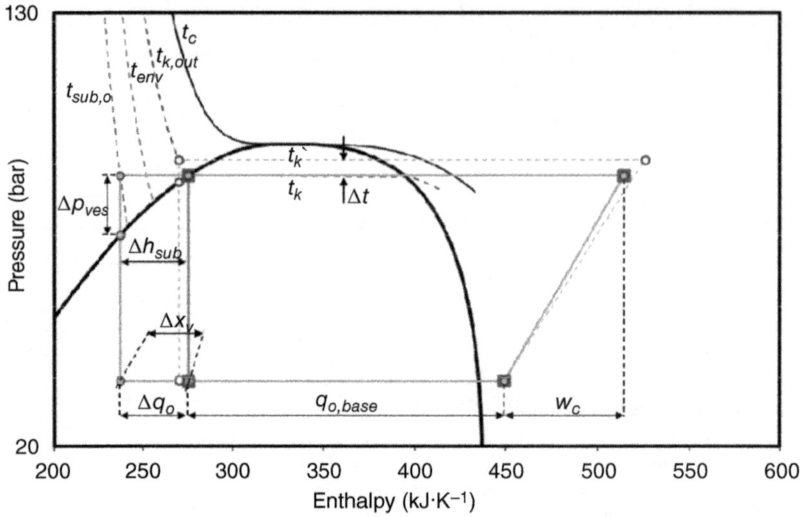

Figure 7.11 p-h diagram of CO_2 cycle and subcooled CO_2 cycle in subcritical.

but in practice for temperatures below the critical [39]), the cycle operates in transcritical conditions. For this last case, the heat exchanger acts as a gas-cooler, with a decreasing temperature profile through the heat rejection process [1]. Since a refrigeration cycle usually alternates its operation between both regimes, the analysis is extended to both.

7.2.2.1 Subcritical CO_2 Subcooling

CO_2 subcooling, when the cycle is in operation in subcritical conditions (if $t_K < t_{crit}$ [39]), is usually performed in cycles with a single-stage expansion device by forcing it to operate at a condensing temperature (t_{K*}) higher than the minimum that the environment will allow ($t_{k,min} = t_{env} + \Delta t$) by control of the total charge of refrigerant in the system [19, 40], as presented in Figure 7.11 in a dashed line. This way, the condenser performs heat rejection at a high temperature and is able to provide a small degree of subcooling by itself. This strategy is common in small capacity refrigeration systems for commercial use working with capillary tubes but is not common in medium or large refrigeration plants, for which control relies on double-stage expansion systems. And obviously, some additional subcooling can be added before.

For systems with double-stage expansion systems, which commonly require rigorous control, CO_2 subcooling is usually performed, as presented in Figure 7.11 (with continuous line). Heat rejection is performed at the minimum possible temperature with floating pressure until reaching saturation at the condenser exit. Then the additional subcooling is provided. It must be highlighted that the subcooling would only be possible if the back-pressure introduces a minimum pressure drop (Δp_{ves}) that guarantees that the vessel is at saturated condition. Since these systems incorporate the back-pressure, a strategy similar to that used for single-stage expansion systems could be used, by forcing the condensing temperature to be higher. However, the results of Nebot-Andrés *et al.* [27] indicate that reaching saturation in the condenser is the best performing condition. Contrasting the operation of a CO_2 cycle (red continuous line with squares) with a subcooled CO_2 cycle (green continuous line

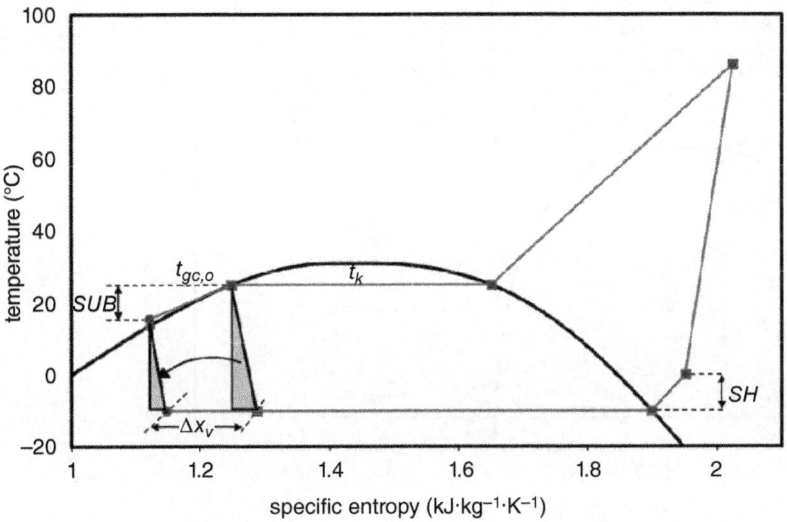

Figure 7.12 t-s diagram of CO_2 cycle and subcooled CO_2 cycle in subcritical.

with circles), the positive effects that the subcooling introduces can be observed: a pressure reduction in the vessel (Δp_{ves}), an increment of the specific refrigerating effect (Δq_o), and a reduction of the vapor quality at the inlet of the evaporator (Δx_v), which can result in a slight increment of the evaporating level [41]. No negative effects are introduced except of the cost of subcooling, which is discussed in subsection 7.2.3.4. Furthermore, as observed in the temperature-entropy diagram in the shaded triangles in Figure 7.12, the introduction of subcooling to the CO_2 cycle also reduces the exergy losses in the throttling processes.

7.2.2.2 Transcritical CO_2 Subcooling

At high heat rejection temperatures, the cycle operates in transcritical conditions and there is only one possible strategy to subcool the CO_2, as schematized in Figure 7.13. Subcooling is done with a subcooling system placed between the gas-cooler and the back-pressure. Since it has been verified that subcooling in CO_2 transcritical cycles allows the optimum high pressure to be reduced [29], the benefits of subcooling are enhanced in transcritical conditions, they being: a reduction of the optimum heat rejection pressure (Δp_{gc}), a reduction of the specific compression work in the compressor (Δw_c), a pressure reduction in the receiver (Δp_{ves}), an increment of the specific refrigerating effect (Δq_o), and a reduction of the vapor quality at the inlet of the evaporators (Δx_v), which can also result in an increment of the evaporating level [41]. Again, the unique drawback is the "cost of subcooling", which is discussed in 7.2.3.4. As observed in Figure 7.14, subcooling is able to reduce the exergy loss especially in the second expansion process, to a larger extent than in subcritical conditions. As analyzed in Section 7.2.1, the subcooling device is subjected to the large variations of CO_2 thermophysical properties in the near-critical region, therefore the heat exchanger used to subcool should follow the same guidelines as gas-coolers.

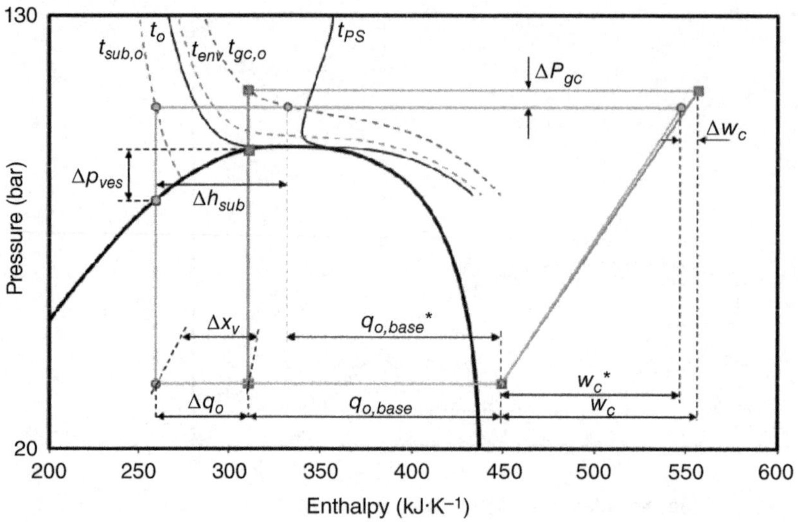

Figure 7.13 p-h diagram of CO_2 cycle and subcooled CO_2 cycle in transcritical.

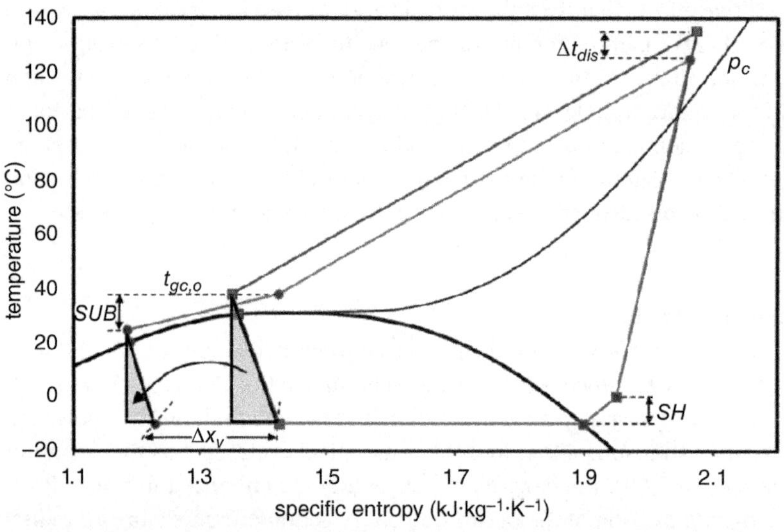

Figure 7.14 t-s diagram of CO_2 cycle and subcooled CO_2 cycle in transcritical.

7.2.3 Benefits of Subcooling

The introduction of subcooling to CO_2 refrigeration and heat pump systems modifies the optimal working conditions of the cycles, as outlined in Section 7.2.2, however, an analysis of the effect on the energy parameters is required to understand the possibilities of this method.

7.2.3.1 Second Law Approach

In order to illustrate the benefits of subcooling, Figures 7.15 and 7.16 present the thermodynamic states of a subcooled CO_2 transcritical cycle operating at an evaporating level of −10°C, outlet temperature of gas-cooler of 35°C, and constant heat rejection pressure of 90 bar, for different subcooling degrees: 0 K representing the cycle without subcooling and 43.9 K corresponding to the maximum subcooling which could be applied to the cycle (at that condition the outlet enthalpy of subcooler coincides with enthalpy of saturated liquid at the evaporating level). As can be observed in Figure 7.15, the introduction of subcooling increases the pressure lift in the back-pressure and reduces the total expansion rate of the second expansion device. The effect, from a second law approach, is highlighted in Figure 7.16, where the shadowed areas represent the exergy losses during the expansion processes. It is observed that for increased subcooling the pressure in the second expansion

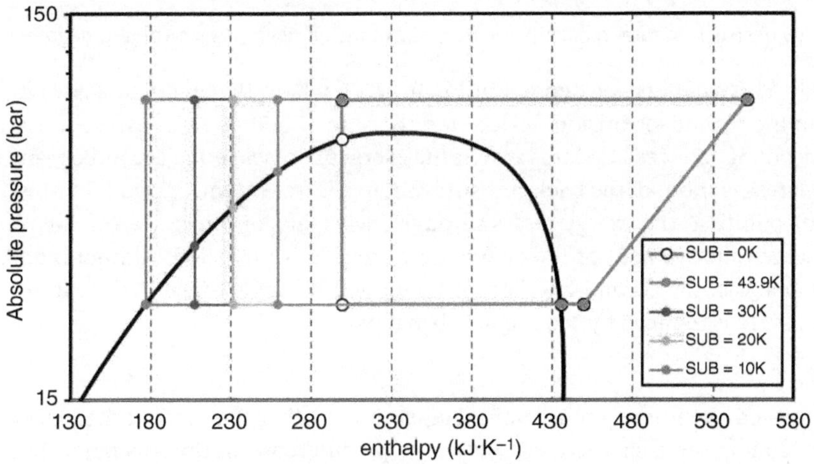

Figure 7.15 Pressure enthalpy diagram of subcooled CO_2 transcritical cycle.

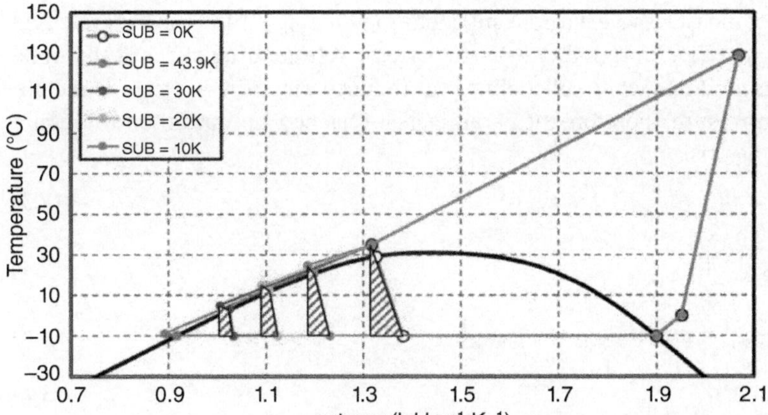

Figure 7.16 Temperature entropy diagram of subcooled CO_2 transcritical cycle.

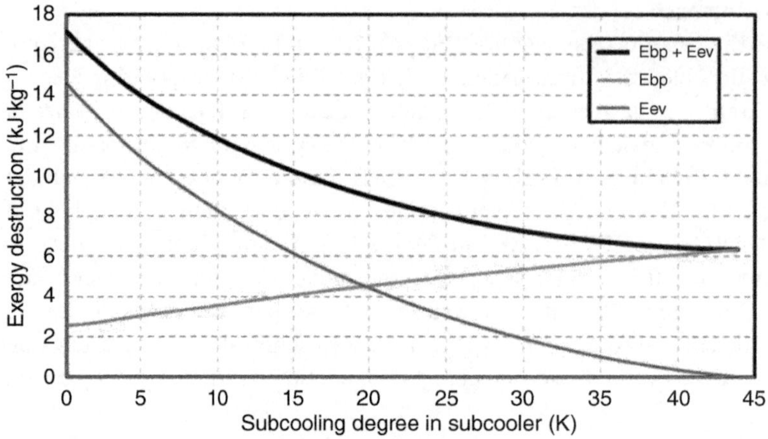

Figure 7.17 Exergy destruction rate in back-pressure and expansion device vs subcooling degree.

device is reduced. This analysis is extended in Figure 7.17, where the energetic losses in back-pressure, in the second expansion device, and the sum of both are represented. It is verified that subcooling reduces to a large extent the exergetic losses in the second expansion valve and increases those of the back-pressure, but to a lesser extent. In overall terms, for the analyzed condition, the exergy losses through the expansion process are reduced by 63%, thus a large improvement of the energy performance is predicted. Although not included in the analysis, the second law improvement will be larger since the optimum heat rejection pressure is reduced by the use of subcooling.

7.2.3.2 Capacity

Subcooling is achieved by reducing refrigerant temperature further than that at the exit of the gas-cooler or condenser, thus there is a net enthalpy difference in the subcooler that increases the specific refrigerating effect of the cycle. Equation (7.3) indicates the cooling capacity of the generic CO_2 cycle with subcooling (Figure 7.10) as the product of refrigerant mass flow rate and specific refrigerating effect in the evaporator. It can be expressed as the sum of the capacity of the CO_2 cycle without subcooling ($\dot{m}_r \cdot q_{o,base}{}^*$) and the heat released in the subcooler ($\dot{Q}_{sub}$) (Eqs. (7.4) and (7.5)). The specific refrigerating effect of the cycle without subcooling is modified when subcooling is introduced, since there is a net reduction of the high rejection pressure, therefore the parameters for the new optimum conditions are represented with an asterisk.

$$\dot{Q}_O = \dot{m}_r \cdot q_o = \dot{m}_r \cdot (q_{o,base}{}^* + \Delta h_{sub}) \tag{7.3}$$

$$\dot{Q}_O = \dot{m}_r \cdot q_{o,base}{}^* + \dot{Q}_{sub} \tag{7.4}$$

$$\dot{Q}_{sub} = \dot{m}_r \cdot \Delta h_{sub} = \dot{m}_r \cdot (h_{gc,out}{}^* - h_{sub,out}) \tag{7.5}$$

$$q_{o,base}{}^* = h_{o,out} - h_{gc,out}{}^* \tag{7.6}$$

To evaluate the increment of capacity due to the introduction of subcooling, Li *et al.* [42] proposed the parameter *RICOSP* to quantify the relation between the increase in capacity of the subcooled system and the heat extracted by the subcooling device. Initially, they concluded [42] that in subcritical cycles RICOSP cannot exceed 1, however in the experimental verification [43] they were able to measure values higher than 1. This effect was outlined in the theoretical approach where authors indicated that RICOSP could exceed the unit at the optimum working conditions. And that is what happens in CO_2 transcritical cycles, since the optimum high pressure is reduced by the use of subcooling, the RICOSP exceeds the unit due to the increase in refrigerant mass flow. For example, using the results of Llopis *et al.* [29], RICOSP values reached 1.19 when working with a DMS single-stage CO_2 cycle at −10 and 40 °C, because the subcooling system reduced the optimum high pressure by 5.2 bar.

7.2.3.3 COP

The COP is expressed as a quotient of the cooling capacity ($\dot{Q}_O$) and the power consumption of the system (P_C), as expressed by Eq. (7.7).

$$COP = \frac{\dot{Q}_O}{P_C} \tag{7.7}$$

If the IHX is considered as a subcooling device, COP modifications are bonded to variations in capacity and power consumption in the cycle due to the thermal coupling between the liquid and suction lines. This element is analyzed in detail in Section 7.3.

However, for an active subcooling method requiring an energy input, the COP can be expressed with Eq. (7.8), where $\dot{Q}_O$ is the cooling capacity, Eq. (7.3); P_C is the power consumption of the CO_2 compressor; and $P_{C,\,sub}$ is the electrical energy used by the subcooling system.

$$COP = \frac{\dot{Q}_O}{P_{C,CO2} + P_{C,sub}} \tag{7.8}$$

Considering a COP_{sub} of the subcooling system as the quotient of the heat transfer in the subcooler and the energy input to activate the subcooling device (Eq. (7.9)), the overall COP can be expressed with Eq. (7.10) using an energy balance in the subcooler. The overall COP depends on the CO_2 enthalpy difference in the subcooler (Δh_{sub}) and on the COP of the subcooler system (COP_{sub}). Subcooling will be positive, from an energy point of view, only if $\frac{\partial COP}{\partial \Delta h_{sub}}$ results are positive, that is when Eq. (7.11) is satisfied. Which is to say that a subcooling system would enhance the performance of a CO_2 cycle as long as $COP_{sub} = f(t_H, t_I)$ is higher than the $COP_{sub} = f(t_H, t_C)$ of the CO_2 cycle. This inequation is generally satisfied for mechanical subcooling systems [29, 44, 45], however, for subcooling systems with reduced COP such as thermoelectric devices [31, 32], their application range is restricted.

$$COP_{sub} = \frac{\dot{Q}_{sub}}{P_{C,sub}} \tag{7.9}$$

$$COP = \frac{q_{o,base} + \Delta h_{sub}}{w_c^* + \frac{\Delta h_{sub}}{COP_{sub}}} \tag{7.10}$$

$$COP_{sub} > COP_{CO2} \tag{7.11}$$

It can be affirmed that a CO_2 subcooled cycle would offer higher COP increments the higher the COP of the subcooling system is, however, the thermodynamic limits of this improvement have not been extensively analyzed.

7.2.3.4 Energy Input

An important aspect in relation to subcooled CO_2 cycles is energy input required by the subcooling system to provide the necessary heat transfer. Equation (7.12) indicates the total energy input to the system, which considers the energy consumption of the CO_2 cycle and of the subcooling system. Equation (7.13) expresses the increment on energy consumption of a subcooled system ('*') in relation to a non-subcooled one, where $\dot{Q}_{sub}$ is the heat transfer rate in the subcooler and COP_{sub} refers to the energy performance of the subcooling system.

$$P_C{}^* = P_{C,CO2} + P_{C,sub} = \dot{m}_r \cdot w_c + P_{C,sub} \tag{7.12}$$

$$\Delta P_C = P_C{}^* - P_C = (\dot{m}_r{}^* \cdot w_c{}^* - \dot{m}_r \cdot w_c) + \frac{\dot{Q}_{sub}}{COP_{sub}} \tag{7.13}$$

If the CO_2 cycle operates in subcritical conditions (Figure 7.11), the optimum working pressure is not modified, thus the increment on energy input in relation to a non-subcooled cycle is the quotient between the heat extracted and the COP of the subcooling system, as detailed by Eq. (7.14). This reasoning is also applicable for conventional refrigerants with subcooled cycles [41, 46]. However, subcooling in transcritical conditions modifies the optimum heat rejection pressure and thus the energy input to the main compressor. At a reduced high pressure, the mass flow rate of the CO_2 cycle is larger than in a non-subcooled cycle ($\dot{m}_r{}^* > \dot{m}_r$) but the specific compression work is lower ($w_{comp}{}^* < w_{comp}$), being the trends opposite. Experimental results with a DMS single-stage plant [47] showed that the CO_2 compressor power consumption was reduced when subcooling the cycle, and the results with a DMS two-stage plant even resulted in decreases of the total system power consumption [48]. Subsequently, it can be affirmed that the increment on energy consumption due to the subcooling system in transcritical conditions will be lower than the one established in subcritical condition, as expressed by Eq. (7.15).

$$\Delta P_C = \frac{\dot{Q}_{sub}}{COP_{sub}} = \dot{m}_r \cdot \frac{\Delta h_{sub}}{COP_{sub}} \tag{7.14}$$

$$\Delta P_C < \frac{\dot{Q}_{sub}}{COP_{sub}} = \dot{m}_r \cdot \frac{\Delta h_{sub}}{COP_{sub}} \tag{7.15}$$

7.2.4 Subcooling Optimization

As mentioned, subcooling in a CO_2 refrigeration system modifies the optimum working conditions, especially in transcritical conditions, where the subcooling is able to reduce the optimum high rejection pressure and modify the behavior of the CO_2 compressor. Obviously, it is necessary for such systems to determine the operating parameters that maximize the COP of the overall system.

COP of the subcooled cycle (Eq. (7.8)) depends on the cooling capacity and on the energy input to the compressor and to the subcooling system. For a fixed operating condition, with fixed t_o, $t_{gc,o}$ and SH at compressor inlet, the power consumption of the CO_2 compressor depends on the high rejection pressure (Eq. (7.16)) [7], and the cooling capacity depends on the high rejection pressure as well as on the subcooling (Eq. (7.17)). Referring to the subcooling system, its cold source at t_I only depends on the subcooling degree, subsequently the energy input to the subcooling system is a function of the subcooling (Eq. (7.18)). Accordingly, it can be affirmed that the COP of the whole system is a function of the heat rejection pressure and of the subcooling degree, as expressed by Eq. (7.19). In subcritical conditions the optimum heat rejection pressure is equal to the condensing pressure and only the subcooling degree needs to be optimized. However, in transcritical conditions the COP of the plant is bounded to two parameters that must be optimized together.

$$P_C = f(p_{gc}) \tag{7.16}$$

$$\dot{Q}_O = f(p_{gc}, SUB) \tag{7.17}$$

$$t_I = f(SUB) \rightarrow P_{C,sub} = f(SUB) \tag{7.18}$$

$$COP = f(p_{gc}, SUB) \tag{7.19}$$

It is important to mention that the classical relations to define the optimum high pressure developed for CO_2 cycles are not suitable [3, 5, 6], since the new optimum conditions depend on the subcooling system.

7.3 Internal Heat Exchanger

7.3.1 Introduction

In this section, the figure of the IHX as the simplest method of subcooling is analyzed and discussed from a theoretical and experimental point of view. Accordingly, it has been divided into three parts. The first part, Section 7.3.2, is focused on explaining the operation of the IHX and how it affects the behavior of a refrigeration facility. The second part, Section 7.3.3, summarizes the most representative analysis from the open literature in subcritical and transcritical conditions. Finally, Section 7.3.4, deals with an experimental analysis with carbon dioxide (CO_2) where the effect of using an IHX in transcritical conditions is presented and discussed. It must be remarked that in this section, the heat exchanger used as IHX operates without phase change in any of the working fluids.

7.3.2 Description and Operation

One of the most extensive subcooling methods used in refrigeration facilities corresponds to the IHX or also named suction-line to liquid-line heat exchanger (SLHX). This internal method consists of installing a heat exchanger at the exit of the condenser/gas-cooler to reduce the temperature of the refrigerant by using the exhausted cold vapor from the evaporator. Figure 7.1 provides a simplified schematic of a refrigeration cycle with IHX in this classical position.

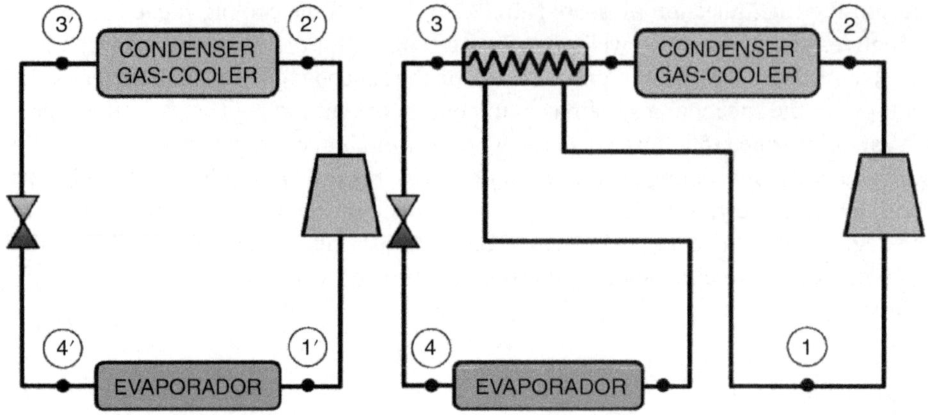

Figure 7.18 Schematic of the refrigeration facility (left) without IHX and (right) with IHX (classical position).

Based on the position showed in Figure 7.18, the IHX provides the following advantages:

a) It prevents flash gas formation at the inlet of the expansion device.
b) It increases the evaporator specific capacity and reduces the vapor quality at the evaporator inlet. The subcooling effect also reduces throttling losses.
c) It ensures the presence of single-phase vapor in the compressor suction line fully evaporating any residual liquid prior to reaching the compressor. This is essential in those systems without useful superheating control.
d) It increases the temperature of the suction line above the dew-point temperature of ambient air, avoiding condensation of the water vapor over the suction line (i.e. small cooling capacity systems).
e) It raises the lubrication oil temperature and, therefore, reduces the refrigerant concentration, improving the oil viscosity. Manufacturers recommend maintaining an oil temperature 15–20 K above suction-side saturation temperature, so the use of this element is strongly recommended for low-temperature applications [49].

Moreover, the use of the IHX may improve the COP of the refrigeration facility and its volumetric capacity due to the effect on the temperature and the pressure at the inlet of the throttling device and the compressor. The variation of both parameters will depend on the heat exchanger used as IHX and the operating conditions of the refrigerating plant.

Figure 7.19 depicts the effects of installing an IHX according to the schematics from Figure 7.18. These effects are described for a basic subcritical cycle, but a similar behavior can be found working in transcritical conditions. The main difference between both operating ranges is the existence of an optimal heat rejection pressure in transcritical conditions, the value of which can be reduced when the IHX is installed.

As it is shown in Figure 7.19, the IHX increases the evaporator capacity ($q > q'$) and the specific volume at the suction port (point 1). The combined effect of both parameters can increase or reduce the volumetric capacity (q_V) (Eq. (7.20)). Additionally, the increment of

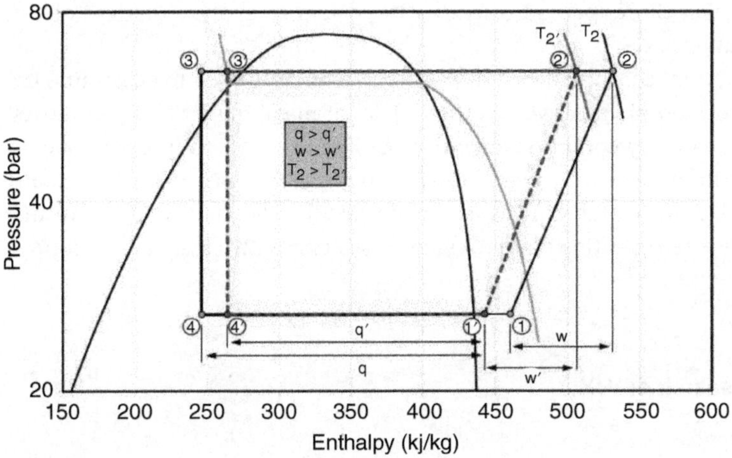

Figure 7.19 Effect of using an IHX in a vapor compression cycle.

the specific volume reduces the mass flow rate driven by the compressor ($\dot{m}_r$) if the compressor volumetric efficiency (η_V) remains almost constant (Eq. (7.21)).

$$q_V = \frac{q}{v_1} \tag{7.20}$$

$$\dot{m}_r = \frac{\eta_V \cdot \dot{V}_G}{v_1} \tag{7.21}$$

In relation to the specific compression work (w), it always increases with the presence of the IHX, affecting negatively the power consumption of the compressor (P_C) and the discharge temperature ($t_2 > t_2$'). However, the reduction of the mass flow rate can compensate for the increment of power consumption Eq. (7.22).

$$P_C = \dot{m}_r \cdot (h_2 - h_1) = \dot{m}_r \cdot w \tag{7.22}$$

The cooling capacity ($\dot{Q}_O$) is affected similarly by the mass flow rate and the evaporator specific capacity. The product of both terms can either improve or reduce the cooling capacity according to Eq. (7.23).

$$\dot{Q}_O = \dot{m}_r \cdot (h_1 - h_4) \tag{7.23}$$

Taking into account the effects over the cooling capacity and the compressor power consumption, the COP of the refrigeration plant can increase or decrease according to Eq. (7.24).

$$COP = \frac{\dot{Q}_O}{P_C} = \frac{q}{w} \tag{7.24}$$

Since the value of $\dot{Q}_O$ and P_C cannot be determined a priori, it is difficult to predict the benefit of using an IHX in a refrigerating plant. To solve this issue, some authors have analyzed which are the key parameters from the refrigerant cycle and from the refrigerant thermophysical properties to state the convenience of using an IHX. However, those methods have been obtained in subcritical conditions, so they are not reliable for transcritical conditions.

Section 7.3.3.1 summarizes those methods, inviting the reader to analyze them to obtain more information about its use.

For transcritical operation, computational models are commonly used to determine the operation of the refrigeration plant as well as the impact of installing an IHX. Moreover, experimental tests in laboratory conditions are also performed to quantify the impact of the IHX and to gather information for modeling. Section 7.3.3.2 summarizes the results published in the open literature attending to its location within the facility. These results correspond to single-stage refrigerating plants or booster systems with a two-stage compression system.

7.3.3 Revision of Research of IHX

7.3.3.1 Predicting Methods

The difficulty of predicting the benefit of using an IHX in a refrigeration facility has been deeply analyzed by several authors attending to different parameters, especially the refrigerant thermodynamic properties.

Domanski *et al.* [50] presented a study where the most influential property was the isobaric heat capacity assuming ideal compression, non-pressure drops and thermal effectiveness equal to 100%. The analysis affirmed that fluids performing poorly in the basic cycle are positively affected by the IHX installation by increasing its COP and volumetric capacity. This conclusion was extended by Domanski [51] using 38 refrigerants including CO_2 in subcritical conditions. From this analysis, the higher the specific heat capacity at the vapor side, the higher the benefit of the IHX. This affirmation is explained taking into account that a high value of the specific heat capacity means a low increment of temperature due to the IHX. However, this affirmation should be completed with a low specific heat capacity at the liquid side, since it would help reduce its temperature before entering the expansion device.

Aprea *et al.* [52] determined a criterion based on the isobaric specific heat to determine whether the adoption of an IHX is a profitable choice. This criterion also assumes ideal isentropic efficiency at the compressor and non-pressure drops. Klein *et al.* [53] also determined a new criterion where the IHX thermal effectiveness, the temperature lift between condensing and evaporating level, the evaporation latent heat and the isobaric specific heat were used as the main parameters. The study analyses the effect of pressure losses in the IHX affirming that they are relevant at high-temperature lifts. Mastrullo *et al.* [54] published a chart for predicting the advantage of using an IHX using 19 different refrigerants, assuming an ideal cycle.

Finally, Hermes [55] analyzed the effect of using an IHX in an ideal refrigeration facility maintaining the cooling capacity as a constraint and the evaporating temperature as a free variable. The results from Hermes corroborated the results obtained by Domanski *et al.* [50] but assuming the cooling capacity as a constraint instead of the evaporating temperature. A second report from Hermes [56] also predicted the effect of a liquid-to-suction heat exchanger over the refrigerant mass charge. The results show a mass charge reduction of almost 15% at -25°C of evaporation level and 5% at 7°C of evaporating temperature

7.3.3.2 Theoretical and Experimental Analysis

Computational models and experimental tests are commonly used to quantify the benefit of using an IHX in a refrigerating plant taking into account not only the operating conditions but also the main characteristics of the refrigerating plant. Focusing on CO_2, the predicting methods stated before are only valid for subcritical conditions, so predictions in transcritical conditions need to be addressed from a computational or experimental point of view. Accordingly, this section is devoted to summarizing some of the studies carried out in different refrigerating facilities, and in different operating conditions. The section is divided into two parts based on the location of the IHX. Thus, the first is focused on the classical position at the exit of the gas-cooler/condenser (Figure 7.18) and those layouts that result from this according to the configuration adopted in the refrigerating plant. The second part covers the combination of the IHX with improved throttling systems as expanders or ejectors.

Classic Vapor Compression Cycle Positions

The use of the IHX as a method to improve the efficiency of the transcritical cycle was first proposed by Lorentzen in 1989 in his Patent no. WO 90/07683 for mobile air conditioning, and later described in 1993 and 1994 as essential for the proper operation of the cycle [57, 58]. From a theoretical point of view, Domanski [51] analyzed theoretically the effect of the IHX over the COP in a single-stage refrigerating facility. Considering ideal conditions, the study demonstrated that in subcritical conditions the IHX penalizes the COP of the system regardless of its thermal effectiveness. Robinson and Groll [59] theoretically determined the improvements of using an IHX in a transcritical cycle assuming a correlation for the isentropic efficiency. Using a gas-cooler outlet temperature of 40°C and several evaporating temperatures, the results revealed improvements between 4% and 7% depending on the thermal effectiveness of the IHX.

Rozhentsev and Wang [60] investigated theoretically the effect of the IHX on the system COP and the optimal pressure. For air conditioning conditions, the effect of using the IHX with different approaches introduced remarkable improvements in COP and a reduction in the optimal heat rejection pressure. Kim *et al.* [61], also theoretically, investigated the performance of a transcritical CO_2 cycle with an IHX for a hot water heater. Varying the length of the IHX, the work analyzed the effect of the IHX length over the optimum discharge temperature, COP, cooling capacity and electrical power compressor.

Chen and Gu [6] presented a deep analysis of the relationship between the optimal heat rejection pressure and IHX thermal effectiveness. This study confirmed the previous assessments commented above and included a polynomial correlation to determine the optimum heat rejection pressure as a function of the environment temperature and the IHX thermal effectiveness. Zhang *et al.* [62] explored the influence of the IHX in subcritical and transcritical conditions from a theoretical point of view. Their analysis concluded that the effect of the IHX over the COP in subcritical conditions is negligible and it only takes relevance in transcritical conditions. Finally, Ituna-Yudonago *et al.* [63] presented a computational fluid dynamics (CFD) numerical investigation to determine the transient behavior of CO_2 in the IHX.

Experimentally, Boewe *et al.* [64] tested the role of the IHX in a mobile A/C system. The maximum benefits registered were 26% for COP and 10% for cooling capacity using an IHX of 2 m length working at 43.3°C gas-cooler air temperature and at 26.7°C evaporator air temperature. Cavallini *et al.* [23, 65] tested in a two-stage transcritical refrigerating plant the effect of the IHX, varying the quality of the vapor at the exit of the evaporator from 0.75 to superheated condition. The results concluded that the use of the IHX increased up to 20% the COP of the system. Similar conclusions were obtained later by Cavallini *et al.* [66] but using also a desuperheater (DSH) (intercooler) between both compression stages.

Cho *et al.* [67] explored the effect of several parameters including the length of the IHX, over a mobile air-conditioning bench-test. The increments registered were up to 9% for the COP at the length of 3 m. Aprea and Maiorino [22] evaluated the performance of a CO_2 transcritical refrigerating plant with and without IHX at the evaporation temperature of 5°C. Varying the gas-cooler air inlet temperature from 25 to 40°C, the increment obtained in terms of COP was ranged between 8.11% and 10.47%. Rigola *et al.* [68] carried out an experimental and numerical study where the possibilities that CO_2 offers for commercial refrigerating cycles where explored. Taking as a reference an evaporating temperature of −10°C, the results clearly showed that at the heat rejection temperatures of 35 and 43°C the COP can be increased up to 30% using an IHX. Torrella *et al.* [21], also experimentally, demonstrated the convenience of using an IHX in CO_2 transcritical cycles. The results evaluated at evaporation temperatures of −5, −10 and −15°C and the heat rejection temperatures of 33.9 and 31°C, revealed increments up to 13% in terms of COP and cooling capacity. However, the discharge temperature experienced increments of up to 10°C which was in accordance with the previously presented analysis. Similar results were obtained by Sánchez *et al.* [69] in a small-capacity refrigerating plant working with a hermetic compressor.

Cabello *et al.* [70] evaluated experimentally the combination of using an IHX and a vapor extraction from the intermediate accumulator tank (also called flash-gas bypass system). The study was performed with an IHX installed after the gas-cooler with three different injection points: before the IHX, after the IHX and just before entering the suction chamber of the compressor. The results concluded that the use of the flash-gas with IHX improves slightly the cooling capacity and the COP (5% and 3.6% on average, respectively), but it allows reduction of the discharge temperature up to 14.7°C.

Sánchez *et al.* [39] analyzed different positions for an IHX in a transcritical refrigerating plant equipped with a two-stage expansion system with an accumulator tank between stages. They compared the classical position at the exit of the gas-cooler with a new one at the exit of the accumulator tank. From the results, they concluded that regardless of position, the use of the IHX was positive in all cases. The best option corresponded to the classical position with enhancements of cooling capacity and COP up to 4.89% and 10.6%, respectively. Moreover, they introduce the option of using both IHX at the same time reaching improvements of 13% in terms of COP but increments of 20 K in the discharge temperature. Karampour and Sawalha [71] published an extended study with nine different layouts for an IHX in a centralized CO_2 booster system. The results demonstrated that the use of the IHX with the only purpose of improving cold COP was negligible. However, if simultaneous refrigeration and heat recovery are proposed, the COP of the global systems can be enhanced up to 12% if a flash-gas bypass system is also included.

Llopis *et al.* [72] tested a brazed-plate IHX in a CO_2 subcritical cycle of a cascade refrigeration facility. The results from this work demonstrated the work presented by Zhang *et al.* [62] since the effect of the IHX hardly affects the COP of the subcritical cycle. The maximum improvements registered were 3.29% at −25°C and 0.45% at −40°C. The same effect was found over the whole cascade facility, for which COP also rises to 3.7% at −35°C and 40°C of evaporating and condensing temperatures, respectively [73].

Finally, Purohit *et al.* [74] carried out an experimental investigation to evaluate the advantages of using an IHX in a transcritical refrigeration cycle, especially at high ambient temperatures. For a heat rejection temperature of 45°C, the experimental tests reported an energy improvement of 5.71% at the evaporation level of −5°C, and 5.01% at the evaporation level of 0°C. Furthermore, the effect of using the IHX allows improvement of the exergy efficiency of the system but affects the discharge temperature with a maximum increment of 24 K.

Taking into account the researchers reported above, it is evident that the use of the IHX in the classical layout (exit of the gas-cooler/exit of the evaporator) is very recommendable in transcritical conditions, since it improves the COP and the cooling capacity at high rejection temperatures. Moreover, it reduces the optimal heat rejection pressure, so it reduces the compressor pressure ratio. In contrast, the IHX increases the compressor discharge temperature, so it compromises the operating conditions of the refrigerating plant. Notwithstanding, focusing on heat pump applications, this aspect could be positive so it should be analyzed deeply to find a balance between the compressor operation and the improvement reached in terms of COP. In subcritical systems, the use of the IHX does not report a significant benefit excepting when it is used in the low-temperature cycle of a cascade refrigerating plant.

Combination of the IHX with Expanders and Ejectors

Use of expanders and ejectors are two different ways to improve the energy efficiency of a CO_2 refrigerating plant, especially when it operates in transcritical conditions (see Section 7.1). The important exergy losses found in the throttling process allows introduction of new expansion devices with the aim of recovering energy from the expansion process. The expander is designed to recover mechanical work that usually is used to assist in driving the main compressor (Figure 7.20), while the ejector is developed to pump vapor (or liquid) from a low pressure side to a higher one (Figure 7.21).

In the open literature, the reader can find extensive documentation about the design and operation of expanders and ejectors, as well as their most common configurations used in a

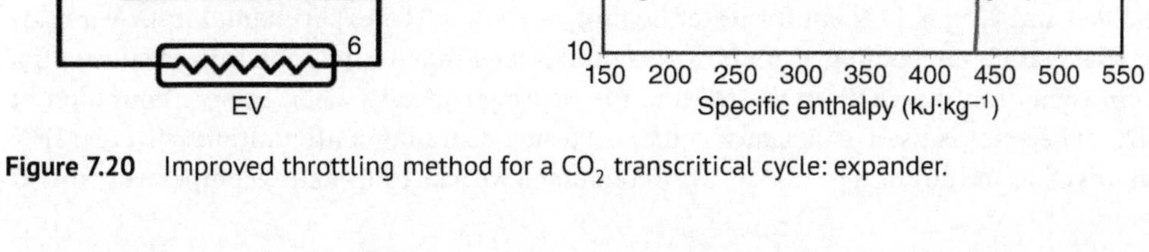

Figure 7.20 Improved throttling method for a CO_2 transcritical cycle: expander.

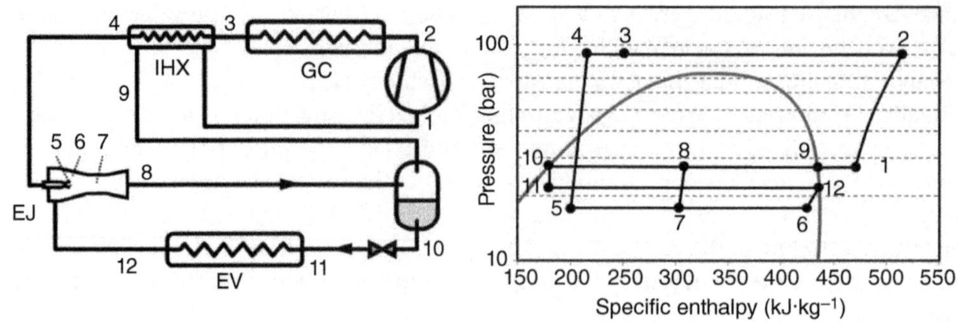

Figure 7.21 Improved throttling method for a CO_2 transcritical cycle: ejector.

refrigeration facility. In this section, the authors would like to show the reader the benefits of combining the IHX with expanders or ejectors in order to improve the operation of the whole system. To achieve this, the section is divided into two parts according to expanders and ejectors.

IHX and Expanders

Robinson and Groll [59] analyzed a modified transcritical cycle equipped with a work recovery turbine and an IHX installed in its classical position. Assuming a constant isentropic turbine efficiency of 60%, the use of the IHX degrades the performance of the work recovery turbine cycle in a range of 6–8%. However, the use of the IHX without a work recovery device increases the COP by up to 7%. Similar results were obtained by Shariatzadeh *et al.* [11] assuming two different isentropic turbine efficiencies of 75% and 65%, and by Zhang *et al.* [75] assuming efficiencies of 60% and 80%.

From the above-mentioned works, it can be highlighted that the use of the IHX is only positive when the isentropic turbine efficiency is very low or no work recovery device is installed.

IHX and Ejectors

Elbel and Hrnjak [76, 77] compared, theoretically and experimentally, four different transcritical mobile air conditioning systems with IHX and ejector. The experimental study, performed at an outdoor temperature of 45°C and an indoor temperature of 27°C, confirmed that the combination of the ejector and the IHX gives better results than those using the classical configuration of IHX. For IHX thermal effectiveness of 60% and optimal operating conditions, the increment of COP and cooling capacity was less than 4% at a constant compressor rotation speed of 1800 rpm. However, if the rotation speed is modified to adjust the cooling capacity to 4.78 kW, the COP could be improved up to 18% with regard to the configuration with the expansion valve.

Xu *et al.* [78] performed an experimental analysis similar to the previous one performed by Elbel and Hrnjak [77] but for water heating purposes. The experimental tests were carried out at different cooling water flow rates and inlet temperatures, obtaining a substantial improvement in the COP up to 16% and the heating capacity when the configuration of IHX and ejector is used. Nakagawa *et al.* [79] tested a similar configuration with two IHX lengths of 30 and 60 cm. The results were conducted with an evaporating temperature of 0°C and a heat rejection temperature of 42°C. Varying the heat rejection pressure and using the combination of ejector and IHX (60 cm), the COP of the system was improved by up to 27% compared to a conventional system with similar IHX. Finally, Zhang *et al.* [80], using a theoretical approach, determined that the positive effect of the IHX in the ejector configuration will depend on the isentropic efficiency level of the ejector. They confirmed that the use of the IHX is only applicable in the cases of lower ejector isentropic efficiencies or higher gas cooler exit/evaporator temperatures.

7.3.4 Experimental Analysis

As an example, this subsection is devoted to showing the experimental results obtained with a CO_2 transcritical refrigeration plant working with and without IHX. The experimental analysis is performed at the heat rejection temperatures of 30 and 35°C and at the evaporating temperatures of 0 and −10°C. Accordingly, the section is split into two subsections that cover a brief description of the experimental setup (Section 7.3.4.1), and a discussion about the experimental results focusing on the discharge temperature, the electrical power consumption, the cooling capacity and the COP (Section 7.3.4.2).

7.3.4.1 Refrigerant System

The experimental facility used to evaluate the performance of the IHX is presented in Figure 7.22 with the corresponding measurement elements. As can be seen, the experimental setup consists of a one-stage vapor compression system with an IHX installed at the exit of the gas-cooler/condenser and the evaporator. The facility has a double-stage expansion system, where the first stage is used to control the heat rejection pressure and the second is installed to maintain the useful superheating at the evaporator. More information about this setup can be found in Sánchez *et al.* [69].

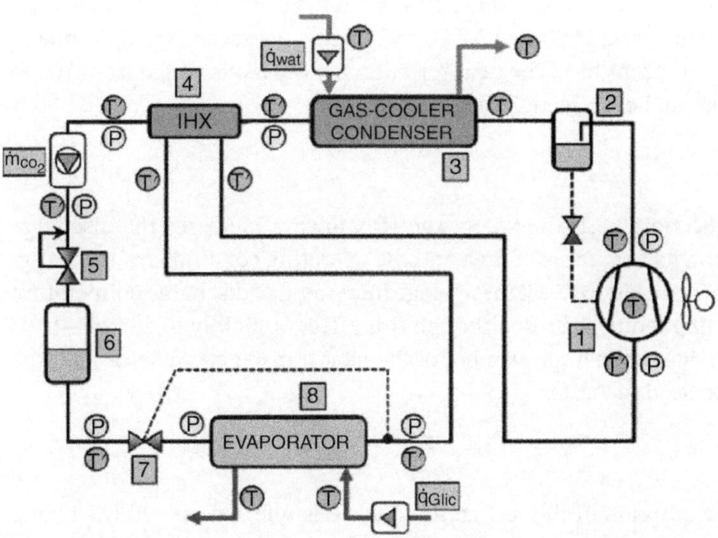

Figure 7.22 Schematic diagram of the experimental refrigeration plant.

According to Figure 7.22, the vapor from the evaporator (8) is compressed with a hermetic compressor (1) to a high-pressure level fixed by an electronic back-pressure valve (5). The compressed refrigerant passes through a coalescing oil separator (2) before entering the gas-cooler/condenser (3) which cools down/condenses the CO_2 depending on the operating conditions. At the exit of the gas-cooler/condenser, the refrigerant is subcooled by a suction-line to liquid-line heat exchanger (IHX) (4) before entering the back-pressure expansion valve (5). This valve expands the refrigerant to an intermediate accumulator tank (6) which feeds the entrance of the thermostatic expansion valve (7) installed at the inlet of the evaporator (8).

The IHX used in this analysis corresponds to a concentric-tube heat exchanger with an inner tube of 12/14 mm of diameter, and an external tube of 20/22 mm of diameter. The inner surface is corrugated, which means a total heat transfer area of 0.022 m^2.

The secondary fluids used in the facility are water for the heat rejection in the gas-cooler/condenser, and a mixture of water and propylene-glycol (70/30% by mass) for the evaporator. In both cases, an external unit is used to maintain the desired conditions of temperature and volumetric flow rate.

The refrigeration plant is fully instrumented with 15 transducers of temperature (T-type), eight pressure transducers, one power consumption and three flow rates for the secondary fluids and the refrigerant. The temperature sensors marked as (T) were placed over pipes and insulated from the environment with the same insulating foam mentioned above. The temperature sensors marked as (T') were installed inside the refrigeration facility with an immersion thermocouple to take more accurate measurements (especially under transcritical conditions).

7.3.4.2 Experimental Results and Discussion

To compare the effects of the IHX, a series of test were obtained maintaining the evaporating temperature level at 0 and −10°C, the useful superheating at the evaporator at 4 K, the inlet temperature of the secondary fluid at 30 and 35°C, and finally, the secondary fluid rate at the gas-cooler/condenser at 0.2 $m^3\ h^{-1}$. The heat rejection pressure was varied from 100 to 75 or 80 bar depending on the heat rejection temperature.

Discharge Temperature

As stated previously in Section 7.3.1, the use of the IHX always increases the discharge temperature, depending on its heat transfer area and the operating conditions of the refrigerating plant. In this case, from Figure 7.23, the average increment of discharge temperature is 9.5 K for both evaporating temperatures, although this effect is slightly higher at −10°C near the critical pressure due to the improvement of the heat transfer coefficients near the pseudocritical region (see Section 7.2.1).

Power Consumption

The electrical power consumption of the refrigeration cycle is referred only to the power consumption of the compressor. As shown in Figure 7.24, the electrical power consumption of the compressor is hardly affected by the use of the IHX, with a maximum deviation of

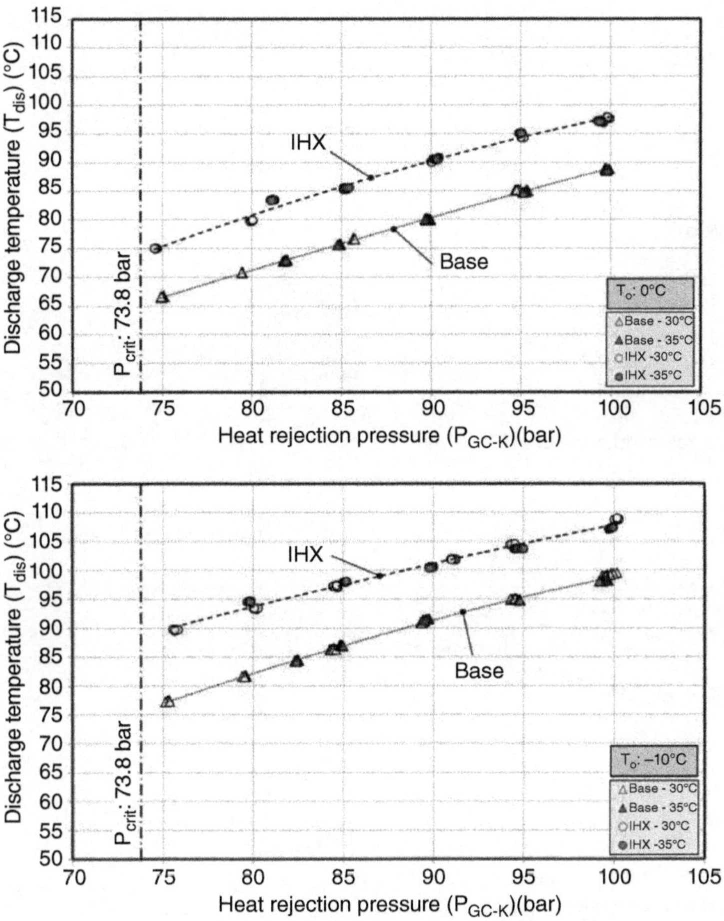

Figure 7.23 Discharge temperature with and without IHX at 0°C (left) and −10°C (right).

2.9%. This behavior means that the effect of the IHX over the mass flow rate offsets the compressor work increment (Eq. (7.22)).

Cooling Capacity

According to Eq. (7.23), the effect of the IHX affects simultaneously the evaporator capacity and the mass flow rate. As both effects are opposite, the cooling capacity can be affected positively or negatively depending on the operating conditions. From Figure 7.25, it is evident that the effect of the IHX at the evaporating temperatures of 0 and −10°C is always positive. Moreover, the enhancement degree depends on the heat rejection temperature and pressure, but it is on average 4%.

COP

The positive effect of the IHX over the cooling capacity combined with the minimal influence on the power consumption enhances the COP of the refrigerating cycle when the IHX

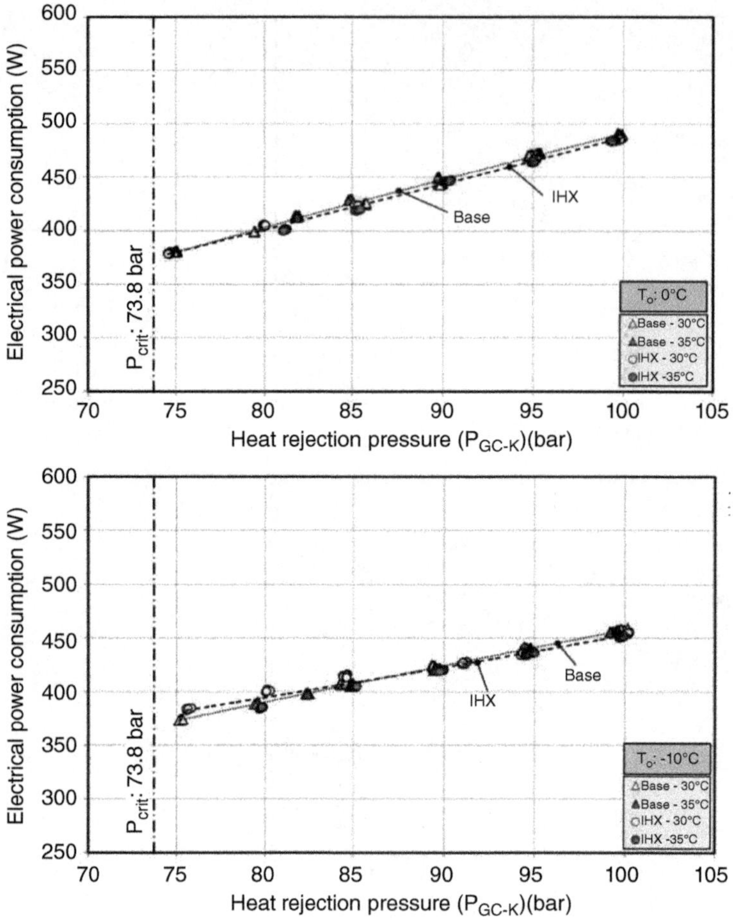

Figure 7.24 Electrical power consumption of the compressor with and without IHX at 0°C (left) and −10°C (right).

is used. This improvement depends on the operating conditions of the refrigerating plant as shown in Figure 7.26. Thus, for the evaporating temperature of −10°C the improvement reached by the IHX is higher than the temperature of 0°C. Moreover, the benefit of the IHX is always greater, the higher the heat rejection temperature is. This behavior matches the reports published in the open literature.

Regarding the optimum operating conditions, Figure 7.26 shows a reduction in the value of the heat rejection pressure that maximizes the COP when the IHX is installed. Table 7.2 presents the values of COP, and cooling capacity power consumption at the optimum operating conditions with and without IHX. As shown, the use of the IHX reduces the optimal heat rejection pressure up to 2 bar regarding the base cycle without IHX.

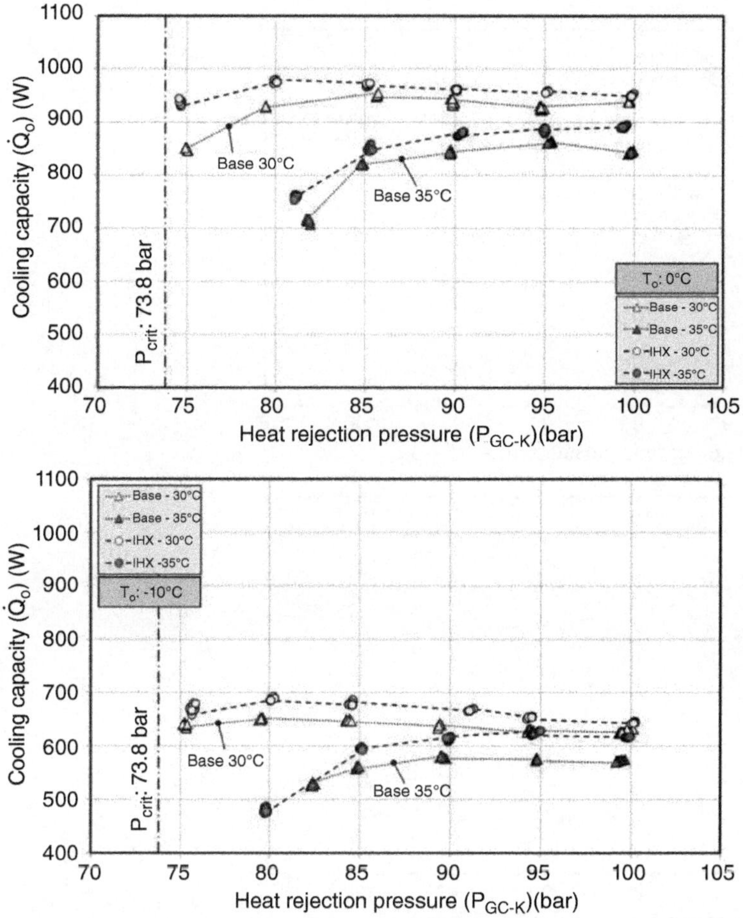

Figure 7.25 Cooling capacity with and without IHX at 0°C (left) and −10°C (right).

7.4 Dedicated Mechanical Subcooling

The DMS system is one of the most commonly used methods to subcool the refrigerant at the exit of the gas-cooler/condenser (GC/K). This auxiliary system is based on a simple vapor compression cycle which is coupled thermally to the CO_2 cycle through a heat exchanger, named subcooler or aftercooler [17]. The DMS, represented in Figure 7.27, performs CO_2 subcooling through evaporation of an auxiliary refrigerant in the subcooler. The refrigerant is compressed by an auxiliary compressor, performing heat rejection to the same hot sink as the CO_2 cycle. The evaporation level of the CO_2 evaporator depends on the cool production demand while the evaporation level in the subcooler is related to the heat transmission

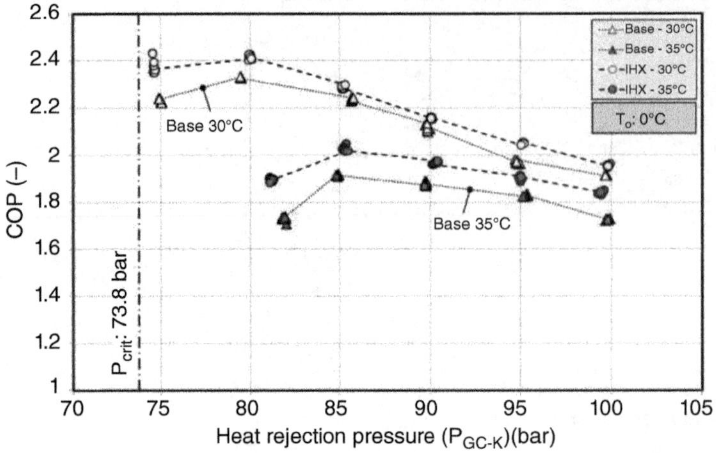

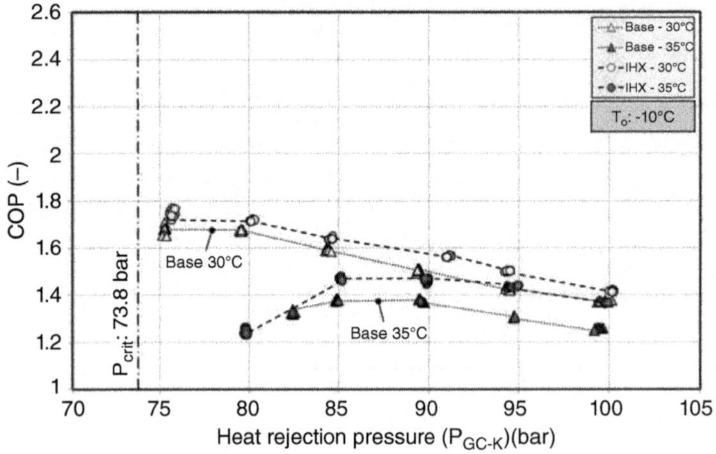

Figure 7.26 COP with and without IHX at 0°C (left) and −10°C (right).

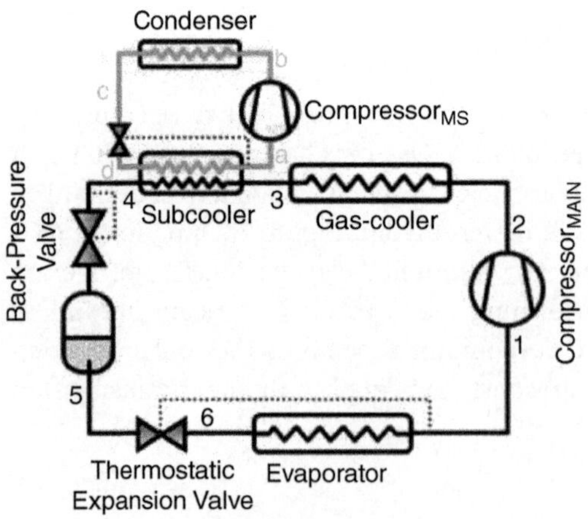

Figure 7.27 CO_2 refrigeration cycle with dedicated mechanical subcooling system.

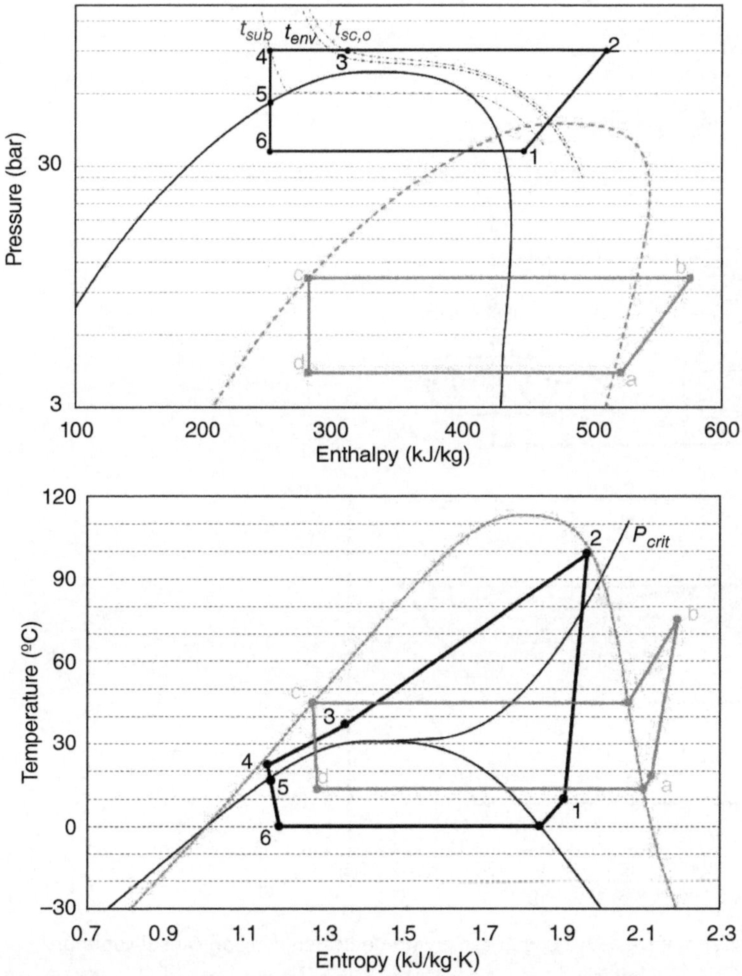

Figure 7.28 p-h and t-s diagram of CO_2 and R-152a DMS cycles.

between both refrigerants, thus it depends on the CO_2 heat rejection level and the degree of subcooling.

In the auxiliary cycle any refrigerant could be used, the effect being positive if the COP of the auxiliary cycle is higher than the COP of the CO_2 cycle [18, 20]. The DMS cycle has to provide the necessary cooling capacity to achieve the optimum subcooling degree, so it has to be designed in order to reach the maximum overall COP values. This maximum COP depends on the operating conditions: heat rejection temperature and evaporating level.

Figure 7.28 illustrates the p-h and t-s diagram for the transcritical single-stage CO_2 refrigeration cycle with DMS. In black, the main points of the CO_2 cycle are represented and in blue those of the DMS cycle, working with R152a.

The same configuration is used in CO_2 booster systems (Figure 7.29), where a two-stage refrigeration system is used to fulfill simultaneously the cooling demands at low temperature (LT) and medium temperature (MT). Generally, this system incorporates an

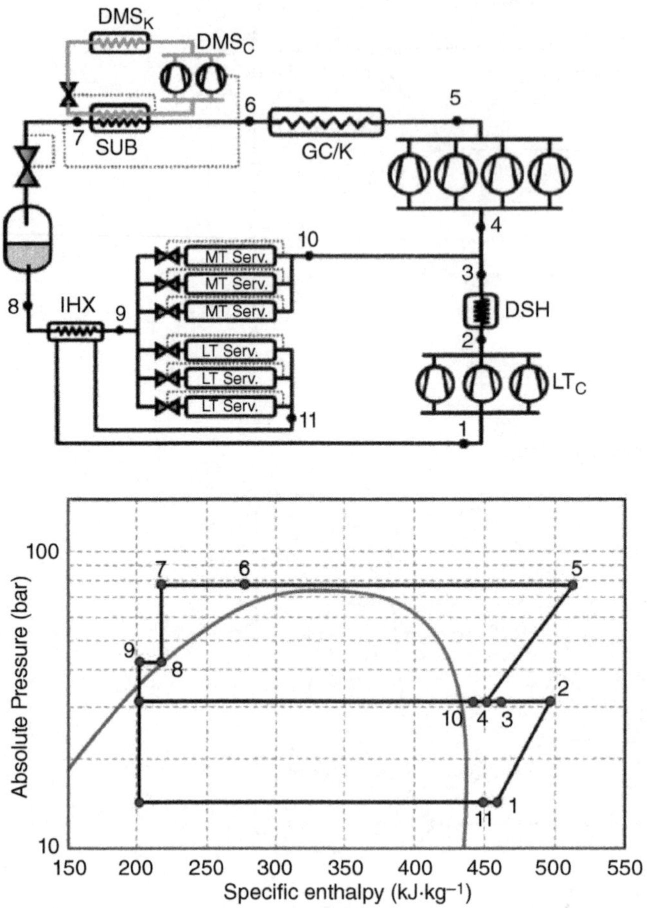

Figure 7.29 Scheme and p-h of the CO_2 booster cycle with dedicated mechanical subcooling.

additional IHX at the receiver exit. It ensures subcooled liquid at the inlet of the expansion valves (point 9, Figure 7.29), guaranteeing a proper operation of the expansion valves due to flash-gas absence, and higher specific cooling capacity in LT evaporators. However, the IHX introduces some superheat at LT compressor suction (point 1, Figure 7.29) [21], it being positive to increase lubricant temperature but unfavorable in relation to the power consumption of the LT compressors. The rise in the discharge temperature of the LT compressors caused by the IHX makes more favorable the use of a DSH between compression stages. The DSH, which is an air-cooled heat exchanger performing heat rejection to the same hot sink as the GC/K, reduces temperature at the inlet of the MT rack, its power consumption and its discharge temperature, thus improving the energy performance of the system [81].

The DMS auxiliary system is also used in heat pumps to enhance the performance of CO_2 cycles, its layout varying from that used for refrigeration purposes, as illustrated in Figure 7.30, which schematizes the cycle proposed by Song *et al.* [82–84] for water heating purposes. The combination is coupled thermally by the use of a water feed system.

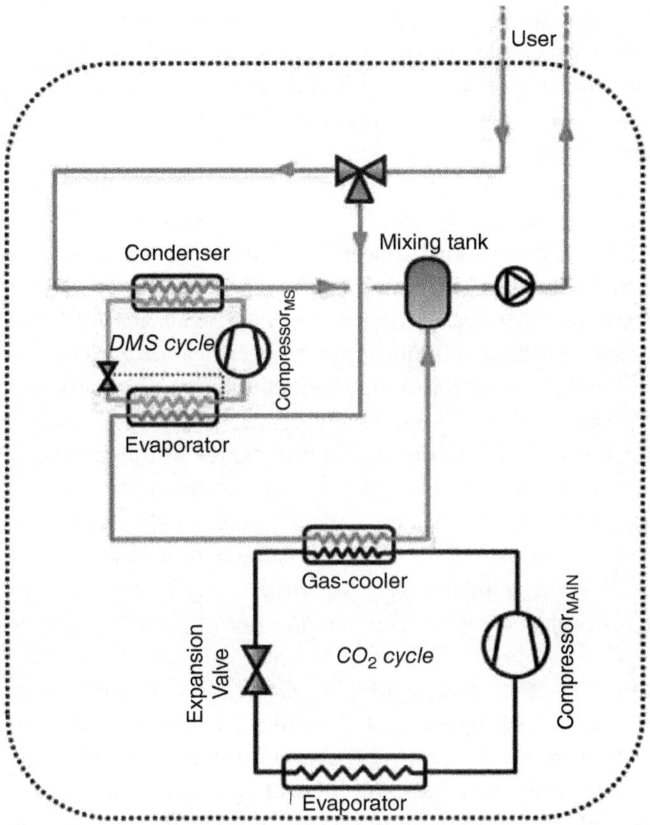

Figure 7.30 CO_2 heat pump with assisted dedicated mechanical subcooling cycle.

Feed water is divided into two currents, one passing through the DMS condenser increasing its temperature and the other going through the DMS evaporator, where it is cooled. This combination looks for two enhancement effects: first, it provides a high heat source level (high evaporating temperature) for the auxiliary system, allowing a higher heating capacity and discharge temperature, and second, the water temperature reduction provides subcooling in the CO_2 cycle, allowing an increase in the CO_2 COP and its specific heating capacity.

7.4.1 Optimum Parameters of the DMS Cycle

7.4.1.1 Subcooling Degree

The subcooling degree (SUB) is the temperature difference between CO_2 at the exit of the gas-cooler and at the exit of the subcooler, Eq. (7.25).

$$SUB = t_{gc,out} - t_{sub,out} \tag{7.25}$$

The SUB directly increments the specific cooling capacity of the main cycle but also influences the power consumption of the DMS auxiliary compressor (see Section 7.2.2.).

When the subcooling degree rises, the specific cooling capacity grows, but the power consumption of the auxiliary compressor rises too and the evaporating level of the DMS cycle becomes lower, thus reducing the individual COP_{DMS}. This combination is the reason for the existence of the optimum subcooling degree (in terms of overall COP), that provides an increment in capacity without penalizing the power consumption.

Figure 7.31 shows the behavior of COP in relation to the SUB for a CO_2 system with DMS at optimized heat rejection pressures. The overall COP rises with incremented SUB, where the increment of capacity of the main cycle is higher than the increment in energy consumption of the DMS cycle, then reaches a maximum, and finally decreases, where the increment in power consumption of the DMS cycle is higher than the rise in capacity. It has been proven that an optimum exists for both subcritical and transcritical operation, and that for refrigeration systems, the optimum SUB is higher when higher the hot sink and lower the evaporating level are [27].

For DMS CO_2 booster architectures, the optimum subcooling degree also depends on the heat rejection level, as analyzed by Catalán-Gil *et al.* [26], but the optimum subcooling degree is higher than in single-stage cycles, since the cycle simultaneously counteracts the low and medium temperature heat loads. Figure 7.32 reflects the optimum subcooling degrees for a wide range of environment temperatures. At temperatures below 8°C subcooling is negative, but as the heat rejection level rises the optimum subcooling degree grows.

In relation to CO_2 heat pumps with DMS, Song *et al.* [84] verified the existence of an optimal intermediate temperature instead of optimum subcooling. This optimal intermediate temperature at the entrance to the gas-cooler (Figure 7.30) is directly related to the subcooling degree, thus it can be derived that an optimum subcooling degree exists for this type of cycle. Also, it was verified that the optimum intermediate level is higher when the ambient temperature is higher.

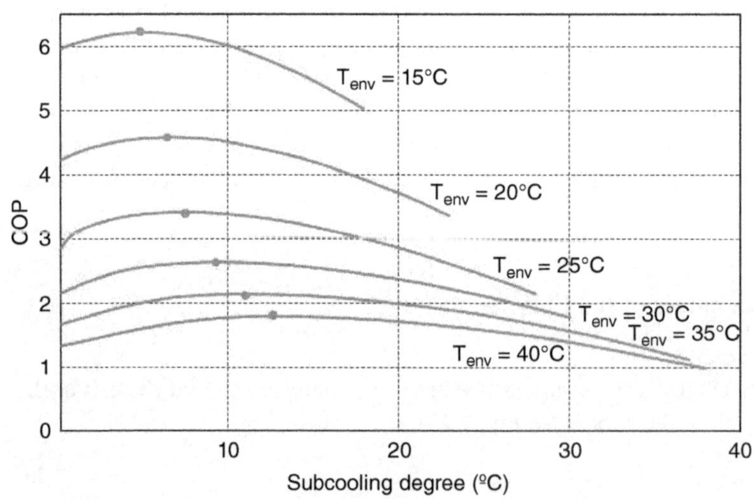

Figure 7.31 Evolution of COP vs subcooling degree for a single-stage CO_2 cycle with DMS at $t_0 = 0$°C and t_{env} with optimized heat rejection pressure [27].

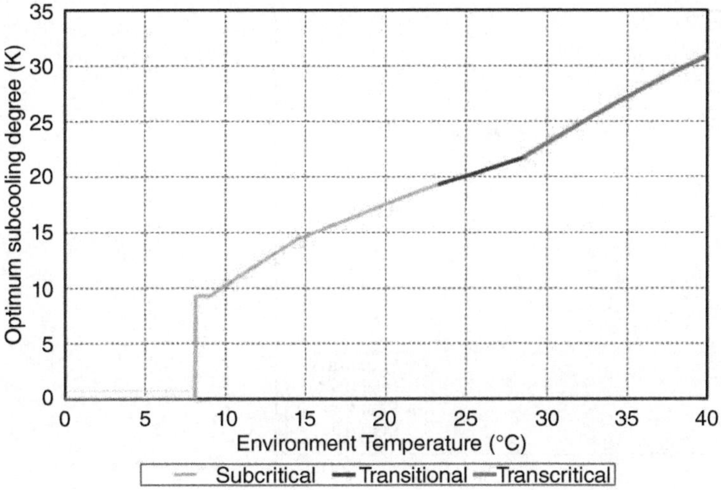

Figure 7.32 Optimum subcooling degree for a DMS subcooled booster with 140 and 41 kW at MT (−6°C) and LT (−32°C).

7.4.1.2 Heat Rejection Pressure

As analyzed in Section 7.2.2, the use of the DMS subcooling system in CO_2 transcritical cycles reduces the optimum high rejection pressure in relation to non-subcooled cycles, allowing a reduction in the effective compression ratio. When CO_2 cycles run in subcritical modes (t_k below t_c) the optimum heat rejection pressure corresponds to the minimum condensing level achievable; however, when the heat rejection level forces the system to operate in transcritical conditions, the high pressure must be optimized in terms of COP. The first correlation for the optimum heat rejection pressure was proposed by Kauf [5] for a single-stage cycle and it is only dependent on $t_{gc,\,o}$ (Eq. (7.26)). Later, Liao *et al.* [3] presented another relation for the optimal heat rejection pressure based on the $t_{gc,\,o}$, t_o and the performance of the compressor (Eq. (7.27)) and Sarkar [4] optimized a single-stage CO_2 transcritical cycle in terms of $t_{gc,\,o}$ and t_o (Eq. (7.28)).

$$p_{opt} = 2.6 \cdot t_{gc,o} - 7.54 \tag{7.26}$$

$$p_{opt} = (2.778 - 0.0157 \cdot t_o) \cdot t_{gc,o} + (0.381 \cdot t_o - 9.34) \tag{7.27}$$

$$p_{opt} = 4.9 + 2.256 \cdot t_{gc,o} - 0.17 \cdot t_o + 0.002 \cdot t_{gc,o}) \tag{7.28}$$

When the CO_2 at the exit of the gas-cooler is subcooled, the optimum pressure of the cycle varies, and for the DMS system, this pressure is lower than the pressure when working without subcooling, leading to a reduction in the compressor's work, helping to improve the behavior of the cycle [29]. Figure 7.33 represents the optimum discharge pressure using the expressions from Eqs. (7.26)–(7.28) and the optimum pressure for the single-stage cycle with DMS presented in Figure 7.27 for $t_o = 0$°C. As can be seen, use of the DMS allows a reduction of the optimum working pressure. The optimum high pressures of the single-stage cycle with R-152a-DMS of Figure 7.33, evaluated with volumetric and

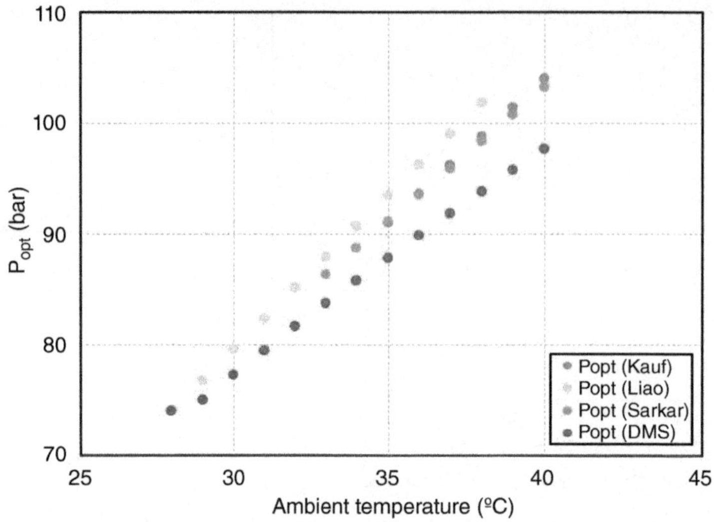

Figure 7.33 Classical relations vs CO_2 with DMS optimum heat rejection pressures.

Table 7.3 Coefficients for the compressor curves for the CO_2 system with DMS.

	CO_2 compressor				**R152a compressor**	
	Transcritical		**Subcritical**			
	η_v	η_g	η_v	η_g	η_v	η_g
a_0	1.0473	0.7634	1.0350	0.4868	0.9926	0.9692
a_1	0.0031	−0.0021	0.0019	−0.0086	−0.0993	−0.1178
a_2	−0.0030	0.0013	−0.0017	0.0115	0.0248	0.0263
a_3	0.0012	−0.0571	−0.0588	−0.2686	−0.0786	−0.0495
a_4	−11.1282	0.5425	−3.6174	20.8432	0.7683	−0.6042

$\eta_V = \eta_G = a_0 + a_1 \cdot P_{suc} + a_2 \cdot P_{dis} + a_3 \cdot \left(\frac{P_{dis}}{P_{suc}}\right) + a_4 \cdot v_{suc}.$

overall compressor's efficiencies (Table 7.3), are detailed in Eq. (7.29). Coefficients of Eq. (7.29) are detailed in Table 7.4.

$$\begin{aligned} p_{opt} &= p_k \quad 15 \le t_{env} < 24^\circ\text{C} \\ p_{opt} &= p_{crit} \quad 24 \le t_{env} < 29^\circ\text{C} \\ p_{opt} &= a \cdot t_{env} + b \quad 29 \le t_{env} \le 40^\circ\text{C} \end{aligned} \tag{7.29}$$

For heat pumps with DMS there is also an optimal pressure, since an increase in the discharge pressure raises the compressor work but also enhances the capacity of the gas-cooler. The existence of an optimal pressure in the subcooler-based system was verified experimentally [82]. In DMS-based refrigeration systems the optimum pressure is reduced when they

Table 7.4 Coefficients for the optimum pressure equation of the DMS.

	$t_o = 0°C$	$t_o = -10°C$
a	2.0523	1.9459
b	15.817	19.177

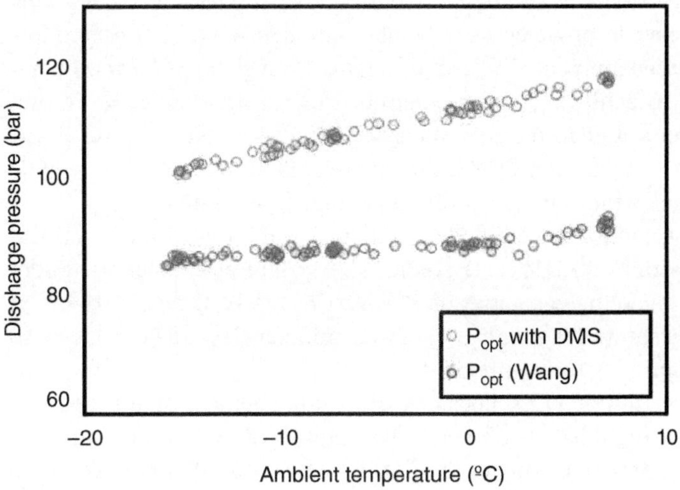

Figure 7.34 Optimum pressures with and without subcooling for CO_2 transcritical heat pumps [83].

are subcooled; however, for heat pump applications the optimum pressures are higher than cycles without subcooling. As an example, Figure 7.34 compares the optimum heat rejection pressures of the cycle described in Figure 7.30 [84] in relation to Wang's correlation [85] for a water heater system, where it can be observed that the use of the DMS system raises the optimum values.

7.4.2 Theoretical Studies

First theoretical studies of the DMS cycle were performed without knowing the existence of the previously mentioned optimum subcooling degrees. Hafner, A. and Hemmingsen, A. K. [28] considered the DMS as a support system with maximum capacity of 30% in relation to the main cycle, stating that the highest improvements were at high heat rejection temperatures. Later, a transcritical cycle was evaluated comparing its performance with and without DMS. Llopis *et al.* [20] simulated a single-stage and a double-stage CO_2 refrigeration system combined with a R290 DMS for three different evaporation conditions (−30, −5 and 5°C) and ambient temperatures from 20 to 35°C. The results showed that increasing the subcooling degree produces an increment in the overall COP but the optimum subcooling degree was not analyzed. The reached COP increments in reference to the base system without DMS were of 18.4% for $t_o = -30°C$, 17.9% for $-5°C$ and 12.3% for 5°C.

Also without optimizing the subcooling degree, Gullo *et al.* [86] presented an energy and environment performance comparison of different CO_2 booster solutions in relation to a cascade architecture for a typical European supermarket (97 kW/−10°C MT, 18 kW/−35°C LT) located in Valencia and Athens, concluding that a CO_2 booster with DMS using R-290 shows the best performance, with maximum COP increments of 32.7% at 30°C of ambient temperature. Later, optimizing pressure and subcooling conditions, Dai *et al.* [44] studied a R152a DMS single-stage system obtaining maximum COPs at the optimum working conditions and most significant improvements for higher ambient temperatures and low evaporation levels. They studied ambient temperatures going from 20 to 40°C for evaporation levels of −30, −5 and 5°C achieving an increment of 25.3% in COP for $t_o = 0°C$ and $t_{env} = 30°C$. After that, authors studied the advantages of using zeotropic mixtures in the DMS [87], concluding that the maximum COP is directly related to the temperature glide of the mixture due to the small heat transfer irreversibility that is generated.

Later, Gullo, P. [88] performed an advanced thermodynamic analysis of a transcritical CO_2 booster supermarket with R290 DMS. He studied the system at a subcooler outlet temperature set to 15°C and cooling capacities of 97 kW at $t_o = -10°C$ and to 18 kW at $t_o = -35°C$. The exergy study showed that only 59% of the inefficiencies can be reduced, so he suggested focusing on improving the components.

Another theoretical comparison of a CO_2 booster with parallel compression in relation to a CO_2 booster with DMS using R290 as refrigerant was presented by Purohit *et al.* [89]. In this case, the reduction in annual energy of the CO_2 booster with DMS relative to the system with parallel compression was from 6.4% to 8.9%.

Recently, Catalán-Gil *et al.* [26] presented an energy analysis of subcooling systems. This theoretical study presents an energy comparison of three advanced CO_2 booster architectures (booster with parallel compression, booster with DMS using R-290 and booster with IMS) in many locations in Europe and Asia, concluding that the DMS system is beneficial for CO_2 boosters for environment temperatures higher than 8.15°C with energy consumption reductions between 1.5% and 3.5% in Southern Europe and in the British Isles and up to 6% in many locations in India.

Referring to heat pump application, other studies have been carried out in the last few years. First, Song *et al.* [84] studied a transcritical CO_2 heat pump combined with a DMS working with R134a. The theoretical study allowed the authors to determine the optimum intermediate water temperature and the optimum discharge pressure that differs from the optimum pressure without subcooling [90]. These theoretical results have allowed the authors to test a prototype experimentally, as will be explained in the next section.

7.4.3 Experimental Studies

Experimental evaluation of the refrigeration system performed by Llopis *et al.* [3] shows the initial results of a R1234yf DMS applied to a CO_2 transcritical plant. This study does not evaluate the plant at subcooling optimum conditions but corroborates the existence of an

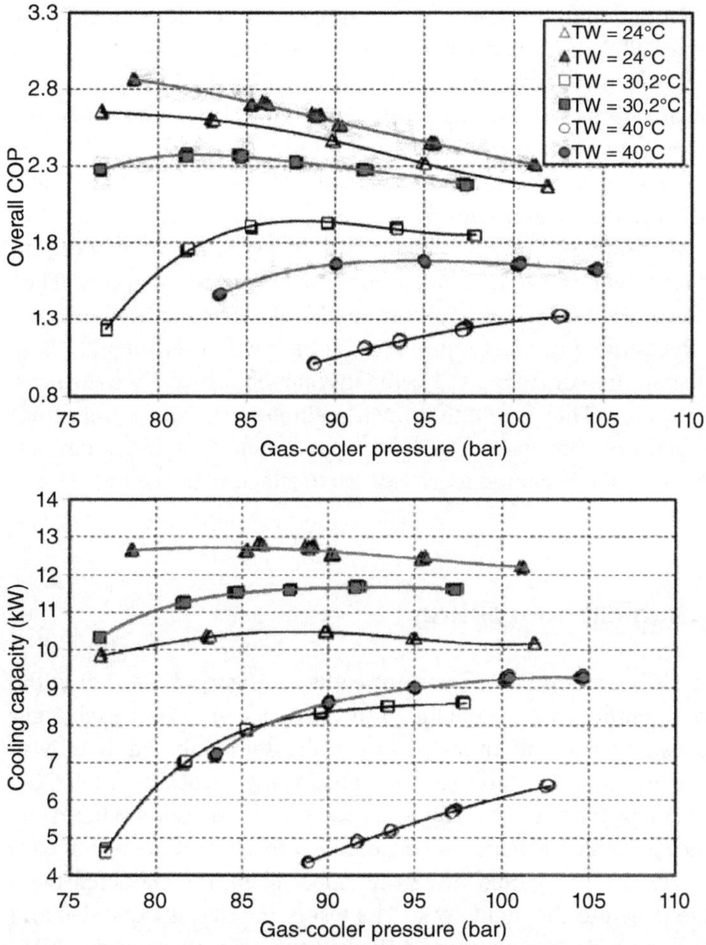

Figure 7.35 Experimental results of a CO_2 refrigeration plan with (red) and without (black) DMS for To = 0°C [47].

optimum discharge pressure. The experimental tests were carried out at two different evaporation levels (−10 and 0°C) and three different water temperatures at the entrance of the gas-cooler: 24.0, 30.2 and 40.0°C. The measured increments in COP at 0°C of evaporation level were 22.8% at 30.2°C water inlet and 17.3% at 40.0°C. In addition, the measured increments in capacity were of 34.9% at 30.2°C and 40.7% at 40.0°C. Authors also corroborated the reduction of the optimal working pressure, being it reduced up to 8 bar (Figure 7.35).

Sanchez *et al.* [69] evaluated a smaller DMS, working with R600a at $t_o = -10$°C at two different rejection levels: 30 and 35°C. They measured an increment of COP of 20.0% for $t_o = -10$°C and $t_{w,gc,in} = 35$°C in relation to the base cycle and of 9.5% with respect to the base cycle with IHX.

A prototype of a CO_2 booster architecture, with an indirect DMS with R134a as refrigerant for supermarket applications, was simulated and experimentally validated by Beshr *et al.* [45] and Bush *et al.* [91]. They evaluated the system at heat rejection levels of 29, 35 and 39°C. They observed a reduction in the optimum pressure of the gas cooler (heat rejection pressure) up to 1.9 bar at 29°C, an increment of cooling capacity up to 37.9% for heat rejection of 35°C and an improvement in the overall COP up to 36.7% at 35°C.

Experimentation with heat pumps was carried out by Song *et al.* [84], who tested the influence of the intermediate water temperature entering the gas-cooler and verified the initial theoretical results. The plant was tested for a feed water temperature of 50°C and a supply water temperature of 70°C varying the ambient temperature from −20 to −7°C. The influence of the feed water flow rate was also tested and the existence of an optimum intermediate temperature for which the COP is maximum has been demonstrated. They also concluded that the effects are different for different ambient air temperatures. Later, Song *et al.* [83] evaluated the optimal discharge pressure in the same experimental plant. Power consumption of both compressors increased as did discharge pressure, and the heating capacity increased much more. They found that the air temperature has an important effect on the performances of the system and the COP decreases when the ambient temperature declined. Further, the optimal discharge pressure also declined with the decrease of air temperature.

7.5 Integrated Mechanical Subcooling

Subcooling in refrigeration CO_2 systems can be performed by the IMS cycle avoiding the use of another auxiliary refrigerant. This cycle adds to the basic layout a heat exchanger called a subcooler, an expansion valve and an auxiliary compressor combined directly to the CO_2 main cycle. This means that the combined system has a single working fluid: CO_2. The structure of this system can be formed in two different ways: the first option is to extract the CO_2 from the exit of the gas-cooler, expand and evaporate it to subcool the rest of CO_2 going through the main cycle, as represented in Figure 7.36a. Then, the extracted mass flow is recompressed and returned to the main cycle. The rest of the CO_2 is expanded and arrives at the evaporator and the main compressor, while the back-pressure valve regulates the high rejection pressure. Before the gas-cooler entrance, both flows are re-mixed. The second option is schematized in Figure 7.36b: the CO_2 is extracted from the vessel, expanded and used to subcool the rest of the CO_2 through the subcooler. This architecture is easier to control, since it avoids the back-pressure to suffer the large variations of CO_2 properties

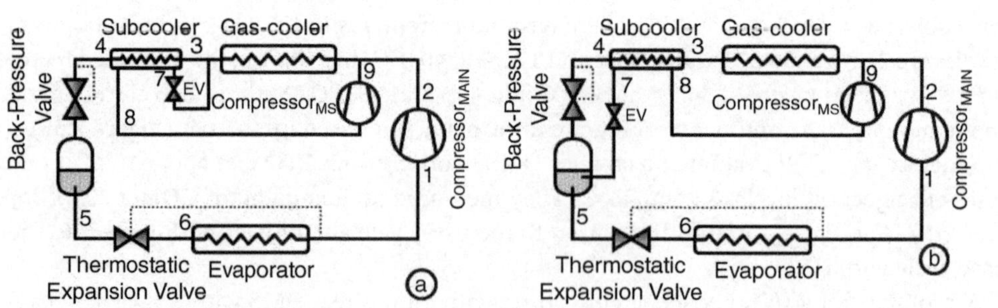

Figure 7.36 CO_2 refrigeration system with IMS: (a) extracting from gas-cooler exit and (b) extracting from receiver.

(Section 7.2.1) and the expansion valve feeding the subcooler will absorb liquid. Although both options have equivalent performances from a thermodynamic point of view for the same t_o in the subcooler, the auxiliary flow in the second option (Figure 7.36b) is larger than in the first (Figure 7.36a). The authors have found no additional research in relation to either scheme.

The benefits provided by this auxiliary system are the same as those of the DMS: an increase of the specific cooling capacity, a reduction of the optimum gas-cooler pressure (compression ratio decreases) and a decrease of the specific compression work, which increase the capacity and the COP of the overall system. Figure 7.37 shows the p-h and t-s

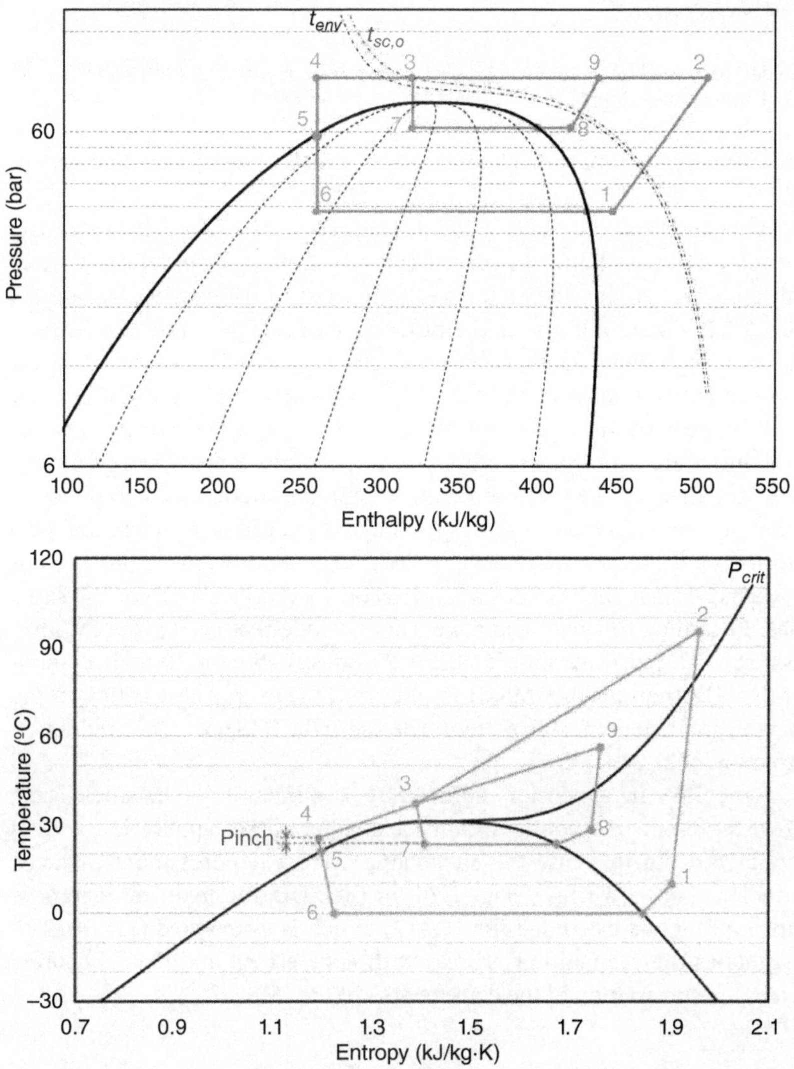

Figure 7.37 p-h and t-s diagram of integrated mechanical subcooling cycle extracting CO_2 from gas-cooler exit.

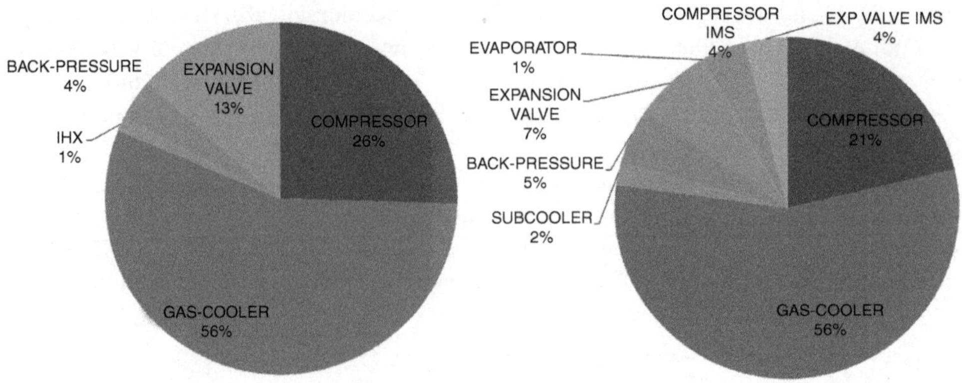

Figure 7.38 Exergy destruction of each component of the base system with IHX (left) and the integrated mechanical subcooling (right).

diagram of the transcritical CO_2 with IMS where the main effects of this subcooling system can be noticed.

In terms of exergy destruction, the main reduction caused by the subcooling is in the evaporator's expansion valve. Subcooling the gas and entering the expansion valve at a lower temperature avoids a big part of the irreversibilities that occur in the expansion stage, as discussed in Section 7.2.3. Figure 7.38 shows the proportion of exergy destruction in each of the components of the IMS (Figure 7.36a) and the single-stage with IHX (Figure 7.18). It can be observed how in the system with IHX the main components that contribute to the irreversibilities of the system are the gas-cooler, the compressor and the expansion valve, whereas for the IMS the compressor irreversibilities represent a smaller part, the expansion valve part is reduced and there are new irreversibilities coming from the new components.

In Figure 7.39, the ratio between the exergy destruction of each component and the cooling capacity of the system for a $t_o = 0°C$ and $t_{env} = 35°C$ can be observed. It allows comparison of both systems because the IMS has a larger capacity than the IHX for the same working conditions. Regarding this parameter, we can conclude that even if the IMS has more components that produce irreversibilities, the exergy destruction in the main components is larger for the IHX than for the IMS. It implies that the exergy destruction in the whole IHX system is bigger than in the IMS system. The use of the IMS produces a reduction in the exergy destruction of 21.34%.

The introduction the IMS in a booster architecture is similar than in single-stage systems. Figure 7.40 represents the configuration for a supermarket application and its pressure-enthalpy diagram. In this case the subcooling system requires more capacity, with a larger subcooler, than in a single-stage [26]. In the SUB, the main refrigerant is cooled (7–8, Figure 7.40) by the expanded flow (7–13) which is evaporated (13–14). The expansion valve regulates the evaporation process with a superheat and the subcooling degree is regulated by speed variation of the compressors of the IMS.

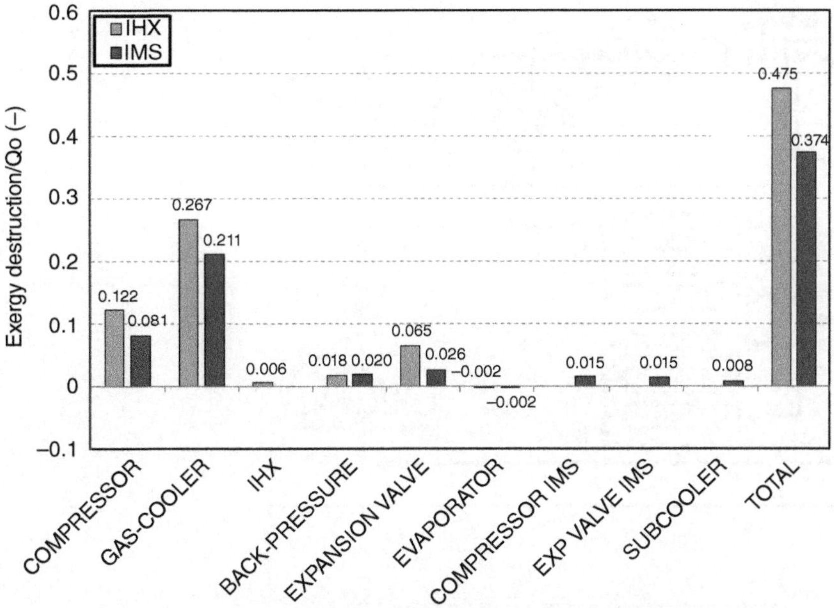

Figure 7.39 Exergy destruction per unit of capacity for the system with IHX and with IMS.

7.5.1 Optimum Parameters of the IMS Cycle

7.5.1.1 Subcooling Degree

As happens with the DMS, variation of the subcooling degree directly affects the specific cooling capacity and the specific compression work of the cycle. Therefore, there is also a subcooling degree for which the COP of the system is maximum. For a system like the one schematized in Figure 7.36a the optimum subcooling degree is defined by Eq. (7.30), which depends on the environment temperature and the evaporation level. Coefficients of Eq. (7.30) are detailed in Table 7.5.

$$SUB_{opt} = a_0 \cdot t_{env}^3 + a_1 \cdot t_{env}^2 + a_2 \cdot t_{env} + a_3 \tag{7.30}$$

The global trend of the optimum subcooling degree for the IMS is to grow when the environment temperature increases, as can be seen in Figure 7.41. The trend is the same for both evaporating conditions even though the values are higher for a lower evaporation temperature and there are three well differentiated zones. First, the subcooling degree rises slightly, but when the temperature at the exit of the gas-cooler is near the CO_2 critical point, the optimum subcooling degree increases dramatically for the decrease until 29–30°C of ambient temperature, where the subcooling degree increases again smoothly. The abrupt change of subcooling degree near the critical point is caused by variation of the approach temperature of gas-cooler/condenser exit with the environment temperature.

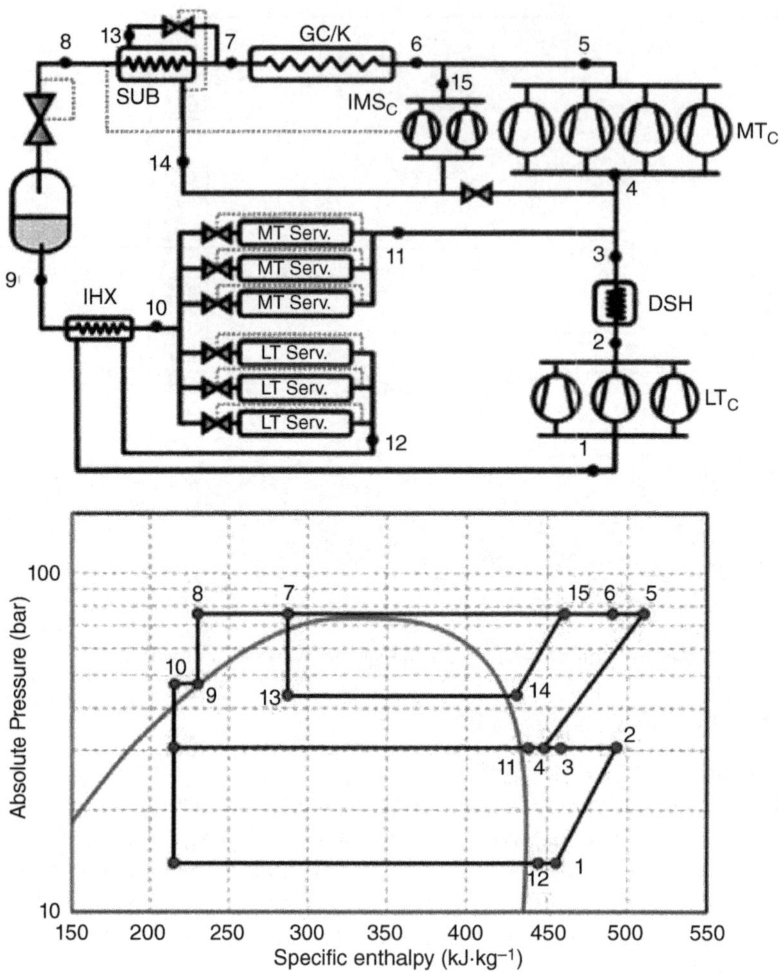

Figure 7.40 Schematic layout of the CO_2 booster system with IMS.

In a booster system with IMS (Figure 7.40), the optimum subcooling degree is lower than in the booster with DMS (Figure 7.29). Figure 7.42 shows the optimum subcooling for environment temperatures from 0 to 40°C with the operation modes depending on the environment temperature, as analyzed in [26]. In this case, subcooling can be applied from 0°C. Another benefit of this system in relation to the DMS is that for temperatures between 11 and 30°C, the subcooling degree is quite similar (around 16 K). Additionally, the use of IMS always improves the efficiency of the CO_2 booster.

7.5.1.2 Heat Rejection Pressure

Subcooling CO_2 at the exit of the gas-cooler modifies the working pressure too, being the optimum discharge pressure different from the optimum pressure of the refrigeration systems working without IMS. When working in subcritical conditions the optimum pressure is equal to the condensing pressure and in transcritical conditions the pressure

Table 7.5 Coefficients for the optimum subcooling degree equation of the single-stage cycle with IMS.

		$15\,°C < t_{env} < 24\,°C$	$24\,°C < t_{env} < 29\,°C$	$29\,°C < t_{env} < 40\,°C$
$t_o = -10\,°C$	a_0	—	−0.0058	—
	a_1	−0.028	0.3368	0.0086
	a_2	1.2729	−5.7531	−0.2026
	a_3	−1.8768	39.7280	12.5280
$t_o = 0\,°C$	a_0	—	−0.0217	—
	a_1	−0.0256	1.6136	0.0070
	a_2	1.1995	−39.916	−0.1182
	a_3	−5.7037	338.7600	7.3554

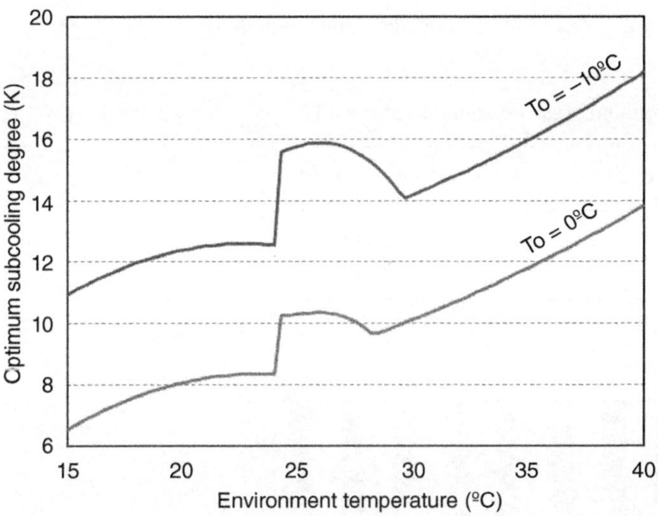

Figure 7.41 Optimum subcooling degree for a CO_2 single-stage with IMS.

goes up linearly as the environment temperature increases, following the line described by equation Eq. (7.29), which coefficients are detailed in Table 7.6 for two evaporating conditions. Between these two zones, when working near the critical point, the optimum pressure is equal to the critical pressure because it is preferable to force condensation with the back-pressure [18].

As has been stated, the optimum pressure of the IMS is different from the discharge pressure of the system working with no subcooling device. This optimal pressure is lower than that of the base system. Figure 7.43 represents the difference between the IMS optimum pressure and the base system with IHX. It can be seen that there is always a reduction in the discharge pressure and this reduction is more important for high ambient temperatures, achieving a reduction of 3.8 bar at 40°C.

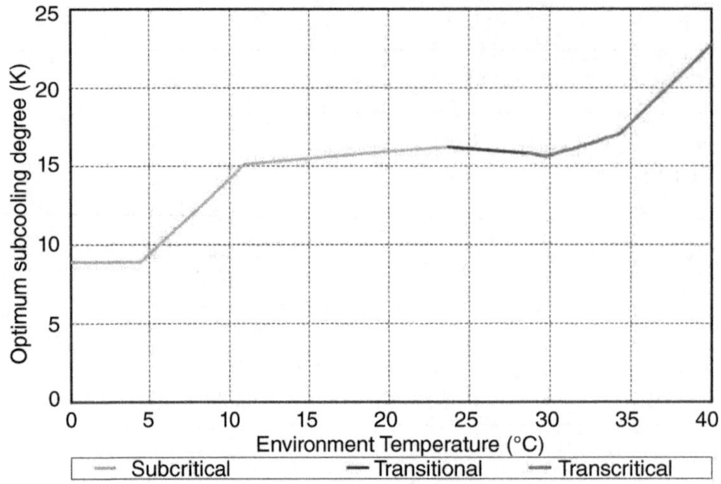

Figure 7.42 Optimum subcooling degree of the CO_2 booster system with IMS.

Table 7.6 Coefficients for the optimum pressure equation of the IMS.

	$t_o = 0°C$	$t_o = -10°C$
a	2.1044	2.1167
b	13.688	13.489

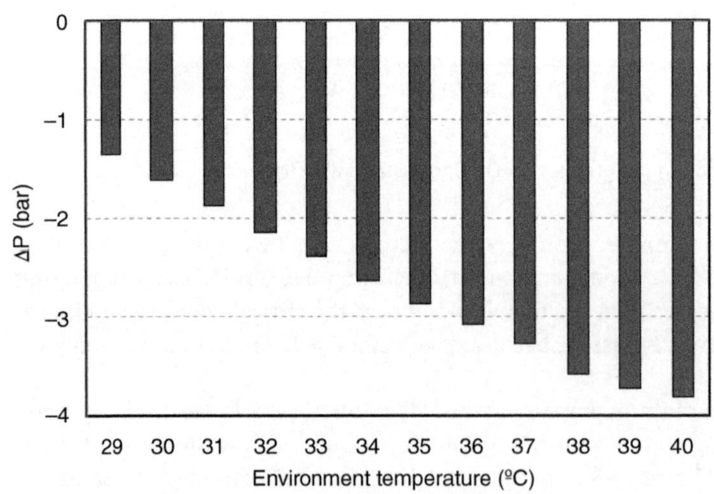

Figure 7.43 Optimum pressure reduction of a single-stage CO_2 cycle with IMS in relation to the base system with IHX.

7.5.2 Theoretical Studies

Although the IMS system appeared in 2007 for any kind of refrigerant [92], the first patent referred to CO_2 applications dates from 2013 [93], where it was considered as a way to reduce power consumption in the main compressors of the cycle and enhance COP and capacity. However, research focused on the application of the IMS in CO_2 cycles is scarce.

Cecchinato *et al.* [24] evaluated a 17.3% increase in energy efficiency in relation to a basic single-stage CO_2 cycle at $t_o = -10°C$ and $t_{gc,o} = 30°C$, concluding that this cycle overcomes the standard double compression cycle, reaching COP increments up to 12%. Later, an exergo-economic analysis of the IMS was made by Gullo and Cortella *et al.* [25]. They compared it with the parallel compression and a gas ejector system for medium temperature services, concluding that IMS achieves a COP improvement regarding parallel compression from 2.8% to 5.5%. However, no more studies have been found by authors in relation to its application in single-stage cycles.

In relation to its use in booster systems, the integration of IMS in a CO_2 booster for a typical European supermarket was analyzed by Catalán-Gil *et al.* [26]. They analyzed the systems with cooling capacities of 140 and 41 kW for medium and low evaporation services, obtaining COP benefits for outdoor temperatures higher than 3.75°C, being a positive solution for all locations in Europe and Asia analyzed with annual energy consumption reductions from 4% to 6% regarding CO_2 boosters with parallel compression, although the IMS system is more suitable to temperate climates.

7.6 Summary

Subcooled CO_2 cycles have been demonstrated to be higher in terms of energy efficiency than classical CO_2 cycles because they are able to reduce the irreversibilities during the expansion processes, which is the main drawback of transcritical cycles. However, since subcooling occurs in the proximities of the critical and pseudocritical regions, where the properties of CO_2 suffer large variations, the benefits of this method depend on the way the cycle is performing in relation to the hot sink source.

This chapter joins the most relevant theoretical and experimental research in relation to the three subcooling systems most used in CO_2 refrigeration and heat pump cycles: the IHX, the DMS system and the integrated subcooling configuration.

The IHX results are mandatory in refrigerating and heat pumping CO_2 cycles, with measured COP increments in relation to cycles without IHX up to 20%. However, the main drawback of the IHX is still the large increment produced in the compressor discharge temperature, which is very relevant for operation at low evaporating levels.

The DMS cycle, based on the use of an additional refrigeration cycle for subcooling purposes, eliminates the problem of increased discharge temperatures, with measured COP increments much larger than with the IHX. This method allows reducing the optimum heat rejection pressure in refrigeration applications, but increases its value for heat pump uses. Nonetheless, this system still relies on the use of another refrigerant.

Finally, the IMS cycle, based on the use of an internal mechanical subcooling device, is designed to only operate with CO_2 as the refrigerant. This configuration, which also solves the compressor discharge temperature problem, is also able to reduce the heat rejection pressure and bring about interesting energy improvements in relation to the IHX cycle.

Although this chapter covers the most relevant research on subcooled CO_2 cycles, it must be highlighted that there are still some important aspects which have not been researched yet, such as the optimum subcooling conditions and their corresponding heat rejection pressures, as well as, the optimum exergo- or thermo-economic layouts for those systems.

Nomenclature

COP	coefficient of performance
D_H	hydraulic diameter, m
DMS	dedicated mechanical subcooling
DSH	desuperheater
h	enthalpy, kJ kg^{-1}
IHX	internal heat exchanger
IMS	integrated mechanical subcooling
L	finite volume length, m
LT	low temperature services
MT	medium temperature services
$\dot{m}_r$	refrigerant mass flow rate, kg s^{-1}
Nu	Nusselt number ($Nu = \alpha \cdot D_H/k$) (–)
p	pressure, bar
Pc	electrical power consumption, kW
Pr	Prandtl number (–)
q	specific enthalpy difference, kJ kg^{-1}
$\dot{Q}$	heat transfer rate, kW
Re	Reynolds number ($Re = v \cdot \rho \cdot D_H/\mu$) (–)
s	entropy, kJ kg^{-1} K^{-1}
SH	degree of superheat, K
SUB	subcooling degree, K
t	temperature, °C
v	fluid velocity, m s^{-1}
$\dot{V}_G$	compressor displacement, m^3 kg^{-1}
w	specific compression work, kJ kg^{-1}
x_v	vapor title, (–)

Greek Symbols

Δ	increment
η	efficiency
ξ	friction factor. Calculated using the Konakov's correlation $\xi = [1.8 \cdot \log_{10}(Re) - 1.5]^{-2}$.
μ	dynamic viscosity, Pa s

κ	thermal conductivity, $W\,m^{-1}\,K^{-1}$
α	convective heat transfer coefficient, $W\,m^{-2}\,K^{-1}$
ρ	density, $kg\,m^{-3}$
ν	specific volume, $m^3\,kg^{-1}$

Subscripts

B	refers to bubble line
base	system considered as reference for comparison
C	cold source level
crit	critical point
env	environment
gc	gas-cooler
H	hot sink level
in	inlet
I	intermediate level
K	condensation
o, out	outlet
O	evaporation
opt	optimum conditions
pc	pseudocritical region
sub	refers to the subcooler
v	volumetric
ves	vessel of refrigeration system
w	refers to wall

References

1 Kim, M.H., Pettersen, J., and Bullard, C.W. (2004). Fundamental process and system design issues in CO_2 vapor compression systems. *Progress in Energy and Combustion Science* 30 (2): 119–174.

2 Groll, E.A. and Kim, J.-H. (2007). Review article: review of recent advances toward transcritical CO_2 cycle technology. *HVAC&R Research* 13 (3): 499–520.

3 Liao, S.M., Zhao, T.S., and Jakobsen, A. (2000). A correlation of optimal heat rejection pressures in transcritical carbon dioxide cycles. *Applied Thermal Engineering* 20 (9): 831–841.

4 Sarkar, J., Bhattacharyya, S., and Gopal, M.R. (2004). Optimization of a transcritical CO_2 heat pump cycle for simultaneous cooling and heating applications. *International Journal of Refrigeration* 27 (8): 830–838.

5 Kauf, F. (1999). Determination of the optimum high pressure for transcritical CO_2-refrigeration cycles. *International Journal of Thermal Sciences* 38 (4): 325–330.

6 Chen, Y. and Gu, J. (2005). The optimum high pressure for CO_2 transcritical refrigeration systems with internal heat exchangers. *International Journal of Refrigeration* 28 (8): 1238–1249.

7 Cabello, R., Sanchez, D., Llopis, R. et al. (2008). Experimental evaluation of the energy efficiency of a CO_2 refrigerating plant working in transcritical conditions. *Applied Thermal Engineering* 28 (13): 1596–1604.

8 Peñarrocha, I., Llopis, R., Tarrega, L. et al. (2014). A new approach to optimize the energy efficiency of CO_2 transcritical refrigeration plants. *Applied Thermal Engineering* 67 (1–2): 137–146.

9 Hu, J., Li, M., Zhao, L. et al. (2015). Improvement and experimental research of CO_2 two-rolling piston expander. *Energy* 93 (Part 2): 2199–2207.

10 Jia, X., Zhang, B., Pu, L. et al. (2011). Improved rotary vane expander for transcritical CO_2 cycle by introducing high-pressure gas into the vane slots. *International Journal of Refrigeration* 34 (3): 732–741.

11 Joneydi Shariatzadeh, O., Abolhassani, S., Rahmani, M. et al. (2016). Comparison of transcritical CO_2 refrigeration cycle with expander and throttling valve including/excluding internal heat exchanger: exergy and energy points of view. *Applied Thermal Engineering* 93: 779–787.

12 Jiang, Y., Ma, Y., Fu, L. et al. (2013). Some design features of CO_2 two-rolling piston expander. *Energy* 55: 916–924.

13 Elbel, S. (2011). Historical and present developments of ejector refrigeration systems with emphasis on transcritical carbon dioxide air-conditioning applications. *International Journal of Refrigeration* 34 (7): 1545–1561.

14 Elbel, S. and Lawrence, N. (2016). Review of recent developments in advanced ejector technology. *International Journal of Refrigeration* 62: 1–18.

15 Hafner, A., Försterling, S., and Banasiak, K. (2014). Multi-ejector concept for R-744 supermarket refrigeration. *International Journal of Refrigeration* 43: 1–13.

16 Pardiñas, Á.Á., Hafner, A., and Banasiak, K. (2018). Novel integrated CO_2 vapour compression racks for supermarkets. Thermodynamic analysis of possible system configurations and influence of operational conditions. *Applied Thermal Engineering* 131: 1008–1025.

17 Mohammadi, S.M.H. (2018). Theoretical investigation on performance improvement of a low-temperature transcritical carbon dioxide compression refrigeration system by means of an absorption chiller after-cooler. *Applied Thermal Engineering* 138: 264–279.

18 Llopis, R., Nebot-Andres, L., Sánchez, D. et al. (2018). Subcooling methods for CO_2 refrigeration cycles: a review. *International Journal of Refrigeration* 93: 85–107.

19 Pottker, G. and Hrnjak, P. (2015). Effect of the condenser subcooling on the performance of vapor compression systems. *International Journal of Refrigeration* 50: 156–164.

20 Llopis, R., Cabello, R., Sánchez, D. et al. (2015). Energy improvements of CO_2 transcritical refrigeration cycles using dedicated mechanical subcooling. *International Journal of Refrigeration* 55: 129–141.

21 Torrella, E., Sánchez, D., Llopis, R. et al. (2011). Energetic evaluation of an internal heat exchanger in a CO_2 transcritical refrigeration plant using experimental data. *International Journal of Refrigeration* 34 (1): 40–49.

22 Aprea, C. and Maiorino, A. (2008). An experimental evaluation of the transcritical CO_2 refrigerator performances using an internal heat exchanger. *International Journal of Refrigeration* 31 (6): 1006–1011.

23 Cavallini, A. et al. (2005). Two-stage transcritical carbon dioxide cycle optimisation: a theoretical and experimental analysis. *International Journal of Refrigeration* 28 (8): 1274–1283.

24 Cecchinato, L., Cecchinato, L., Corradi, M. et al. (2009). Thermodynamic analysis of different two-stage transcritical carbon dioxide cycles. *International Journal of Refrigeration* 32 (5): 1058–1067.

25 Gullo, P. and G. Cortella (2016). Comparative Exergoeconomic Analysis of Various Transcritical R744 Commercial Refrigeration Systems. *ECOS 2016 – The 29th International Conference on Efficiency, Cost, Optimization, Simulation and Environmental Impact of Energy Systems.*, U.O. Ljubljana (Ed.). Portoroz, Slovenia.

26 Catalán-Gil, J., Llopis, R., Sánchez, D. et al. (2019). Energy analysis of dedicated and integrated mechanical subcooled CO_2 boosters for supermarket applications. *International Journal of Refrigeration* 101: 11–23.

27 Nebot-Andrés, L., Llopis, R., Sánchez, D. et al. (2017). CO_2 with mechanical subcooling vs CO_2 cascade cycles for medium temperature commercial refrigeration applications thermodynamic analysis. *Applied Sciences* 7 (9): 955.

28 Hafner A. and Hemmingsen A. K. (2015). R744 refrigeration technologies for supermarkets in warm climates. *24th International Congress of Refrigeration*, IIR (Ed.) Yokohama, Japan.

29 Llopis, R., Nebot-Andrés, L., Cabello, R. et al. (2016). Experimental evaluation of a CO_2 transcritical refrigeration plant with dedicated mechanical subcooling. *International Journal of Refrigeration* 69: 361–368.

30 Schoenfield, J., Muehlbauer, J., Hwang, Y., et al. (2008). Integration of a thermoelectric subcooler into a carbon dioxide transcritical vapor compression cycle refrigeration system. International Refrigeration and Air Conditioning Conference. Paper 903. http://docs.lib.purdue.edu/iracc/903 (accessed 5 Spetember 2020).

31 Schoenfield, J., Hwang, Y., and Radermacher, R. (2012). CO_2 transcritical vapor compression cycle with thermoelectric subcooler. *HVAC&R Research* 18 (3): 297–311.

32 Sarkar, J. (2013). Performance optimization of transcritical CO_2 refrigeration cycle with thermoelectric subcooler. *International Journal of Energy Research* 37 (2): 121–128.

33 Arora, A., Singh, N.K., Monga, S. et al. (2011). Energy and exergy analysis of a combined transcritical CO_2 compression refrigeration and single effect H2O-LiBr vapour absorption system. *International Journal of Exergy* 9 (4): 453–471.

34 Salajeghe, M. and Ameri, M. (2016). Effects of further cooling the gas cooler outlet refrigerant by an absorption chiller, on a transcritical CO_2-compression refrigeration system. *International Journal of Exergy* 21 (1): 110–125.

35 Liao, S.M. and Zhao, T.S. (2002). Measurements of heat transfer coefficients from supercritical carbon dioxide flowing in horizontal mini/micro channels. *Journal of Heat Transfer* 124 (3): 413–420.

36 Lemmon E. W., Bell, I.H., Huber M. L., et al. (2018). NIST standard reference database 23: reference fluid thermodynamic and transport properties-REFPROP, Version 10.0. National Institute of Standards and Technology.

37 Hendricks, R.C., Simoneau, R.J., and Smith, R.V. (1995). *Survey of Heat Transfer to Near-Critical Fluids*. Boston, MA: Springer US.

38 Gnielinski, V. (1976). New equations for heat and mass transfer in turbulent pipe and channel flow. *International Chemical Engineering* 16: 359–368.

39 Sánchez, D., Patiño, J., Sanz-Kock, C. et al. (2014). Energetic evaluation of a CO_2 refrigeration plant working in supercritical and subcritical conditions. *Applied Thermal Engineering* 66 (1–2): 227–238.

40 Pisano, A., Martínez-Ballester, S., Corberán, J.M. et al. (2015). Optimal design of a light commercial freezer through the analysis of the combined effects of capillary tube diameter and refrigerant charge on the performance. *International Journal of Refrigeration* 52: 1–10.

41 Qureshi, B.A., Inam, M., Antar, M.A. et al. (2013). Experimental energetic analysis of a vapor compression refrigeration system with dedicated mechanical sub-cooling. *Applied Energy* 102: 1035–1041.

42 Li, Z., Chen, E., Jing, Y. et al. (2017). Thermodynamic relationship of subcooling power and increase of cooling output in vapour compression chiller. *Energy Conversion and Management* 149: 254–262.

43 Chen, E., Li, Z., Yu, J. et al. (2019). Experimental research of increased cooling output by dedicated subcooling. *Applied Thermal Engineering*. 154: 9–17.

44 Dai, B., Liu, S., Sun, Z., et al. (2017). Thermodynamic performance analysis of CO_2 transcritical refrigeration cycle assisted with mechanical subcooling. Energy Procedia.

45 Beshr, M., Bush, J., Aute, V., et al. (2016). Steady state testing and modeling of a CO_2 two-stage refrigeration system with mechanical subcooler. Refrigeration Science and Technology.

46 Zubair, S.M. (1994). Thermodynamics of a vapor-compression refrigeration cycle with mechanical subcooling. *Energy* 19 (6): 707–715.

47 Nebot-Andrés, L., Llopis, R., Sánchez, D., et al. (2016). Experimental evaluation of a dedicated mechanical subcooling system in a CO_2 transcritical refrigeration cycle. Refrigeration Science and Technology.

48 Bush, J., Beshr, M., Aute, V. et al. (2017). Experimental evaluation of transcritical CO_2 refrigeration with mechanical subcooling. *Science and Technology for the Built Environment*: 23 (6): 1013–1025.

49 Bitzer GmbH (2017). Bitzer operating instructions. (https://www.bitzer.de/shared_media/documentation/kb-104-7.pdf) (accesed March 2019).

50 Domanski, P.A., Didion, D.A., and Doyle, J.P. (1994). Evaluation of suction-line/liquid-line heat exchange in the refrigeration cycle. *International Journal of Refrigeration* 17 (7): 487–493.

51 Domanski, P.A. (1995). Theoretical evaluation of the vapor compression cycle with a liquid-line/suction-line heat exchanger, economizer, and ejector. National Institute of Science and Technology, Interagency Report 5606. Gaithersburg, MD: National Institute of Standards and Technology.

52 Aprea, C., Ascani, M., and De Rossi, F. (1999). A criterion for predicting the possible advantage of adopting a suction/liquid heat exchanger in refrigerating system. *Applied Thermal Engineering* 19 (4): 329–336.

53 Klein, S.A., Reindl, D.T., and Brownell, K. (2000). Refrigeration system performance using liquid-suction heat exchangers. *International Journal of Refrigeration* 23 (8): 588–596.

54 Mastrullo, R., Mauro, A.W., Tino, S. et al. (2007). A chart for predicting the possible advantage of adopting a suction/liquid heat exchanger in refrigerating system. *Applied Thermal Engineering* 27 (14–15): 2443–2448.

55 Hermes, C.J.L. (2013). Alternative evaluation of liquid-to-suction heat exchange in the refrigeration cycle. *International Journal of Refrigeration* 36 (8): 2119–2127.

56 Hermes, C.J.L. (2015). Refrigerant charge reduction in vapor compression refrigeration cycles via liquid-to-suction heat exchange. *International Journal of Refrigeration* 52: 93–99.

57 Lorentzen, G. and Pettersen, J. (1993). A new, efficient and environmentally benign system for car air-conditioning. *International Journal of Refrigeration* 16 (1): 4–12.

58 Lorentzen, G. (1994). Revival of carbon dioxide as a refrigerant. *International Journal of Refrigeration* 17 (5): 292–301.

59 Robinson, D.M. and Groll, E.A. (1998). Efficiencies of transcritical CO_2 cycles with and without an expansion turbine: Rendement de cycles transcritiques au CO_2 avec et sans turbine d'expansion. *International Journal of Refrigeration* 21 (7): 577–589.

60 Rozhentsev, A. and Wang, C.C. (2001). Some design features of a CO_2 air conditioner. *Applied Thermal Engineering* 21 (8): 871–880.

61 Kim, S.G., Kim, Y.J., Lee, G. et al. (2005). The performance of a transcritical CO_2 cycle with an internal heat exchanger for hot water heating. *International Journal of Refrigeration* 28 (7): 1064–1072.

62 Zhang, F.Z., Jiang, P.X., Lin, Y.S. et al. (2011). Efficiencies of subcritical and transcritical CO_2 inverse cycles with and without an internal heat exchanger. *Applied Thermal Engineering* 31 (4): 432–438.

63 Ituna-Yudonago, J.F., Belman-Flores, J.M., Elizalde-Blancas, F. et al. (2017). Numerical investigation of CO_2 behavior in the internal heat exchanger under variable boundary conditions of the transcritical refrigeration system. *Applied Thermal Engineering* 115: 1063–1078.

64 Boewe, D., Yin, J., Park, Y. C., et al. (1999). The Role of Suction Line Heat Exchanger in Transcritical R744 Mobile A/C Systems. SAE International.

65 Cavallini, A., Cecchinato, L., and Corradi, M., et al. (2006). Experimental investigation on the effects of internal heat transfer in a two-stage transcritical carbon dioxide cycle. *7th IIR-Gustav Lorentzen Conference on Natural Working Fluids (GL2006).* I.-. IIF (Ed.). Trondheim, Norway.

66 Cavallini, A., M. Corradi, and E. Fornasieri, (2007). Experimental investigation on the effect of the Internal Heat Exchanger and Intercooler effectiveness of the energy performance of a two-stage transcritical carbondioxde cycle. *22nd International Congress of Refrigeration*, IIR (Ed.) Beijin, China.

67 Cho, H., Ryu, C., and Kim, Y. (2007). Cooling performance of a variable speed CO_2 cycle with an electronic expansion valve and internal heat exchanger. *International Journal of Refrigeration* 30 (4): 664–671.

68 Rigola, J., Ablanque, N., Pérez-Segarra, C.D. et al. (2010). Numerical simulation and experimental validation of internal heat exchanger influence on CO_2 transcritical cycle performance. *International Journal of Refrigeration* 33 (4): 664–674.

69 Sánchez, D., Catalán-Gil, J., Llopis, R., et al. (2016). Improvements in a CO_2 transcritical plant working with two different subcooling systems. Refrigeration Science and Technology.

70 Cabello, R., Sánchez, D., Patiño, J. et al. (2012). Experimental analysis of energy performance of modified single-stage CO_2 transcritical vapour compression cycles based on vapour injection in the suction line. *Applied Thermal Engineering* 47: 86–94.

71 Karampour, M. and S. Sawalha (2014). Investigation of using Internal Heat Exchangers in CO_2 Trans-critical Booster System. *11th IIR Gustav Lorentzen Conference on Natural Refrigerants*, I.-. IIF (Ed.). Hanzhou, China.

72 Llopis, R., Sanz-Kock, C., Cabello, R. et al. (2015). Experimental evaluation of an internal heat exchanger in a CO_2 subcritical refrigeration cycle with gas-cooler. *Applied Thermal Engineering* 80: 31–41.

73 Llopis, R., Sanz-Kock, C., Cabello, R. et al. (2016). Effects caused by the internal heat exchanger at the low temperature cycle in a cascade refrigeration plant. *Applied Thermal Engineering* 103: 1077–1086.

74 Purohit, N., Gupta, D.K., and Dasgupta, M.S. (2018). Experimental investigation of a CO_2 trans-critical cycle with IHX for chiller application and its energetic and exergetic evaluation in warm climate. *Applied Thermal Engineering* 136: 617–632.

75 Zhang, Z., Tian, L., Chen, Y. et al. (2014). Effect of an internal heat exchanger on performance of the transcritical carbon dioxide refrigeration cycle with an expander. *Entropy* 16 (11): 5919.

76 Elbel, S. and P. Hrnjak (2004). Effect of internal heat exchanger on performance of transcritical CO_2 Systems with Ejector. *International Refrigeration and Air Conditioning Conference*. Paper 708. Purdue University, EEUU.

77 Elbel, S. and Hrnjak, P. (2008). Experimental validation of a prototype ejector designed to reduce throttling losses encountered in transcritical R744 system operation. *International Journal of Refrigeration* 31 (3): 411–422.

78 Xu, X.-x., Chen, G.-m., Tang, L.-m. et al. (2011). Experimental evaluation of the effect of an internal heat exchanger on a transcritical CO_2 ejector system. *Journal of Zhejiang University Science A* 12 (2): 146–153.

79 Nakagawa, M., Marasigan, A.R., and Matsukawa, T. (2011). Experimental analysis on the effect of internal heat exchanger in transcritical CO_2 refrigeration cycle with two-phase ejector. *International Journal of Refrigeration* 34 (7): 1577–1586.

80 Zhang, Z.-y., Ma, Y.-t., Wang, H.-l. et al. (2013). Theoretical evaluation on effect of internal heat exchanger in ejector expansion transcritical CO_2 refrigeration cycle. *Applied Thermal Engineering* 50 (1): 932–938.

81 Karampour, M. and Sawalha, S. (2018). State-of-the-art integrated CO_2 refrigeration system for supermarkets: a comparative analysis. *International Journal of Refrigeration* 86: 239–257.

82 Song, Y. and Cao, F. (2018). The evaluation of the optimal medium temperature in a space heating used transcritical air-source CO_2 heat pump with an R134a subcooling device. *Energy Conversion and Management* 166: 409–423.

83 Song, Y., Ye, Z., Wang, Y. et al. (2018). The experimental verification on the optimal discharge pressure in a subcooler-based transcritical CO_2 system for space heating. *Energy and Buildings* 158: 1442–1449.

84 Song, Y., Li, D., Cao, F. et al. (2017). Investigation of the optimal intermediate water temperature in a combined r134a and transcritical CO_2 heat pump for space heating. *International Journal of Refrigeration* 79: 10–24.

85 Wang, S., Tuo, H., Cao, F. et al. (2013). Experimental investigation on air-source transcritical CO_2 heat pump water heater system at a fixed water inlet temperature. *International Journal of Refrigeration* 36 (3): 701–716.

86 Gullo, P., Elmegaard, B., and Cortella, G. (2016). Energy and environmental performance assessment of R744 booster supermarket refrigeration systems operating in warm climates. *International Journal of Refrigeration* 64: 61–79.

87 Dai, B., Liu, S., Li, H. et al. (2018). Energetic performance of transcritical CO_2 refrigeration cycles with mechanical subcooling using zeotropic mixture as refrigerant. *Energy* 150: 205–221.

88 Gullo, P. (2018). Advanced thermodynamic analysis of a transcritical R744 booster refrigerating unit with dedicated mechanical subcooling. *Energies* 11 (11):.

89 Purohit, N., Gullo, P., and Dasgupta, M.S. (2017). Comparative assessment of low-GWP based refrigerating plants operating in hot climates. *Energy Procedia* 109: 138–145.

90 Yang, D., Song, Y., Cao, F. et al. (2016). Theoretical and experimental investigation of a combined R134a and transcritical CO_2 heat pump for space heating. *International Journal of Refrigeration* 72: 156–170.

91 Bush, J., Beshr, M., Aute, V. et al. (2017). Experimental evaluation of transcritical CO_2 refrigeration with mechanical subcooling. *Science and Technology for the Built Environment* 23 (6): 1013–1025.

92 Shapiro, D. (2007). Refrigeration system with mechanical subcooling. Google Patents. US7628027B2

93 Lesage, G. and J. Kantchev (2013). Mechanical subcooling of transcritical R-744 refrigeration systems with heat pump heat reclaim and floating head pressure. Google Patents. CA2807643C

8

High Temperature CO_2 Heat Pump System and Optimization

Lin Chen[1,2] *and Dipankar N. Basu*[3]

[1] *Institute of Engineering Thermophysics, Chinese Academy of Sciences, Beijing 100190, China*
[2] *School of Aeronautics and Astronautics, University of Chinese Academy of Sciences, Beijing 100049, China*
[3] *Department of Mechanical Engineering, Indian Institute of Technology Guwahati, Guwahati, Assam 781039, India*

8.1 Background

Heat pump is an efficient energy saving device due to the fact that heating energy capacity can be several times larger than that which is consumed. Heat pump could be a very effective instrument to make use of waste heat and low-grade energy and upgrade it into higher temperature heat [1–5]. A high temperature heat pump applies the same technology in a relatively higher temperature region. The common sources of such heat pump systems include ambient air, water, geothermal, and also industrial waste heat which is abundant in many industrial processes [6–9].

According to the operating principle, heat pump can be classified into vapor compression heat pump, absorption heat pump, chemical heat pump and steam jet heat pump. Vapor compression heat pump, which is also called mechanical heat pump, is the most mature and widely implemented device nowadays. Due to the harmful effect on ozone layer depletion and global warming, conventional refrigerants such as chlorofluorocarbons (CFCs), hydrochlorofluorocarbons (HCFCs) and hydrofluorocarbons (HFCs) are due to be phased out [10, 11]. This situation has led to the renaissance of natural refrigerants like ammonia, water, and carbon dioxide. Among these refrigerants, ammonia is toxic and flammable [12], and water cannot be used in vapor compression refrigeration cycles because of low density and low working pressure [13]. Besides, water has low COP and is not cost-effective [14]. On the other hand, CO_2 has several advantages over other refrigerants, such as zero ozone depletion potential (ODP) and zero effective GWP, compatibility with normal lubricants and common machine construction materials, non-flammability, non-toxicity, easy availability and very low cost [12, 15–20].

CO_2 has a relatively low critical temperature of 31.1°C, so it is very easy for a system to be operated in the transcritical and/or supercritical region. This characteristic point leads to several unique characteristics for CO_2 heat pumps [21]: (i) in many cases the system is operated under transcritical conditions; high-side pressure is determined by refrigerant charge and not by saturation pressure, which affects the total COP and capacity; (ii) the

Transcritical CO_2 Heat Pump: Fundamentals and Applications,
First Edition. Xin-Rong Zhang and Hiroshi Yamaguchi.

lower compression ratio of CO_2 compared to fluorocarbons results in higher isentropic efficiency and high volumetric capacity; (iii) large refrigerant temperature glide during heat rejection. With proper heat exchanger design the refrigerant can be cooled to a few degrees above the entering coolant (air, water) temperature, and this contributes to high COP of the system; (iv) High temperature yield. Water up to 90° can easily be produced, making it possible for direct use in the food and beverage industry, hotels, restaurants, and hospitals requiring sterilization, etc.; (v) Downsizing. Despite the drawbacks, CO_{2}- based heat pumps still offer extensive possibilities in both heating and cooling applications due to their outstanding advantages.

In this chapter, recent progress in the basic operation system design, key equipment development (compressors, heat exchangers, etc.), as well as system application studies are summarized. New concepts and optimization analysis for the high temperature CO_2 heat pump systems are also introduced and analyzed. Several representative applications for CO_2 heat pump in water heater, space heating, air conditioning, and drying, are also included in this chapter.

8.2 Basic System Design

8.2.1 Key Features in High Temperature CO_2 Heat Pump

A high temperature heat pump adopts the same technology as a conventional heat pump, but it yields at a high temperature. Nellissen and Wolf's analysis [22] had suggested that in applications such as pasteurization, drying, distillation, sterilization, pressurized hot water production and other industrial processes, there is an urgent need for high temperature heat pumps to offer a temperature range of 80–150°C. Many heat pump technologies with CO_2 as the refrigerant for high temperature heat pump have been discussed. Commercially available products with temperatures up to 120°C have been presented using CO_2 as the working fluid [23]. Eikevik *et al.* [24, 25] investigated high temperature heat pump using CO_2 for fish, fruits, vegetables and dairy products drying. White *et al.* [26] developed a CO_2 heat pump prototype for up to 65°C heat delivery. The thermodynamic properties of most refrigerants used in the low temperature heating process does not apply to high temperature applications. It seems that the only technological limitation is the compressor, as other components are commercially available [27]. In recent years, the study of CO_2 heat pumps has become a hot field, R.U. Rony *et al.* [28] summarized the research works by categorizing them by different applications (water/air/ground source heat pump, hybrid solar system, hybrid geothermal, etc.). In earlier chapters of this book, the basic principles of the CO_2 heat pump system have also been introduced. In this chapter, the high temperature features are discussed.

In a CO_2 based transcritical cycle, the evaporation process takes place in the subcritical region while the heat rejection process happens in the supercritical region, as shown in Figure 8.1 [29]. This is different form a subcritical cycle whose evaporator and heat exchanger both operate below the critical point. During the heat rejection process, refrigerant gives off heat by phase transition at constant temperature in the subcritical process whereas, in the transcritical cycle, the temperature of the refrigerant decreases continuously without a phase change. Although the subcritical cycle has an efficient heat transfer path,

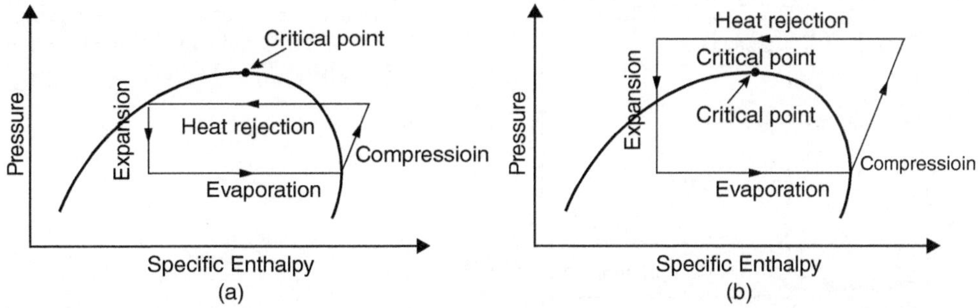

Figure 8.1 P-h diagrams for CO_2 heat pump cycles [29]: (a) subcritical cycle and (b) transcritical cycle.

the supercritical properties of CO_2 still makes it unmatchable in a heat pump system. The pressure-enthalpy diagram also shows that in the transcritical cycle, the compression ratio is 2–3 times lower than that in subcritical cycles (in the compression process, the conventional subcritical cycle operates at a pressure ratio up to eight, whereas the transcritical cycle operates at a pressure ratio within the range of three to four [30]. This low compression ratio can contribute to high system efficiency (which may be important for COP enhancement in related systems [31]).

8.2.2 Overall System Design

In Figure 8.2, the basic system design and operation process of a transcritical CO_2 heat pump system are shown. In Figure 8.2a the schematic picture of system layout is shown. It can be seen in Figure 8.2 that from the internal heat exchanger the fluid is heated from saturated vapor (6) to superheated state (1) and then by compressing process the fluid pressure and temperature is further increased to the supercritical region (state 2). Then the supercritical state CO_2 is cooled to state (3) through a gas cooler (there is no phase change, thus so-called gas cooler). The typical process of a "gas cooler" in this section is very critical for the system, as the fluid goes through supercritical state toward liquid state or near the saturation state in the critical region. Very large density and flow property changes happen in this section. The CO_2 is then further cooled through the internal heat exchanger, which is specially designed for higher thermal efficiency of the system. The fluid is then expanded to the two-phase region (state 5) and is led to the low temperature side evaporator. In Figure 8.2b the respective T-s curves are shown. From Figure 8.2b, in the gas to supercritical process (state 1 to state 2), a very steep gradient in the temperature curve can be seen, which indicates that in the heat pump system heat output process (state 2 to state 3) a very high temperature can be obtained. That is one typical process for supercritical state and could be utilized in applications such as water heating, space heating and others.

8.2.3 Real System Construction

Figure 8.3 [32] shows an industrial CO_2 heat pump water heater. The gas cooler is composed of four helical-type heat exchanger units and adopts a macro tube with 4.8 mm inner

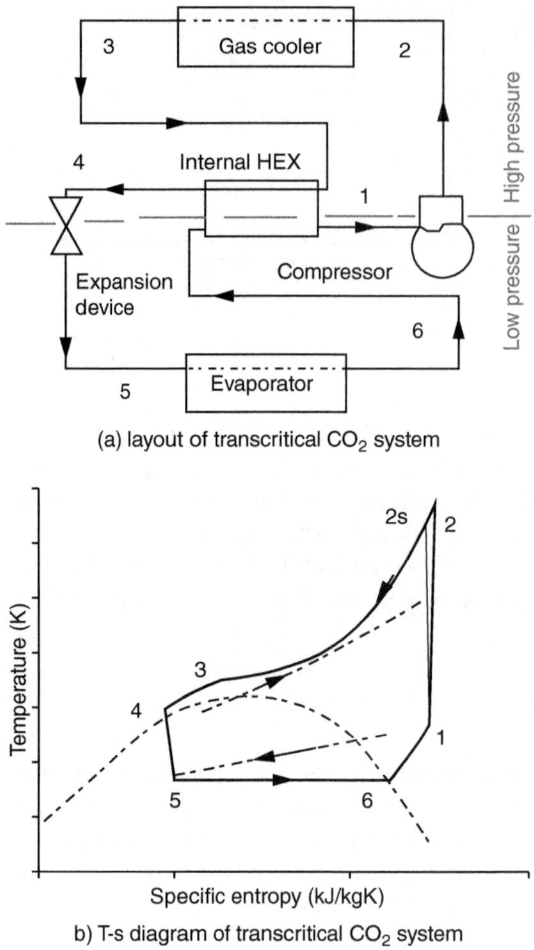

Figure 8.2 Schematic of CO_2 heat pump system with internal heat exchanger. (a) system layout; (b) T-s diagram.

diameter. The evaporator has two compartments, each containing four heat exchanger units. Two fans between the compartments are used to supply air. A reciprocating-type compressor with two cylinders is selected for this system.

8.3 High Temperature Operation and Key Equipment

8.3.1 Basic High Temperature CO_2 Heat Pump Operations

The high temperature CO_2 heat pump system has different operating principles for different applications, though the basic system design and general mechanisms are similar. In this section, several key processes for different kinds of high temperature CO_2 heat pump systems are introduced and compared.

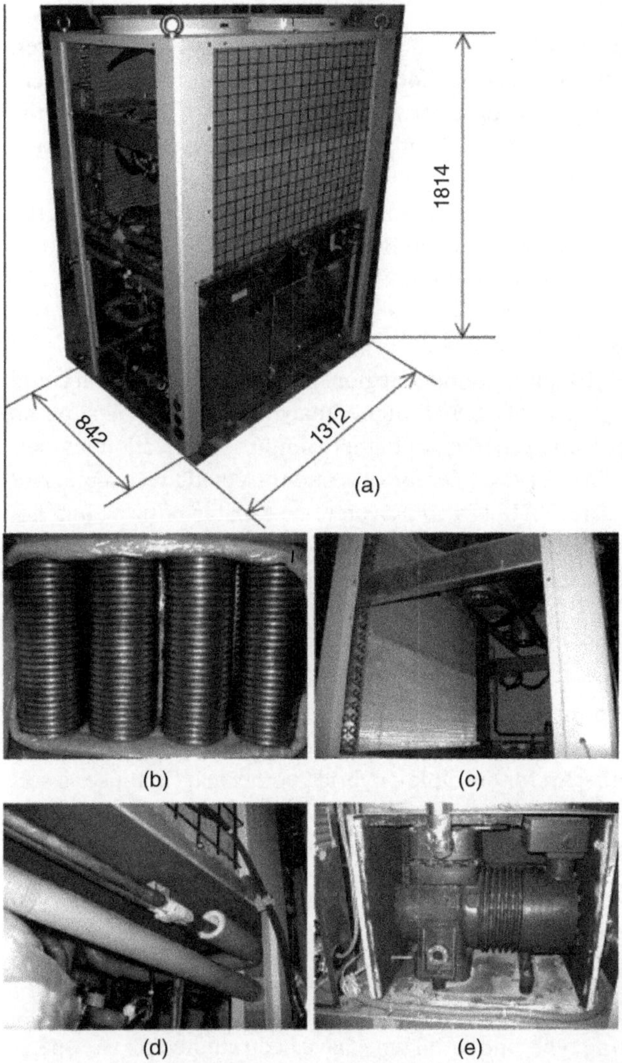

Figure 8.3 Pictures of the actual CO_2 heat pump water heater and its components [32]: (a) exterior, (b) gas cooler, (c) evaporator, (d) internal heat exchanger, and (e) compressor.

8.3.1.1 Water Source Heat Pump

Water source heat pump (WSHP) could be one good choice when simultaneous heating and cooling is required. An internal heat exchanger can be used to transfer the heat rejected from the cooling unit to the heating unit to increase the overall COP. But in cold climates the water source would be limited due to freezing. One example of such a water source system is from Mayekawa Company (Australia) [33], where the CO_2-based WSHP is used to supply both hot water at 90°C and chilled water at −9°C with a total COP of 8.0.

8.3.1.2 Air Source Heat Pump

The air source heat pump (ASHP) has been widely used for its energy saving features. CO_2-based air source heat pump is advantageous for its good operation data in cold climate regions. According to Hu *et al.* [34], a large part (~35%) of the energy is used for the frost-melting process in a CO_2 driven air source heat pump system. Due to the relatively high compressor outlet temperature, CO_2-based heat pump could be operated in cold climate regions while traditional water-based or air-based ones have to be adjusted below the freeze temperature. Usually, a hot gas bypass defrosting method should be applied in cold climates, and a discharge pressure control is proposed for COP optimization [34].

8.3.1.3 Ground Source Heat Pump

The ground source heat pump (GSHP) can be found in regions where the ground heat could be well utilized. The installation expenses of GSHP is higher than other kinds of heat pump systems. Compared with other types, the GSHP has a higher compressor rejection temperature and larger temperature drop in the gas cooler for CO_2 and thus could recover a great amount of heat from the ground. The GSHP is also proven to be capable of providing hot water at higher than 90°C while traditional heat pumps can only give 60°C. Indeed, this is not typical for ground source-based CO_2 heat pumps, but typical for CO_2-based water heater systems.

8.3.1.4 Hybrid Heat Pump

The hybrid heat pump (HHP) system indicates the combination of two or more energy sources to be utilized in CO_2 heat pump systems. Typical systems may include hybrid solar (to connect with flat plate solar collector for example), hybrid geothermal (that uses a secondary source such as solar, air, water, etc.) and other kinds. All such systems could be established when the overall system capacity and COP could be improved.

8.3.2 Compressors

CO_2 heat pump needs a relatively higher discharge pressure in the compressor (90–130 bar) than conventional CFC/HFC working fluids (10–40 bar). Different kinds of compressors, such as the rotary type, reciprocating type, and scroll type, have been studied for the operation of CO_2 heat pump systems. Table 8.1 shows a summary of CO_2 compressors in Japan. It can be seen from this table that different types of compressors have been commercially developed and applied in real applications. Two-stage design is the main type. Representative pictures of those types of compressors are shown in Figure 8.4. Typical displacement of the CO_2 compressors changes from 3.3 to 4.5 cm^3 [36–39]. The revolutions change between 1800 and 7200 min^{-1} [40]. The conditions are designed with a suction pressure around 4.0–4.5 MPa, a discharge pressure around 9.0–10.0 MPa and a gas cooler outlet temperature around 22°C.

Dorin Company has produced reciprocating-type compressors for supercritical CO_2 working fluid. It is found that such a type of compressor discharge pressure will always have higher value than the critical pressure (7.377 MPa for CO_2), which is affected by the gas cooler outlet temperature [41]. Mitsubishi Company developed one single rotary-type compressor for commercial heat pump system [42], which has a COP of 4.5 for residential

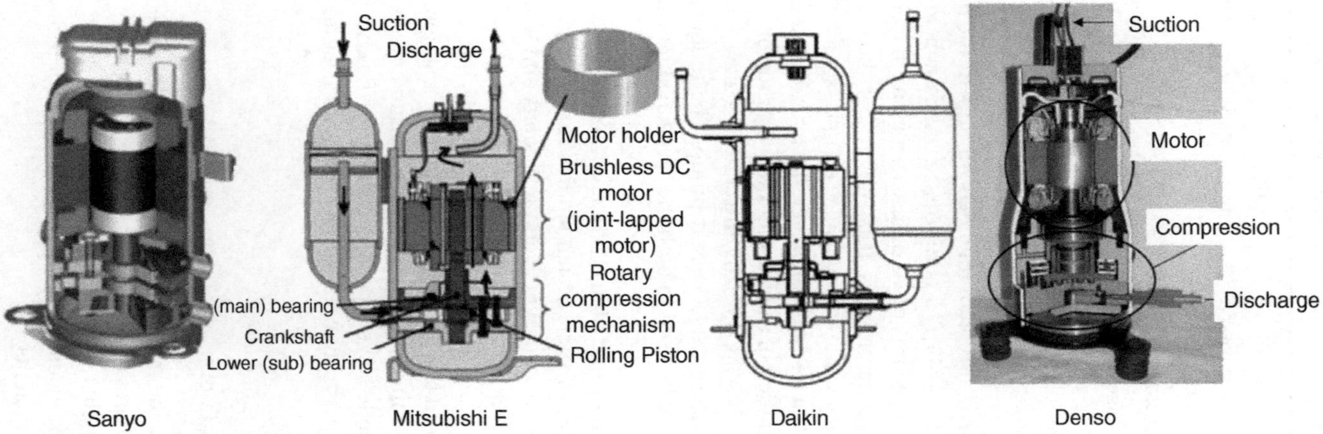

Figure 8.4 Schematic of the compressor layout for various companies [36–39].

Table 8.1 Basic specifications for CO_2 compressors in Japan [35].

Manufacture	Sanyo	Mitsubishi E	Daikin	Denso
Reference	[35]	[36]	[37]	[38]
Compressor type	Two-stage rotary	Single rotary	Swing	Scroll
Motor type	DC motor			
Size(mm)	Φ118 × 217			Φ137 × 285
Displacement(cm3)	1st stage:3.33	—	—	
2nd stage:2.33	4.5	4.2	4.3	

use. A scroll-type compressor is not recommended for supercritical CO_2 systems due to its low COP in system operation [43]. In addition, hermetic and semi-hermetic types of system configuration have also become areas of research focus in recent years [44–46]. For the single-stage and two-stage compressors, it is shown that the single-stage compressor has higher volumetric efficiency while the two-stage compressors show higher value when compression ratio is higher than 2.8 [46]. However, the overall system efficiency would be largely affected by the combination of a gas cooler and compressor system and its detailed designs. Recently, the Boost HEAT Company introduced a new and original concept of a thermal compressor for transcritical CO_2 heat pump. The uniqueness of the system is that the same working fluid can be used in both the thermal engine and the heat pump to enhance thermal efficiency.

To obtain high heating capacity and high efficiency, novel compressors have been developed by researchers. Examples of such design studies can be found as the work of Sato *et al.* [42], who developed a compressor which employs rotary and scroll mechanisms in the first and second stage respectively. Yokoyama *et al.* [47] developed a two-stage rotary compressor with refrigerant injection. Compared with single type, the compressor efficiency of the two-stage type is superior in the low rotational speed or high pressure-ratio as shown in Figure 8.5. This can lead to much less gas leaking during compression. The two-stage type is also superior to the single-stage type in compressor efficiency and heating capacity because it can improve these performances with refrigerant injection during high pressure ratio operation.

8.3.3 Heat Exchanger/Gas Cooler

For CO_2 heat pump systems, the gas cooler undertakes the heat rejection task from the supercritical temperature and pressure conditions. A great number of tests can be found in literature for the heat transfer characteristics of the supercritical CO_2 heat transfer process [15–20]. The current section is focused on the current status of heat exchanger design for high temperature heat pump systems.

Three types of the water-CO_2 heat exchangers (double tube, smooth tube, and dimple tube) were investigated and compared by Taira [48] and others (see Figure 8.6 and related explanations in refs [48–50]). In a double tube heat exchanger, leakage is detectable, but

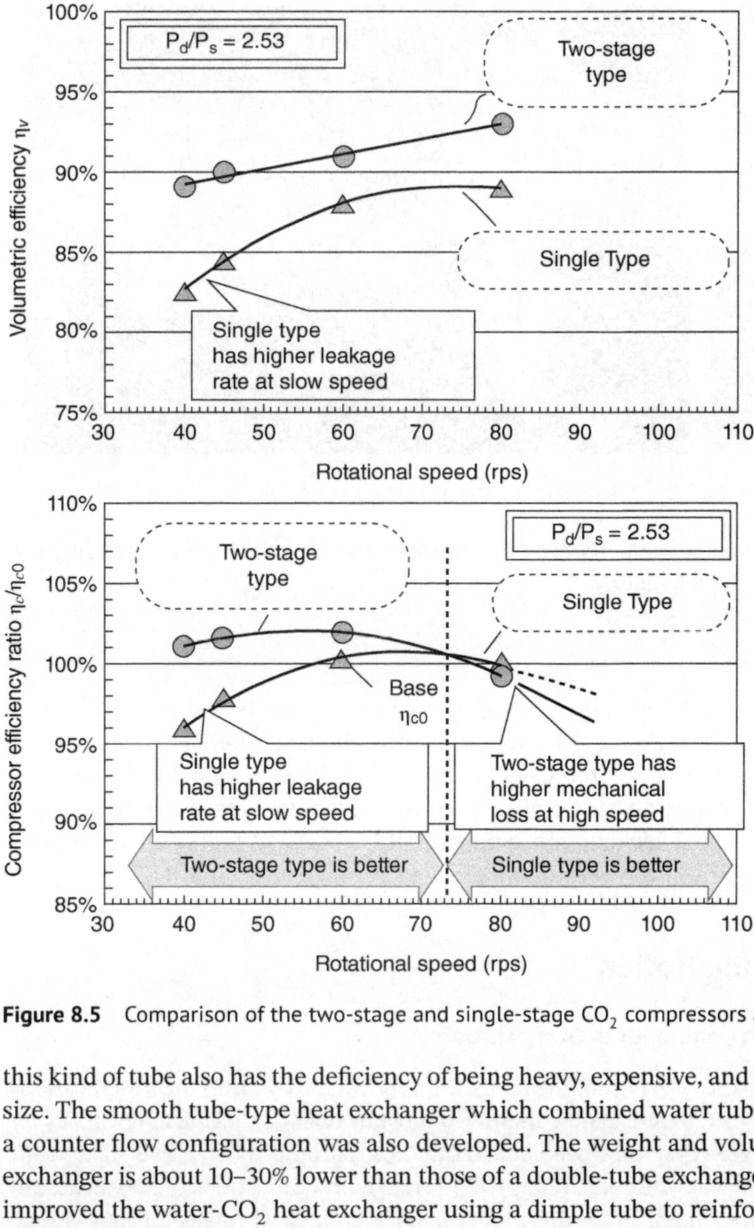

Figure 8.5 Comparison of the two-stage and single-stage CO_2 compressors at Pd/Ps¼2.53 [47].

this kind of tube also has the deficiency of being heavy, expensive, and difficult to reduce in size. The smooth tube-type heat exchanger which combined water tube and CO_2 tube with a counter flow configuration was also developed. The weight and volume of the new heat exchanger is about 10–30% lower than those of a double-tube exchanger [48]. Taira further improved the water-CO_2 heat exchanger using a dimple tube to reinforce the heat transfer augmentations [48, 49].

Several new types of heat exchanger/gas cooler have been developed. For example, a "twist and spiral gas cooler" features a twisted pipe with three lines of spiral grooves for a water pipe. This aims to improve transfer area by increasing the contact surfaces. This arrangement can also accentuate the turbulence effect and reduce the pressure loss in the refrigerant. Research on a capillary tube heat exchanger (proposed in Ref. [50]) illustrated that the set mode of the water tube and CO_2 tube inner diameter were the most important factors affecting the efficiency. Tests on a tube-in-tube heat exchanger [50] revealed that

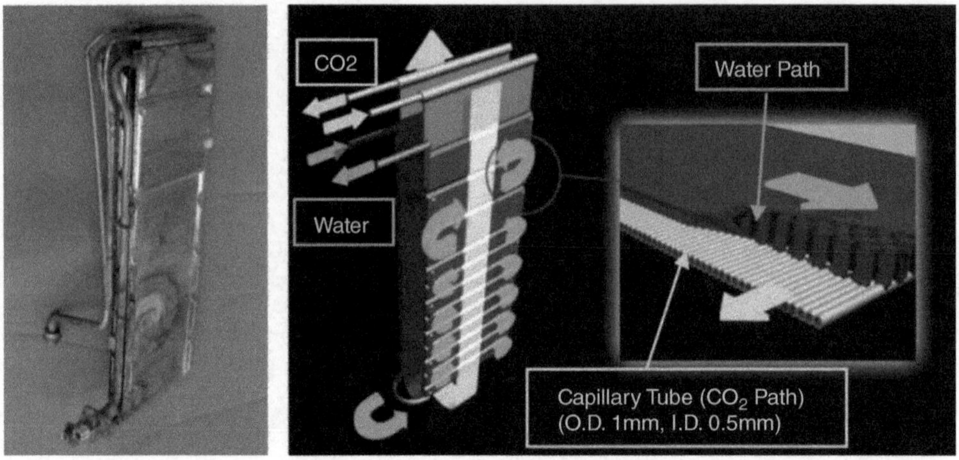

Figure 8.6 Developed a capillary tube heat exchanger developed by Sakakibara *et al.* [38].

the water side heat transfer coefficient can be improved by more than double by using the design of a dimple heat exchanger.

8.3.4 Expander

An expansion device in a high temperature CO_2 heat pump system is used to distribute CO_2 working fluid into the evaporator. In this process the pressure difference is maintained in the gas cooler and the evaporator to generate continuous flows. Generally, the working fluid will become two-phase and has a highly compressible feature in the ejector/expander, which process could be modeled to improve the overall system COP. More detailed designs and discussions of expanders can be found in previous review articles [28, 35].

8.4 System Optimization

8.4.1 Basic System Components Optimization

Sarkar [31] studied the cycle modifications in typical air-conditioning applications (t_{ev} = 50°C and t_{co} = 40°C). COP can be improved most by using a turbine and least by an internal heat exchanger (COP improvement: IHX, 7.5%; Turbine, 17.5%; Two-stage, 25%; PCE, 17.5%; Ejector, 16%; Vortex generator, 10% [31]). And modifications of two-stage and parallel compression economization are more effective compared to ejector and vortex. Optimum discharge pressure reduction can be achieved by using two-stage, but turbine will be more cost effective (reduction of optimal discharge pressure by each kind: IHX, 2%; Turbine, 6%; Two stage, 6.5%; PCE, 5.2%; Ejector, 3.6%; Vortex generator, 3.5% [31]).

8.4.2 Discharge Pressure Optimization

According to previous researchers [28, 35], the optimum discharge pressure and system COP rely on the key temperatures of such as the evaporator, gas cooler exit and compressor

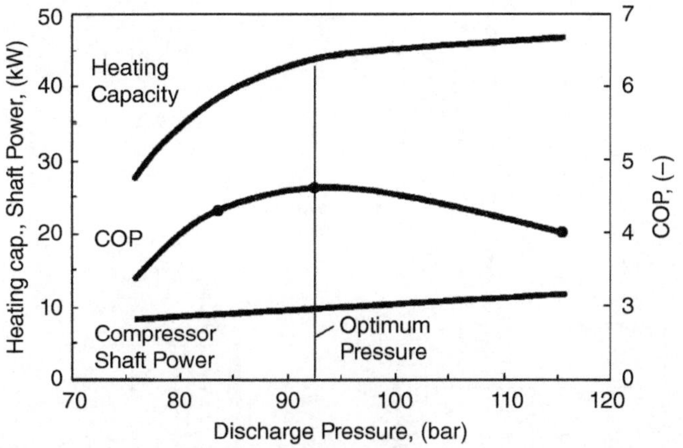

Figure 8.7 Variation of heating capacity, heating-COP and compressor shaft power with the discharge pressure for a CO_2 heat pump [53].

outlet. And it is generally recommended that the gas cooler and evaporator temperatures are the most important ones to determine the optimal discharge pressure: it will increase with gas cooler exit temperature and decrease with evaporator temperature. Based on the experimental study [51], Sarkar *et al.* [52] concluded a reasonable correlation by performing regression analysis on data obtained from cycle simulation. The correlation is as follows:

$$P_{d,opt} = 4.9 + 2.256t_{co} - 0.17t_{ev} + 0.002t^2{}_{co} \tag{8.1}$$

8.4.3 System Optimization

Figure 8.7 [53] shows the effect of the compressor discharge pressure on heating capacity, heating-COP and compressor shaft power for heat pump. At low discharge pressure capacity appears to show a steep increase, but with the increasing of discharge pressures it flattens out, while the compressor shaft presents a linear increase with discharge pressure. COP increases at the beginning, reaching the maximum point and then decreases with the increase of discharge pressures. Besides, the variation in COP around the optimum may be rather flat which depends on the shape of the compressor isentropic efficiency curve and the length of the internal heat exchanger [54].

8.5 Applications and Challenges

8.5.1 Heating and Cooling

Heating and cooling is the biggest sector of high temperature CO_2 heat pump applications. Figure 8.8 is a combined space and water heating system design schematic with a T-s diagram for space heating application. In order to fix a very low return temperature from the heating system, radiator and air heating are connected in series. In such a representative system, hot water supply is achieved by the exchanging of heat from the compressor outlet gas cooler and also from the space heating system. The temperature of the hot water supply

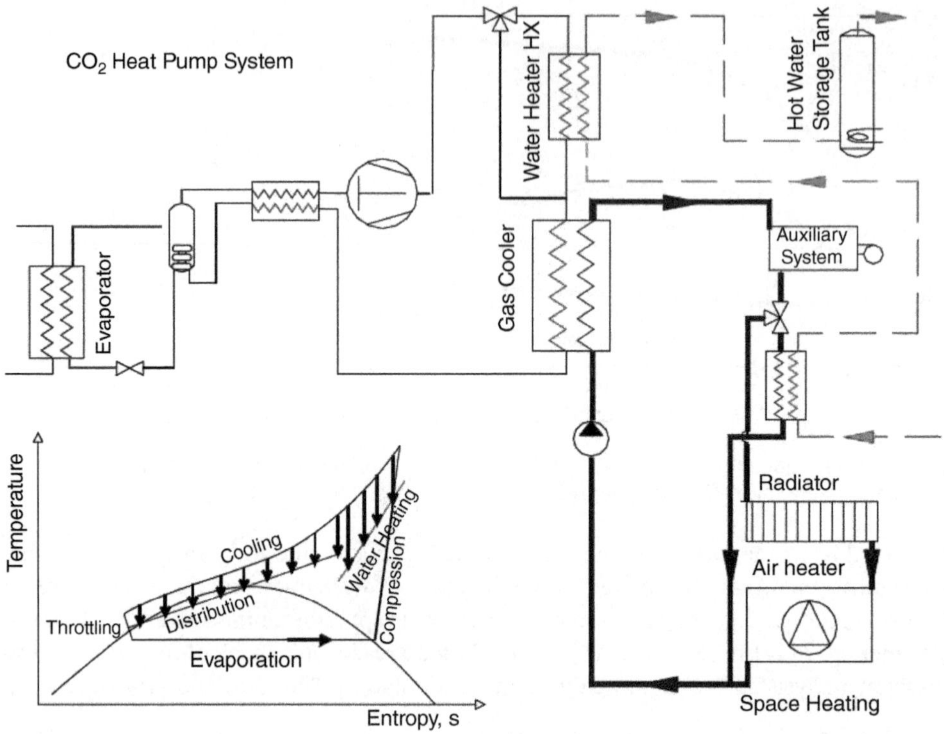

Figure 8.8 System design for a combined space and water heating system. The process is also illustrated in the T-s diagram [53].

can be adjusted by a hot water tank and by changing the flow rate of the gas side and the water side. Such a combined heating and cooling system has recently been tested in several countries, but the efficiency and functionality of the system is still in discussion.

8.5.2 Other Industrial Sectors

In real industrial sectors, many other kinds of CO_2 heat pump systems can be found. For example, the air-cycle and refrigerant-cycle of a one-stage air-source heat pump dryer has been proposed [55]. The system model consists of five major modules (compressor, capillary tube, condenser, evaporator, and tumbler) and a recirculating fan. In the air re-circulation cycle, air exits the drum then enters the evaporator and condenser, then finally returns to the drum. Fin-and-tube type heat exchangers are utilized as evaporator and condenser/gas-cooler. While the expansion device uses a capillary tube, the compressor is the semi-hermetic reciprocating type and the drum is the tumbler type. Such a dryer system is much simpler than the water/space heating system but the well control of the CO_2 side flow is still a problem.

8.5.3 COP Analysis and Comparison

In Figure 8.9, the COP comparison between CO_2 and R22 working fluids are shown for different operation modes. It can be seen that in AC mode the COP values (cooling COP) of the CO_2 unit are slightly lower than those of the HCFC-22 unit, whereas in heat pump mode, the COP (heating COP) of the CO_2 unit is slightly higher than those of the HCFC-22 unit. For cooling mode, CO_2 could provide similar output capacity while for heating mode CO_2 is advantageous in heat pump operation. Such results agree with recent experiments and comparisons for CO_2 heat pumps [28, 35]. The key feature for a CO_2 heat pump working in winter or cold regions is its relative higher efficiency and the ability to provide higher compressor outlet temperature and higher heating capacity.

COP variation with time for both CO_2 and R-134a systems in a heat pump system is depicted in Figure 8.10 [56]. For these two systems, the suction and discharge pressure both increase with time, therefore the COP presents an overall decreasing trend. Generally speaking, the simulated results are in very good agreement with trends in experiment research. Besides, it is worth noting that the work consumption at the compressor also increases as pressure increases, thus deteriorating the COP.

Seasonal performances of CO_2 heat pump systems are quite unique compared with other kinds of heat pumps. Minetto *et al.* [57] studied the monthly dated energy consumption for each sector of domestic use and compared the monthly COP, as shown in Figure 8.11. It is reported that the overall system COP, the COP_{tot} which is defined as the ratio between

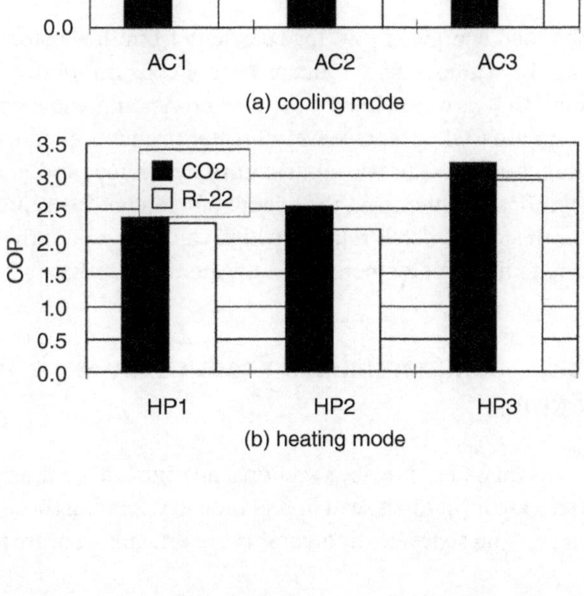

Figure 8.9 System COP measured for different modes of operation [53]. (a) cooling mode, Ambient temperatures are 28, 35 and 46°C for AC1, AC2 and AC3, respectively. (b) heating mode, Ambient temperatures are 2, 7, and 14 for heat pump1, heat pump2, and heat pump3, respectively.

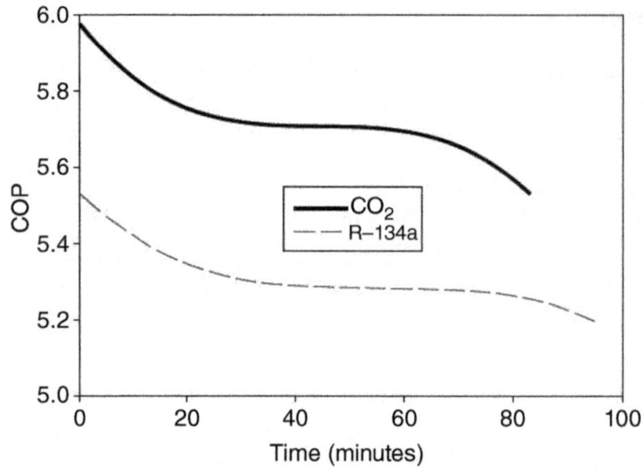

Figure 8.10 Simulation results for CO_2 and R-134a for drying model [56].

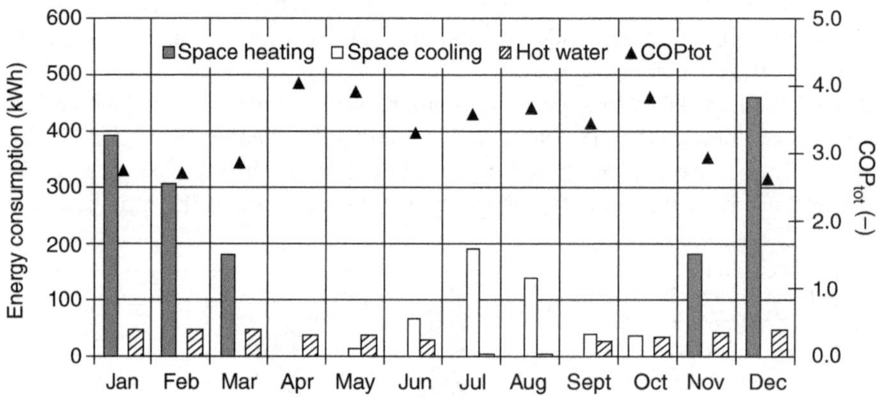

Figure 8.11 Simulated COP_{tot} and energy consumption for space heating, cooling and hot water production [57].

the total energy output (load capacity from the customer side) and the energy input. During the wintertime, the major energy consumption was space heating; In April and May, high COP_{tot} was achieved because no heating and significant cooling were required, so in this case COP_{tot} represents hot water production efficiency. In the summertime (July and August) tap water was almost entirely heated, and heat recovery from space cooling had significantly increased the overall efficiency. The annual system COP averaged around 3.2 for the test, which is relatively high value. In winter the system COP can be higher than 2.5, due to the high temperature hot water output that can be realized by this system.

8.6 Commercialized Products by High Temperature CO_2 Heat Pump

As discussed in previous sections, heating and cooling applications of CO_2 heat pump systems are most often seen in real industry. Among those applications, CO_2 heat pump water heater and space heating/cooling are the most mature products that have been accepted by the market. Indeed, it was the early 2000s when the CO_2 heat pump water heater appeared for the first time in NTNU (Norwegian University of Science and Technology). At the same time, CRIEPI (Central Research Institute of Electric Power Industry, Japan), TEPCO (Tokyo Electric Power Company, Japan) and DENSO company in Japan collaborated on the CO_2 heat pump water heater system and confirmed that the system could provide hot water at a temperature higher than 90°C, even when the ambient temperature was below −20 °C. After that, major industrial companies in Japan started to design and sell CO_2 heat pump water heater units from 2001. Currently, Mitsubishi, Daikin, Sanyo, Hitachi, Matsushita, Toshiba, Denso, Chofu, and other companies are selling such units in Japan. It is reported that by 2018, sales of the famous "Eco-Cute" unit had exceeded six million [58]. Most recently in 2017, Denso and Stiebel Eltron put into market one Stiebel Air System (using Denso CO_2 heat pump) for air conditioning. The system is designed for the construction industry in Germany, which fits the concept of "near-zero energy houses" to be built by 2020 [59]. In the US, there are also such kinds of CO_2-based heat pump water heaters in the market (for example, the Sanden product [60]). In China, several makers (Dongqi, Kaide, Gaoli, Arco, Haier, etc.) have developed new types of CO_2 heat pump water heaters [61]. It has been proved that CO_2 heat pump systems could achieve relatively high COP in different seasons and could have advantages over traditional kinds when emission and environmental problems are considered. Besides the commercial application in heating and cooling, real CO_2 heat pump systems have also been tested in real house heating/cooling, or combined water heating and space heating/cooling [62], which indicates a promising future market around the world. Readers could also refer to other chapters of this book for more details of CO_2 heat pump system applications in very recent years.

8.7 Summary

In this chapter, the basic concepts, system construction, key equipment, operations, efficiency analysis and optimization of high temperature CO_2 heat pumps are summarized. Recent developments and the optimization of CO_2 heat pump systems are also introduced and compared in this chapter. The results show that CO_2 heat pumps yield a performance that is comparable with traditional working fluid cycles, while for low temperature (winter season) it has advantageous system behaviors. In summary, though, there are still problems in the theoretical and mechanical design of systems and components for high temperature CO_2 heat pump systems, such as high efficiency heat exchangers and compressors, ejectors. The way toward commercial use of CO_2 heat pump systems in high temperature applications and also multi-functional systems appears very promising. It is hoped that this summary study of high temperature CO_2 heat pump systems could be useful as a section for researchers and engineers in this field.

Acknowledgments

This chapter is a short review analysis on the high temperature CO_2 heat pump system and its applications, system components, commercialized products and challenges. The support from the Young Professionals Program (Chinese Academy of Sciences), the Chinese Academy of Sciences Key Research Program of Frontier Sciences (No.ZDBS-LY-JSC018) and the NSFC-JSPS International Cooperation Program (No. 51961145201) are gratefully acknowledged by the authors. The authors are also grateful for the discussions with Prof. Haisheng Chen (Chinese Academy of Sciences, China), Prof. Hiroshi Yamaguchi (Doshisha University, Japan), Prof. Yuhiro Iwamoto (Nagoya Institute of Technology, Japan), and Prof. Xin-Rong Zhang (Peking University, China).

Nomenclature

CFC	chlorofluorocarbon
HCFC	hydrochlorofluorocarbon
HFC	hydrofluorocarbon
COP	coefficient of performance
ODP	ozone depletion potential
GWP	global warming potential
P	pressure, bar
h	enthalpy, $kJ \cdot kg^{-1}$
HEX	heat exchanger
T	temperature, °C
s	entropy, $kJ \cdot kg^{-1}$
WSHP	Water source heat pump
ASHP	air source heat pump
GSHP	ground source heat pump
HHP	hybrid heat pump
DC	direct current
AC	alternating current
Pd	discharge pressure
Ps	suction pressure
IHX	internal heat exchanger
HP	heat pump

Greek Symbols

Φ	size, mm
η	efficiency

Subscripts

v	volumetric
c	compressor
c0	compressor base value

ev	evaporator
co	cooler
opt	optimum
tot	total

References

1 Lorentzen, G. (1994). Revival of carbon dioxide as a refrigerant. *International Journal of Refrigeration* 17 (5): 292–301.

2 Neksa, P., Hrnjak, P., Pettersen, J. et al. (1999). CO_2 as a working fluid, short course SC3. *20th International Congress of Refrigeration*. Sydney.

3 Kim, M.H., Pettersen, J., and Bullard, C.W. (2004). Fundamental process and system design issues in CO_2 vapor compression systems. *Progress Energy Combustion Science* 30 (2): 119–174.

4 Sarkar, J., Bhattacharyya, S., and Ramgopal, M. (2005). Transcritical CO_2 heat pump systems: exergy analysis including heat transfer and fluid flow effects. *Energy Conversion Management* 46 (13–14): 2053–2067.

5 Hepbasli, A. and Kalinci, Y. (2009). A review of heat pump water heating systems. *Renewable and Sustainable Energy Reviews* 13 (6–7): 1211–1229.

6 Lund, J.W., Freeston, D.H., and Boyd, T.L. (2005). Direct application of geothermal energy: 2005 worldwide review. *Geothermics* 34 (6): 691–727.

7 Omer, A.M. (2008). Ground-source heat pumps systems and applications. *Renewal and Sustainable Energy Reviews* 12 (2): 344–371.

8 Omojaro, P. and Breitkopf, C. (2013). Direct expansion solar assisted heat pumps: a review of applications and recent research. *Renewal and Sustainable Energy Reviews* 22: 33–45.

9 Johnson, I., Choater, B., and Dillich, S. (2008). Waste heat recovery: opportunities and challenges. In: *Proceedings of: The Minerals, Metals & Materials (TMS) Society Annual Meeting*, 47–52.

10 Bolaji, B.O. and Huan, Z. (2013). Ozone depletion and global warming: case for the use of natural refrigerant – a review. *Renewal and Sustainable Energy Reviews* 18: 49–54.

11 Varotsos, C. (2008). The 20th anniversary of the Montreal protocol and the unexplainable 60% of ozone loss. *Environtal Science and Pollution Research* 15 (6): 448–449.

12 ASHRAE Standard. (2019) ASHRAE 15 and 34 safety standard for refrigeration systems and designation and classification of refrigerants ISO 5149 mechanical refrigerating systems used for cooling and heating-safety requirements. https://www.ashrae.org/technical-resources/bookstore/standards-15-34 (accessed 12 November 2019).

13 Lachner, B.F., Nellis, G.F., and Reindl, D.T. (2007). The commercial feasibility of the use of water vapor as a refrigerant. *International Journal of Refrigeration* 30 (4): 699–708.

14 ASHRAE (2014). ASHRAE position document on natural refrigerants. American Society of Heating, Refrigerating and Air-Conditioning Engineers, Inc. Atlanta, GA, USA.

15 Chen, L., Zhang, X.R., Yamaguchi, H., and Liu, Z.S. (2010). Effect of heat transfer on the instabilities and transitions of supercritical CO_2 flow in a natural circulation loop. *International Journal of Heat and Mass Transfer* 53: 4101–4111.

16 Chen, L. and Zhang, X.R. (2014). Experimental analysis on a novel solar collector system achieved by supercritical CO_2 natural convection. *Energy Conversion and Management* 77: 173–182.

17 Chen, L., Zhang, X.R., and Jiang, B. (2014). Effects of heater orientations on the natural circulation and heat transfer in a supercritical CO_2 rectangular loop. *Transactions of ASME - Journal of Heat Transfer* 136 (5): 052501-1–052501-12.

18 Yu, S.C., Chen, L., Zhao, Y. et al. (2015). Thermodynamic analysis of representative power generation cycles for low-to-medium temperature applications. *International Journal of Energy Research* 39: 84–97.

19 Chen, L. (2016). *Microchannel Flow Dynamics and Heat Transfer of near-Critical Fluid*. Singapore: Springer.

20 Chen, L. and Iwamoto, Y. (2017). *Advanced Applications of Supercritical Fluids in Energy Systems*. Hershey, Pennsylvania, USA: IGI Global.

21 Stenstadvold, K. and Neksa, P. (2005), Transcritical CO_2 heat pumping system. In Proc. of: The 8th IEA Heat Pump Conference Conference, May 30 - June 2, 2005, Las Vegas, USA.

22 Nellissen, P. and Wolf, S. (2014). Heat pumps in non-domestic applications in Europe: potential for an energy revolution. Presentation given at the 8th EHPA European Heat Pump. Retrieved from Germany.

23 IEA (2014). Application of industrial heat pumps, final report, Part 1, report No. HPPAN35-1.

24 Eikevik, T., Alves-Filho, O., and Strømmen, I. (2004). Heat pump drying – new technologies and operational modes for production of a new generation of high-quality dried products. Paper presented at the International Workshop and Symposium on Industrial Drying (IWSID). Mumbai, India

25 Eikevik, T., Strømmen, I., and Alves-Filho, O. (2005). Heat pump fluidised bed dryer with CO_2 as refrigerant e measurements of COP and SMER. Paper presented at the 3rd Nordic Drying Conference. Karlstad, Sweden:

26 White, S.D., Yarrall, M.G., Cleland, D.J., and Hedley, R.A. (2002). Modelling the performance of a transcritical CO_2 heat pump for high temperature heating. *International Journal of Refrigeration* 25 (4): 479–486.

27 Bamigbetan, O., Eikevik, T.M., Neksa, P. et al. (2017). Extending hydrocarbon heat pumps to higher temperatures: predictions from simulations. Paper presented at the 30th International Conference on Efficiency, Cost, Optimization, Simulation and Environmental Impact of Energy Systems. San Diego, California.

28 Rony, R.U., Yang, H.J., Krishnan, S., and Song, J.C. (2019). Recent advances in transcritical CO_2 (R744) heat pump system: a review. *Energies* 12 (3): 457.

29 Austin, B.T. and Sumathy, K. (2011). Trans-critical carbon dioxide heat pump systems: a review. *Renew. Sustainable Energy Review* 15: 4013–4029.

30 Staub, J., Rasmusen, B.D., and Robinson, M. (2004). CO_2 as refrigerant: the transcritical cycle. ACHR News.

31 Sarkar, J. (2010). Review on cycle modifications of transcritical CO_2 refrigeration. *Journal of Advanced Research in Mechanical Engineering* 1 (1): 22–29.

32 Seiichi, Y., Daisuke, K., Kiyoshi, S., and Sunao, K. (2011). Development and validation of static simulation model for CO_2 heat pump. *International Journal of Heat and Mass Transfer* 54 (9): 1896–1906.

33 Mayekawa Co. Ltd. (2019). Water source heat pump system. www.mayekawa.com.au/products/heat-pumps/heatcom-water-heat-source (accessed 12 November 2019).

34 Hu, B., Li, Y., Cao, F., and Xing, Z. (2015). Extremum seeking control of COP optimization for air-source transcritical CO_2 heat pump water heater system. *Applied Energy* 147: 361–372.

35 Zhang, J.F., Qin, Y., and Wang, C.C. (2015). Review on CO_2 heat pump water heater for residential use in Japan. *Renewable and Sustainable Energy Reviews* 50: 1383–1391.

36 Hideaki, M. and Shinichi, T. (2007). Rotary compressor for a CO_2 heat pump water heater. *Mitsubishi Electric Advance* 120: 9–12.

37 Masahide, H., Hideki, M., Hiroyuki, T. et al. (2006). Development of the high efficiency and low noise swing compressor for CO_2 heat pump water heater. *Proceedings of the 18th International Compressor Engineering Conference*. Paper C100, USA.

38 Sakakibara, H., Kato, H., Akiyama, Y. et al. (2002). Development of natural refrigerant (CO_2) hot water supplier for residential use. *Denso Technology Review* 7: 81–90.

39 Masahiro, K. (2008). CO_2 heat pump heating and water heater system for cold area. *Sanyo Technology Review* 38: 10–15.

40 Hiroshi, M. (2003). Development and performance evaluation of CO_2 heat pump water heater for cold climates. *IEA Heat Pump Centre Newsletter* 21: 9–12.

41 Jiang, Y., Ma, Y., Li, M., and Fu, L. (2013). An experimental study of trans-critical CO_2 water-water heat pump using compact tube-in-tube heat exchangers. *Energy Conversation and Management* 76: 92–100.

42 Sato, H., Kimata, Y., Hotta, Y. et al. (2012). Development of two-stage compressor for CO_2 heat-pump water heaters. *Mitsubishi Heavy Industries Technology Review* 49: 92–97.

43 Hiwata, A., Iida, N., Futagami, Y., et al. (2002). Performance investigation with oil-injection to compression chambers on CO_2-scroll compressor. *Proceedings of the International Compressor Engineering Conference*. Paper 1577. West Lafayette, IN, USA: Purdue University.

44 Rozhentsev, A. and Wang, C.C. (2001). Some design features of a CO_2 air conditioner. *Applied Thermal Engineering* 21: 871–880.

45 Cavallini, A., Cecchinato, L., Corradi, M. et al. (2005). Two-stage transcritical carbon dioxide cycle optimization: a theoretical and experimental analysis. *International Journal of Refrigeration* 28: 1274–1283.

46 Zhang, L., Yang, M., and Huang, X., (2016). Performance Comparison of Single-stage and Two-stage Hermetic Rotary CO_2 Compressor. *Proceedings of the International Compressor Engineering Conference*. Paper 2473. West Lafayette, IN, USA.

47 Yokoyama, T., Sasaki, K., Sekiya, S., and Maeyama, H., 2008. Developing a two-stage rotary compressor for CO_2 heat pump systems with refrigerant injection. *Proceedings of the 19th International Compressor Engineering Conference*. Paper 1193. Purdue, USA.

48 Taira, S. (2010). The development of heat pump water heaters using CO_2 refrigerant. *Paper presented in Proceedings of the International Symposium on Next Generation Air Conditioning and Refrigeration Technology*.

49 Wang, C.C., Yang, K.S., Liu, Y.P., and Chen, I.Y. (2011). Effect of cannelure fin configuration on compact air cooling heat sink. *Applied Thermal Engineering* 31: 1640–1647.

50 Andou, T., Machida, K., Imagawa, T., and Yamamoto, T. (2010). Development of energy saving technology for CO_2 heat pump water heater. *Panasonic Technology Journal* 56: 27–32.

51 Cabello, R., Sanchez, D., Llopis, R., and Torrella, E. (2008). Experimental evaluation of the energy efficiency of a CO_2 refrigerating plant working in transcritical conditions. *Applied Thermal Engineering* 28 (13): 1596–1604.

52 Sarkar, J., Bhattacharyya, S., and Ramgopal, M. (2004). Optimization of a transcritical CO_2 heat pump cycle for simultaneous cooling and heating applications. *International Journal Refrigeration* 27 (8): 830–838.

53 Neksa, P. (2002). CO_2 heat pump systems. *International Journal of Refrigeration* 25: 421–427.

54 Neksa, P., Rekstad, H., Zakeri, G.R., and Schiefloe, P.A. (1998). CO2-heat pump water heater: characteristics, system design and experimental results. *International Journal of Refrigeration* 21: 172–179.

55 Lee, B.H., Sian, R.A., and Wang, C.C. (2019). A rationally based model applicable for heat pump tumble dryer. *Drying Technology* 37 (6): 691–706.

56 Sian, R.A. and Wang, C.C. (2019). Comparative study for CO_2 and R-134a heat pump tumble dryer - a rational approach. *International Journal of Refrigeration* 106: 474–491.

57 Minetto, S., Cecchinato, L., Brignoli, R. et al. (2016). Water-side reversible CO_2 heat pump for residential application. *International journal of refrigeration* 63: 237–250.

58 Japan eco-cute heat pump water heater. https://accelerate24.news/regions/japan/japan-eco-cute-and-c-stores-lead-natrefs-growth/2019 (accessed 5 September 2020).

59 Denso Co. Ltd (2019). Denso CO_2 heat pump. https://www.denso-am.eu/products/life-energy/co2-heat-pump (accessed 12 November 2019).

60 Sanden Co. Ltd (2019). Sanden Water Heater. www.sandenwaterheater.com (accessed 12 November 2019).

61 Several major makers of CO_2 heat pump systems. https://kmdqkjgf.cn.biz72.com/offerlist; http://info.hp.hc360.com/2014/07/011755248341-4.shtml; http://www.kaorinb.com/products4.aspx?cateid=88;http://www.arcochina.com/cpjj1.asp (accessed 12 November 2019).

62 Dongqi Co. Ltd (2019). Dongqi CO_2 heat pumps. https://www.dqbest.net/ (accessed 12 November 2019).

9

Performance Analysis and Optimization of a CO_2 Heat Pump Water Heating System

Ryohei Yokoyama

Department of Mechanical Engineering, Osaka Prefecture University, 1-1 Gakuen-cho, Naka-ku, Sakai, Osaka 599-8531, Japan

9.1 Introduction

In Japan, energy consumption in the residential sector accounts for about 16% of the total in all sectors, and the energy consumption for hot water supply accounts for about 28% of that in the residential sector. This is because hot water is used not only for washing, cooking, and shower but also for bath. Thus, the energy saving in hot water supply has been an important issue in the residential sector. Under this situation, water heating systems, each of which is composed of a heat pump using CO_2 as a natural refrigerant and a hot water storage tank, called "ECO CUTE," have been developed and commercialized widely [1]. More than seven million units have been installed into residential houses during the period from 2001 to the present.

The performance of CO_2 heat pumps has been enhanced dramatically during the period through the technological development of their components such as compressors and gas coolers. On the other hand, importance has also been given to the performance of water heating systems in case they are operated under a daily change in hot water demand. The performance of the CO_2 heat pump only, or coefficient of performance (COP) is affected by the air temperature as well as the inlet and outlet water temperatures, while the performance of the water heating system is affected by many conditions. The ambient conditions such as air and feed water temperatures, the hot water demand, and operating conditions such as startup time, shutdown time, and outlet water temperature during operation of the CO_2 heat pump affect the inlet water temperature and resultantly the COP through the temperature distribution in the storage tank. In addition to the COP, the storage and system efficiencies, and the volumes of stored and unused hot water are considered as system performance values, and these are also affected by the aforementioned various conditions through the temperature distribution in the storage tank. As a result, the system performance is affected by the operational history of the past several days, and changes complexly with days. Therefore, in order to attain the maximum system performance, it is necessary to analyze the system performance under the aforementioned various conditions, estimate daily changes in system performance values accurately in relation to the conditions, and determine operating conditions optimally based on them.

Transcritical CO_2 Heat Pump: Fundamentals and Applications,
First Edition. Xin-Rong Zhang and Hiroshi Yamaguchi.

Many theoretical and experimental studies have been conducted for the performance analysis on CO_2 heat pumps. However, few studies have been conducted for the performance analysis on CO_2 heat pump water heating systems [2–5]. It takes much time to conduct the performance analysis on water heating systems under the aforementioned various conditions by experiment, and thus it is very difficult to optimize the operating conditions. On the other hand, it is expected that numerical simulation enables one to conduct the performance analysis very efficiently, which may lead to optimization of the operating conditions. We have conducted many studies for the performance analysis on water heating systems by numerical simulation [6–14]. In addition, we have estimated daily changes in system performance values based on the results obtained by numerical simulation, and have optimized the operating conditions [15].

In this chapter, we present performance analysis and optimization of a CO_2 heat pump water heating system by numerical simulation. First, a summary of the system modeling and numerical solution is described. Second, some studies on performance analysis are described. They are related with analysis under periodically steady state for daily repeated hot water demand, analysis for performance enhancement by extracting tepid water from the middle of the storage tank, and analysis under unsteady state for daily change in hot water demand. Finally, performance estimation and optimal operation are described. Daily changes in system performance values are estimated by neural network models based on the results obtained by performance analysis under unsteady state. In addition, operating conditions are determined optimally based on the system performance values obtained by the estimation.

9.2 System Configuration

A unifunctional CO_2 heat pump water heating system only with the function of hot water supply is investigated in this chapter. Figure 9.1 shows the configuration of the CO_2 heat pump water heating system investigated here. This system is composed of a CO_2 heat pump and a hot water storage tank. The CO_2 heat pump is composed of a compressor, a gas cooler, an expansion valve, and an evaporator. The system is equipped with a fan, a pump, and motors M1 to M3 as auxiliary machinery. Here, inlet and outlet water are defined as water at the inlet and outlet of the gas cooler, respectively. In the charging mode, the system heats water using the refrigeration cycle of the CO_2 heat pump and stores hot water in the storage tank. In the tapping mode, hot water stored in the storage tank is retrieved and supplied to a tapping site.

In the conventional system, hot water stored in the storage tank is retrieved from its top. In such a system, the gradient of the vertical temperature distribution in the storage tank with water temperature stratified becomes small because of heat conduction in the vertical direction during storage for a long time. Namely, the temperature in the lower part rises while that in the upper part drops, and the area of tepid water in the middle part expands. As a result, the temperature of the inlet water entering the CO_2 heat pump rises, which leads to a decrease in the heat pump COP, and resultantly system efficiency also decreases. On the other hand, the volumes of hot water stored after heat pump operation in the charging mode and unused after hot water supply in the tapping mode decrease, which may result in

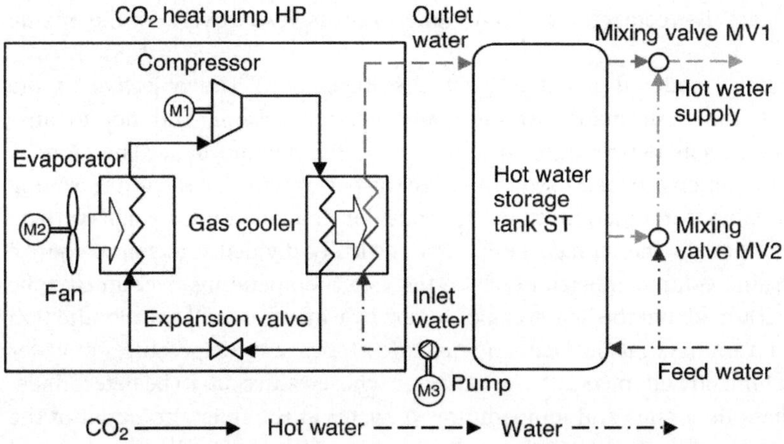

Figure 9.1 Configuration of CO_2 heat pump water heating system.

a shortage in the hot water supply. In order to overcome these defects of the conventional system simultaneously, it is necessary to restore the large gradient of the temperature distribution in the storage tank. For this purpose, it is considered to be effective to extract tepid water from the side of the storage tank. In fact, such revised systems have been developed and commercialized. However, it is never clarified how the performance enhancement of a water heating system can be attained. Here, the revised system is also investigated in addition to the conventional one.

9.3 System Modeling [6]

The system modeling is conducted as follows.

A simplified static model is adopted for the CO_2 heat pump. Although the CO_2 heat pump includes the aforementioned four components, they are not taken into account explicitly, and it is expressed by one model. The mass flow rates and temperatures of water at the inlet and outlet, heat output, and power consumption are adopted as basic variables whose values are to be determined. The mass and energy balance relationships as well as the energy input and output relationship are adopted as basic equations to be satisfied. The remaining equations to be considered are approximate functions of the power consumption and heat pump COP, and they are expressed in relation to the ambient air and inlet/outlet water temperatures. Although detailed explanation about this modeling is omitted here, the modeling results in a set of nonlinear algebraic equations, which is expressed by:

$$\boldsymbol{f}_{\mathrm{HP}}(\boldsymbol{y}_{\mathrm{HP}}(t), t) = \mathbf{0} \tag{9.1}$$

at each time t, where $\boldsymbol{f}_{\mathrm{HP}}$ is the vector for the aforementioned equations, and $\boldsymbol{y}_{\mathrm{HP}}$ is the vector for the aforementioned variables.

Similarly, a set of nonlinear algebraic equations for the mixing valve is expressed by:

$$\boldsymbol{f}_{\mathrm{MV}}(\boldsymbol{y}_{\mathrm{MV}}(t), t) = \mathbf{0} \tag{9.2}$$

where $\boldsymbol{f}_{MV}$ and $\boldsymbol{y}_{MV}$ are the vectors for equations and variables, respectively, for the mixing valve.

A detailed dynamic model is adopted for the storage tank. The objective of the study is to analyze the performance of the water heating system, and not to analyze the three-dimensional temperature distribution in the storage tank. Therefore, a one-dimensional simulation model is used to analyze the performance of the water heating system in a reasonable computation time. To consider the one-dimensional vertical temperature distribution in the storage tank, it is vertically divided into many control volumes with the same volume, in each of which the water temperature is assumed to be uniform. It is also assumed that the heat transfer occurs by water flow and heat conduction as well as heat loss from the tank surface. The mass flow rates and temperatures of water for each control volume are adopted as basic variables whose values are to be determined. In addition, the mass flow rates and temperatures of water at the inlet and outlet of the top and bottom of the storage tank are adopted as variables. The mass and energy balance relationships for each control volume are adopted as basic equations to be satisfied. Although detailed explanation about the modeling is omitted here, the modeling results in a set of nonlinear differential algebraic equations, which is expressed by:

$$\boldsymbol{f}_{ST}(\boldsymbol{x}_{ST}(t), \dot{\boldsymbol{x}}_{ST}(t), \boldsymbol{y}_{ST}(t), t) = \boldsymbol{0} \tag{9.3}$$

where $\boldsymbol{f}_{ST}$ is the vector for the aforementioned equations, $\boldsymbol{x}_{ST}$ is the vector for the variables with their derivatives, i.e. the temperatures of water for all the control volumes, $\boldsymbol{y}_{ST}$ is the vector for all the other variables without their derivatives, $\dot{\boldsymbol{x}}_{ST}$ is the derivative of $\boldsymbol{x}_{ST}$ with respect to time t.

At the connection points among the CO_2 heat pump, mixing valves, and storage tank, connection conditions are taken into account to equalize the values of the corresponding variables. The outlet water temperature is given as an operating condition. The feed water temperature and the mass flow rate and temperature of hot water from the storage tank to the tapping site are given as boundary conditions. The ambient air temperature is given as an ambient condition.

9.4 Numerical Solution [16]

The aforementioned modeling for the performance analysis by numerical simulation is conducted by a building block approach as follows: The component models for the CO_2 heat pump, mixing valve, and storage tank, and the substance model for water are defined independently; The system model is composed of the component and substance models as well as the connection, operating, boundary, and ambient conditions.

The equations for the CO_2 heat pump and mixing valve are static, while those for the storage tank are dynamic. Therefore, the modeling of the system results in a set of nonlinear differential algebraic equations, which is expressed by:

$$\left.\begin{array}{l} \boldsymbol{f}(\boldsymbol{x}(t), \dot{\boldsymbol{x}}(t), \boldsymbol{y}(t), t) = \boldsymbol{0} \\ \boldsymbol{x}(t_0) = \boldsymbol{x}_0 \end{array}\right\} \tag{9.4}$$

where $\boldsymbol{f}$ is the vector for all the equations composed of $\boldsymbol{f}_{HP}$ of Eq. (9.1), $\boldsymbol{f}_{MV}$ of Eq. (9.2), and $\boldsymbol{f}_{ST}$ of Eq. (9.3) as well as the connection, operating, boundary, and ambient conditions, $\boldsymbol{x}$

is the vector for the variables with their derivatives composed of $\boldsymbol{x}_{ST}$ in Eq. (9.3), $\boldsymbol{y}$ is the vector for all the other variables without their derivatives composed of $\boldsymbol{y}_{HP}$ in Eq. (9.1), $\boldsymbol{y}_{MV}$ in Eq. (9.2), and $\boldsymbol{y}_{ST}$ in Eq. (9.3), $\dot{\boldsymbol{x}}$ is the derivative of $\boldsymbol{x}$ with respect to time t, and $\boldsymbol{x}_0$ is the initial value of $\boldsymbol{x}$ at the initial time t_0.

The set of nonlinear differential algebraic equations expressed by Eq. (9.4) is solved numerically by a hierarchical combination of the Runge–Kutta and Newton–Raphson methods. A concrete solution algorithm is shown briefly here. For a value of the sampling time interval Δt, the Runge–Kutta method is used to derive the values of $\boldsymbol{y}(t)$ and $\boldsymbol{x}(t+\Delta t)$ from that of $\boldsymbol{x}(t)$ at any time t. A common formula for this purpose is as follows:

$$\boldsymbol{f}(\boldsymbol{x}(t)+\dot{\boldsymbol{x}}_{[r]}k_{[r+1]}\Delta t, \dot{\boldsymbol{x}}_{[r+1]}(t), \boldsymbol{y}_{[r+1]}(t), t+k_{[r+1]}\Delta t)=\boldsymbol{0}$$
$$(r=0,1,2,\cdots) \quad (9.5)$$

where k is the constant, and the subscript $[r]$ denotes the number of applications of the formula. For example, $k_{[1]}=0$, $k_{[2]}=1/2$, $k_{[3]}=1/2$, and $k_{[4]}=1$ according to the Runge–Kutta formula considering the fourth order of Δt. For each application, the values of $\dot{\boldsymbol{x}}_{[r+1]}$ and $\boldsymbol{y}_{[r+1]}$ are derived using the following equation based on the Newton–Raphson method:

$$\begin{aligned}
\begin{Bmatrix} \dot{\boldsymbol{x}}_{[r+1](s+1)}(t) \\ \boldsymbol{y}_{[r+1](s+1)}(t) \end{Bmatrix} &= \begin{Bmatrix} \dot{\boldsymbol{x}}_{[r+1](s)}(t) \\ \boldsymbol{y}_{[r+1](s)}(t) \end{Bmatrix} \\
&- \left[\partial \boldsymbol{f}\left(\boldsymbol{x}(t)+\dot{\boldsymbol{x}}_{[r]}k_{[r+1]}\Delta t, \dot{\boldsymbol{x}}_{[r+1](s)}(t), \boldsymbol{y}_{[r+1](s)}(t), t+k_{[r+1]}\Delta t\right)/\partial \dot{\boldsymbol{x}}_{[r+1]}, \right. \\
&\left. \partial \boldsymbol{f}\left(\boldsymbol{x}(t)+\dot{\boldsymbol{x}}_{[r]}k_{[r+1]}\Delta t, \dot{\boldsymbol{x}}_{[r+1](s)}(t), \boldsymbol{y}_{[r+1](s)}(t), t+k_{[r+1]}\Delta t\right)/\partial \boldsymbol{y}_{[r+1]}\right]^{-1} \\
&\times \boldsymbol{f}(\boldsymbol{x}(t)+\dot{\boldsymbol{x}}_{[r]}k_{[r+1]}\Delta t, \dot{\boldsymbol{x}}_{[r+1](s)}(t), \boldsymbol{y}_{[r+1](s)}(t), t+k_{[r+1]}\Delta t)
\end{aligned}$$
$$(s=0,1,2,\cdots) \quad (9.6)$$

where the subscript (s) denotes the number of repeats for the convergence calculation. After the first application of the formula, the value of $\boldsymbol{y}(t)$ is derived as $\boldsymbol{y}_{[1]}$. In addition, after all the applications, the value of $\boldsymbol{x}(t+\Delta t)$ is derived from those of $\boldsymbol{x}(t)$ and $\dot{\boldsymbol{x}}_{[r+1]}$ ($r=0$, 1, 2, ...). For example, $\boldsymbol{x}(t+\Delta t)=\boldsymbol{x}(t)+(\dot{\boldsymbol{x}}_{[1]}+2\dot{\boldsymbol{x}}_{[2]}+2\dot{\boldsymbol{x}}_{[3]}+\dot{\boldsymbol{x}}_{[4]})\Delta t/6$ according to the Runge–Kutta formula considering the fourth order of Δt.

9.5 Conditions for Performance Analysis and Optimization

Table 9.1 shows the specifications of the CO_2 heat pump water heating system used commonly in the performance analysis and optimization. The values of model parameters included in the equations are estimated based on measured data for existing devices. The rated heat output of the CO_2 heat pump is set at 4.5 kW. As an example, Figure 9.2 shows measured values and approximate functions for the power consumption, COP, and the resultant heat output of the CO_2 heat pump in relation to the inlet water temperature for the air and outlet water temperatures of 16 and 85°C, respectively. Here, each value is relative to its rated one for the air and inlet/outlet water temperatures of 16, 17, and 65°C, respectively. The volume of the storage tank is set at 370L.

Table 9.1 Specifications of a CO_2 heat pump water heating system.

Equipment	Specification	Value
CO_2 heat pump	Rated heat output	4.50 kW
Hot water storage tank	Volume	370 L
	Height	1.43 m
	Diameter	0.56 m
	Overall heat transfer coefficient	0.80 W m^{-2} °C^{-1}

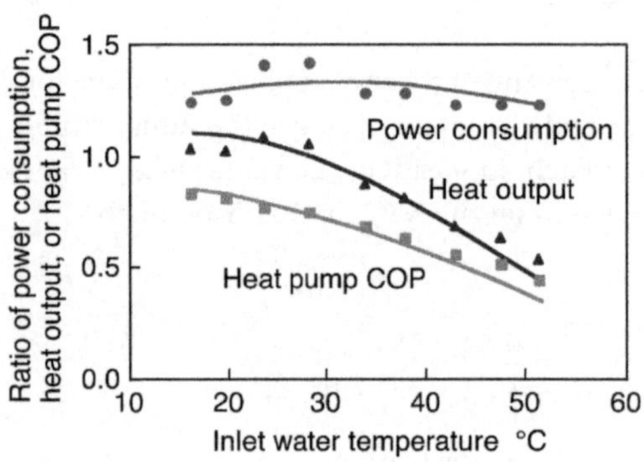

Figure 9.2 Performance characteristics of CO_2 heat pump.

Table 9.2 Ambient conditions.

		Unit (°C)
Season	**Ambient air**	**City water**
Summer	25	24
Mid-season	16	17
Winter	7	9

As the ambient conditions, the ambient air and city water temperatures in three seasons are set as shown in Table 9.2, which are prescribed by the Japan Refrigeration and Air Conditioning Industry Association [17]. These values are assumed to be constant throughout the days.

To analyze the performance under a periodically steady state, the hourly changes in the flow rate and temperature of the standardized hot water demand are given as shown in Figure 9.3, which is also prescribed by Japan Refrigeration and Air Conditioning Industry Association [17]. Here, the height of each vertical line means the flow rate, as indicated. The temperature is shown above each vertical line. In addition, the thickness of each vertical line means the duration. The heat for the total hot water demand is 46.15 MJ d^{-1} for the aforementioned city water temperature in mid-season.

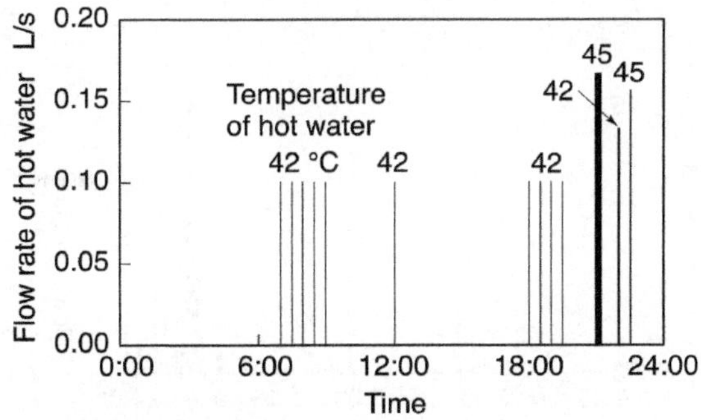

Figure 9.3 Hourly change in hot water demand.

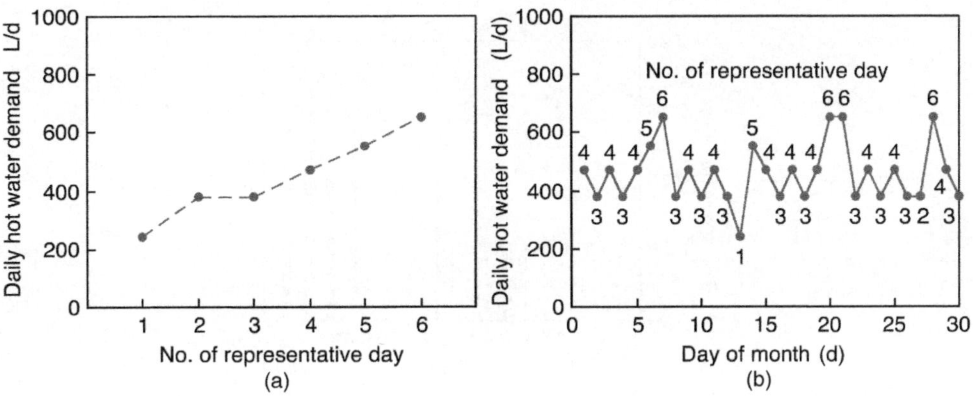

Figure 9.4 Daily change in hot water demand: (a) daily demands on six representative days; (b) daily change on 30 consecutive days.

To analyze the performance under unsteady state, a month composed of 30 consecutive days which are categorized into six representative days is set [18]. On each representative day, an hourly change in a simulated hot water demand is prescribed. Figure 9.4a, and b show the daily hot water demands on the six representative days and the daily change in the hot water demand on the 30 consecutive days, respectively. The 1st and 2nd representative days correspond to holidays with smaller and larger hot water demands, respectively, on which days residents are out of the house. The 3rd and 4th representative days correspond to weekdays with smaller and larger hot water demands, respectively. The 5th and 6th representative days correspond to holidays with smaller and larger hot water demands, respectively, on which days residents are in the house. As examples, Figures 9.5a–c show the hourly changes in the hot water demand on the 3rd, 4th, and 6th representative days, respectively. Here, the height and thickness of each vertical line means the flow rate and duration, respectively. The temperature is set at 42°C.

The system is assumed to be operated in the charging and tapping modes independently during the night time and daytime, respectively. In the charging mode, the outlet water temperature of the CO_2 heat pump is set at an appropriate value selected from 65 to 85°C. In addition, the CO_2 heat pump is started up at an appropriate time during the night. On the other hand, it is shut down with a shutdown condition satisfied at an appropriate time before starting hot water supply. The shutdown condition is that the inlet water temperature

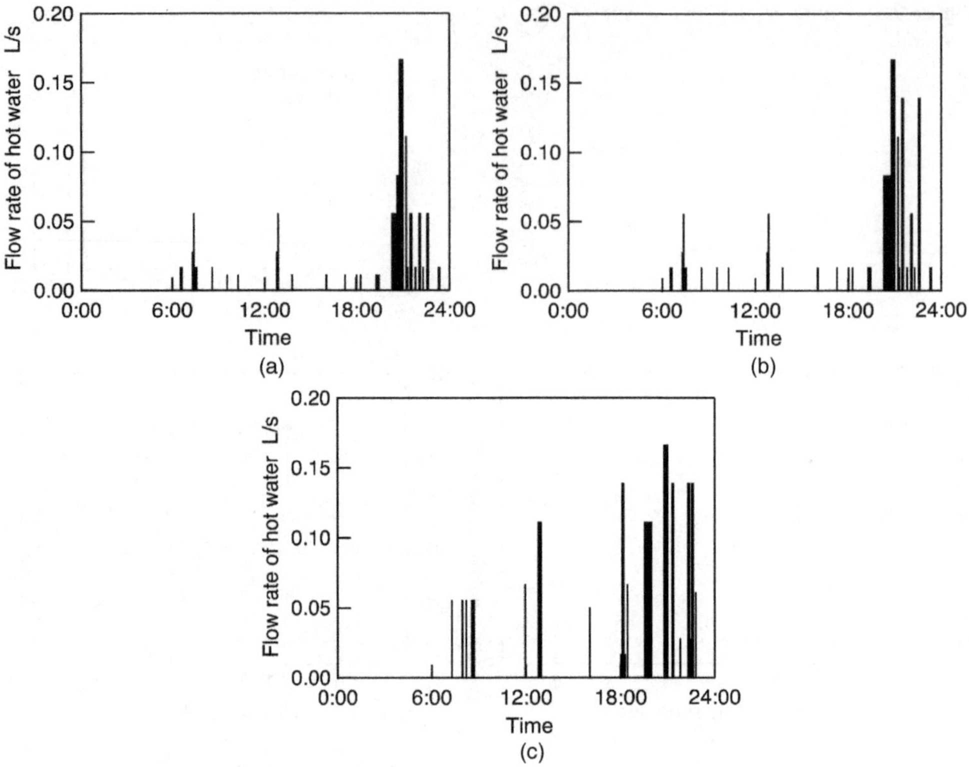

Figure 9.5 Hourly change in hot water demand on three representative days: (a) 3rd representative day; (b) 4th representative day; (c) 6th representative day.

of the CO_2 heat pump attains an appropriate value selected from 30 to 50°C. As the initial condition, the temperature distribution of water in the storage tank at 0:00 on the 1st day is set appropriately.

The number of control volumes for the storage tank is set at 200, and the sampling time interval for the Runge–Kutta method is set at 10 and 180 seconds for the cases with and without water flow, respectively.

9.6 Performance Analysis Under Periodically Steady State [7, 9]

In this section, as a fundamental analysis, performance analysis is conducted under periodically steady state when a constant energy demand arises every day. Here, the standardized hot water demand prescribed by the Japan Refrigeration and Air Conditioning Industry Association shown in Figure 9.3 is used for this purpose. The CO_2 heat pump is started up at 02:00 every day.

First, it is investigated how long it takes to get to the periodically steady state when the CO_2 heat pump water heating system starts to be used from an initial state. Here, it is

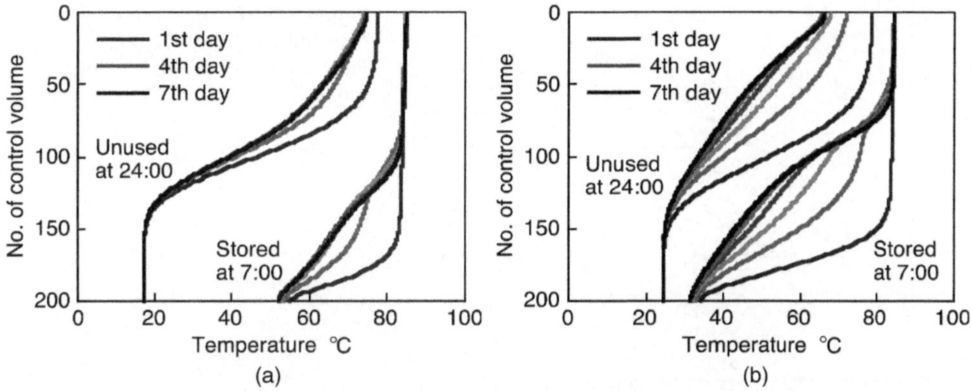

Figure 9.6 Daily changes in temperature distributions in storage tank: (a) 07:00 after heat pump operation; (b) 24:00 after hot water supply.

assumed that the temperature of all the control volumes of the storage tank is equal to that of the city water. Figure 9.6 shows the daily changes in the temperature distributions of hot water in the storage tank at 07:00 after heat pump operation and at 24:00 after hot water supply. Figures (a) and (b) correspond to the condition with outlet water temperature during heat pump operation of 85°C. Figure (a) corresponds to the conditions with the inlet water temperature for heat pump shutdown of 50°C in mid-season, while Figure (b) corresponds to the conditions with the inlet water temperature for heat pump shutdown of 30°C in summer. In Figure (a), since the heat for the daily hot water demand in mid-season is relatively large, and the inlet water temperature for heat pump shutdown is relatively high, the temperature distributions converge in three days. In Figure (b), since the heat for the daily hot water demand in summer is relatively small, and the inlet water temperature for heat pump shutdown is relatively low, it takes five days until the temperature distributions converge. In the following all the performance analyses under periodically steady state, the numerical simulation is conducted for eight days.

Second, as the base case, performance analysis is conducted using the following conditions. The ambient conditions in mid-season are used, and the outlet water temperature for heat pump operation and the inlet water temperature for heat pump shutdown are set at 85 and 50°C, respectively. Figure 9.7a, and b show the changes in the temperature distribution in the storage tank in the charging and tapping modes, respectively, obtained on the 8th day. Since the outlet water with a temperature of 85°C enters the top of the storage tank and the inlet water temperature rises up to 50°C in the charging mode, the temperature gradient changes with the vertical position, but its change is gradual. The temperature gradient becomes small gradually in the tapping mode, and the area of tepid water expands up to about two thirds of the height of the storage tank. Figure 9.8a, and b show performance characteristics of the system. Figure (a) shows the temperature distributions in the storage tank at 07:00 after heat pump operation and at 24:00 after hot water supply, and these are the same with those shown in Figure 9.7, while Figure (b) shows the hourly change in heat pump COP. During the heat pump operation, water in the storage tank is extracted from the bottom and enters the CO_2 heat pump; the hourly change in the heat pump COP depends on the temperature distribution at a lower part of the storage tank at 24:00. From Figures

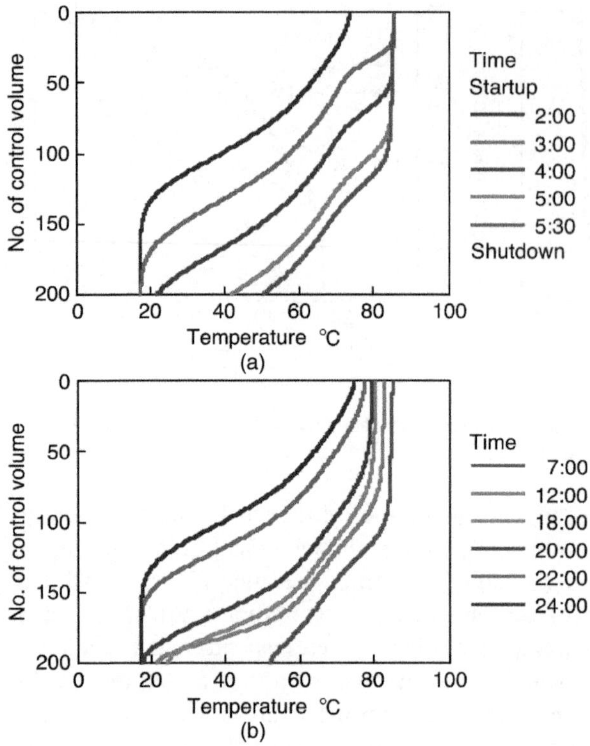

Figure 9.7 Change in temperature distribution in storage tank: (a) charging mode; (b) tapping mode.

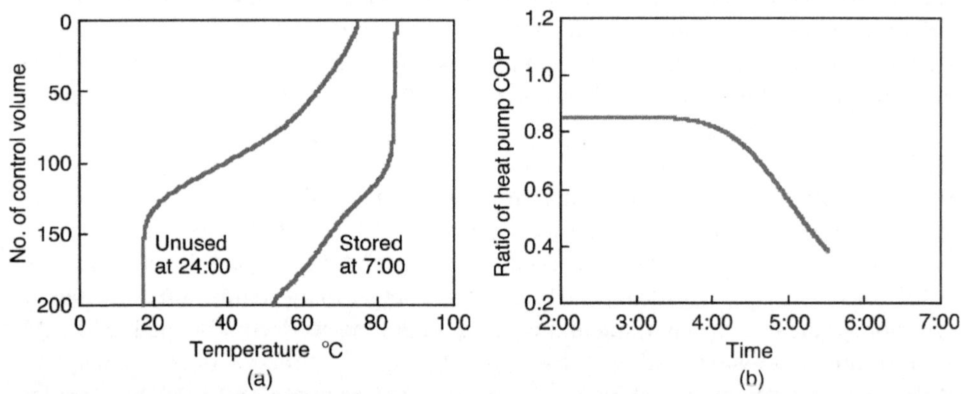

Figure 9.8 Performance characteristics in base case: (a) hot water stored at 07:00 and unused at 24:00; (b) hourly change in heat pump COP.

(a) and (b), the volumes of hot water stored at 07:00 and unused at 24:00, and the heat pump COP can be evaluated.

Third, the influences of ambient and operating conditions on the performance characteristics are investigated. Figure 9.9 shows performance characteristics of the system when the season changes, namely the ambient air and city water temperatures change. The volume of hot water stored at 07:00 in winter is the largest, and that in summer is the smallest. On

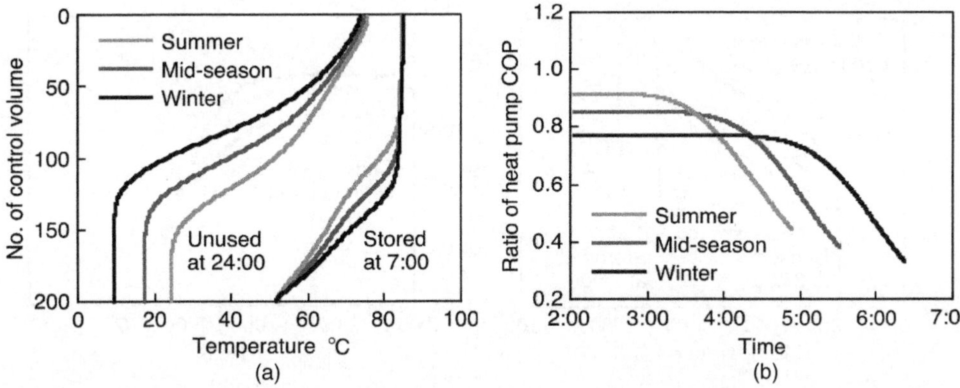

Figure 9.9 Influence of season on performance characteristics: (a) hot water stored at 07:00 and unused at 24:00; (b) hourly change in heat pump COP.

the other hand, the volume of hot water unused at 24:00 in summer is the largest, and that in winter is the smallest. The temperature gradient in summer is the smallest, and that in winter is the largest. As aforementioned, the hourly change in the heat pump COP depends on the temperature distribution at a lower part of the storage tank at 24:00. Generally, the heat pump COP increases with an increase in the ambient air temperature, and decreases with an increase in the inlet water temperature. In this case, the influence of the ambient air temperature is greater than that of the city water temperature. Thus, the heat pump COP in summer is the highest, and that in winter is the lowest. Figure 9.10 shows performance characteristics of the system when the outlet water temperature during heat pump operation changes as 85, 75, and 65°C. The volumes of hot water stored at 07:00 and unused at 24:00 with the outlet water temperature of 85°C is the largest, and that with the outlet water temperature of 65°C is the smallest. The temperature gradient does not depend significantly on the outlet water temperature. Generally, the heat pump COP increases with any decrease in the outlet water temperature. Thus, the heat pump COP with the outlet water temperature of 65°C is the highest, and that with the outlet water temperature of 85°C is the lowest. Figure 9.11 shows performance characteristics of the system when the inlet water temperature for heat pump shutdown changes as 50, 40, and 30°C. The volumes of hot water stored at 07:00 and unused at 24:00 with the inlet water temperature for heat pump shutdown of 50°C is the largest, and that with the inlet water temperature for heat pump shutdown of 30°C is the smallest. The temperature gradient increases with the inlet water temperature for heat pump shutdown. The heat pump COP with the inlet water temperature for heat pump shutdown of 30°C is the highest, and that with the inlet water temperature for heat pump shutdown of 50°C is the lowest.

Finally, the following daily system performance values are evaluated based on the results shown in Figures 9.9–9.11: heat pump COP, storage and system efficiencies, and volumes of stored and unused hot water. Here, the heat pump COP is defined as the ratio of the daily heat output to daily power consumption, the storage efficiency is defined as the ratio of daily heat supply to daily heat output, and the system efficiency is defined as the ratio of daily heat supply to daily power consumption, which is equal to the product of the heat

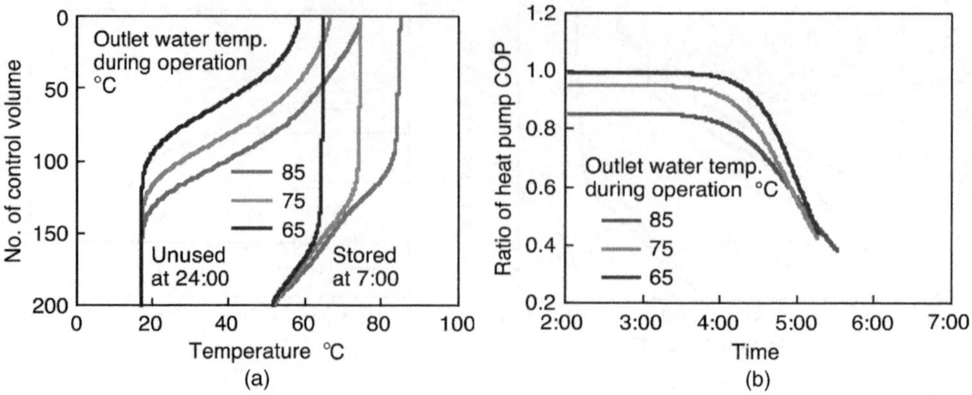

Figure 9.10 Influence of outlet water temperature during heat pump operation on performance characteristics: (a) hot water stored at 07:00 and unused at 24:00; (b) hourly change in heat pump COP.

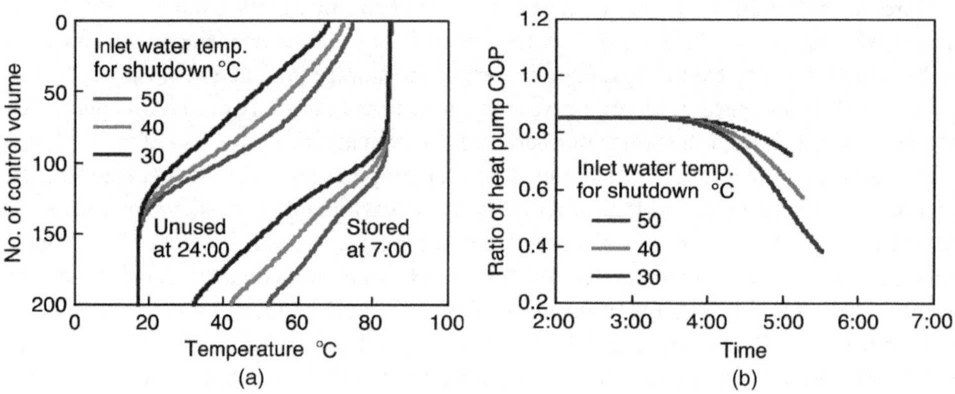

Figure 9.11 Influence of inlet water temperature for heat pump shutdown on performance characteristics: (a) hot water stored at 07:00 and unused at 24:00; (b) hourly change in heat pump COP.

pump COP and storage efficiency. The volumes of stored and unused hot water are evaluated as the volumes of hot water with a temperature of 42°C obtained by mixing the hot water with temperatures higher than 42°C and the city water. Among these system performance values, the system efficiency and the volume of unused hot water are the most important, because the former and latter are criteria for energy saving and reliable hot water supply, respectively. Figure 9.12 shows the relationship between the system efficiency and the volume of unused hot water for all the combinations of the values for the outlet water temperature during heat pump operation and the inlet water temperature for shutdown in all the seasons. These system performance values are relative to those with the outlet water temperature during heat pump operation of 85°C and the inlet water temperature for heat pump shutdown of 50°C in mid-season.

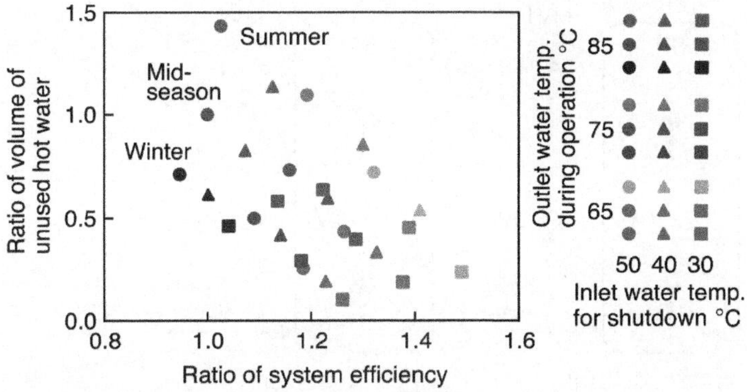

Figure 9.12 Relationship between system efficiency and volume of unused hot water.

9.7 Performance Enhancement by Extracting Tepid Water [13]

In this section, the performance of a CO_2 heat pump water heating system with extracting tepid water from the side of the storage tank is analyzed by numerical simulation. A performance analysis for the conventional and revised systems is conducted under periodically steady state, and their system performance values are compared. Through this analysis, the effect of extracting tepid water from the side of the storage tank on the performance enhancement is investigated.

It is necessary to determine the strategy as to how tepid hot water is extracted. Here, the position for water extraction is set at the nth control volume CVn. On the assumption that the temperature of hot water supplied to the tapping site is prescribed exactly, the following strategy for hot water supply is adopted: If the water in CVn can be used for hot water supply by mixing it with feed water or water extracted from the top, it has priority over water extracted from the top. Based on this strategy, the following three modes for hot water supply shown in Figure 9.13 are set using the temperature T_{STn} of water in CVn, and thresholds $\overline{T}$ and $\underline{T}$, which are higher and lower, respectively, slightly than the temperature for hot water supply in consideration of the error in measuring temperature:

- Mode A: In case that $T_{STn} > \overline{T}$, the heat of water in CVn can be used for hot water supply by mixing it with water with a lower temperature, and water in CVn mixed with feed water is supplied to the tapping site.
- Mode B: In case that $\underline{T} < T_{STn} \leq \overline{T}$, the temperature of water in CVn may not be suitable for hot water supply in modes A and C in consideration of the error in measuring temperature, and water in CV1 mixed with feed water is supplied to the tapping site. This mode denotes a conventional one without water extraction.
- Mode C: Even in case that $T_{STn} \leq \underline{T}$, the heat of water in CVn can be used for hot water supply by mixing it with water with a higher temperature, and water in CVn mixed with water in CV1 is supplied to the tapping site.

Modes A to C are switched to another one using the mixing valves MV1 and MV2 based on the temperature T_{STn}. When the system is operated in the charging mode without hot water demand during the night time, the temperature T_{STn} rises with time, and the mode

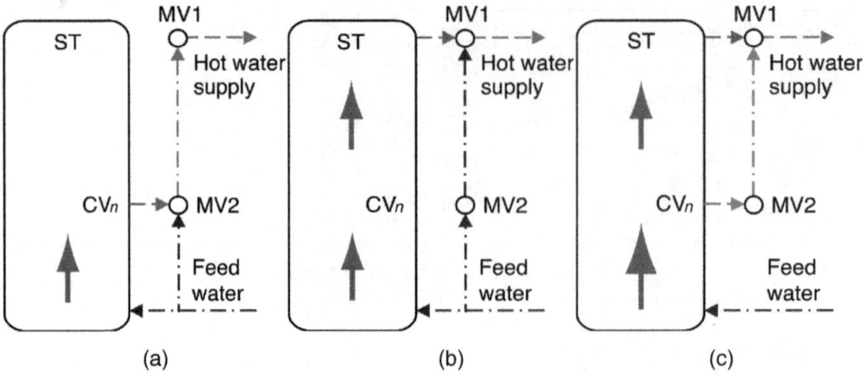

Figure 9.13 Modes for hot water supply: (a) mode A; (b) mode B; (c) mode C.

changes from C to A. On the other hand, when the system is operated in the tapping mode without additional heat pump operation during the daytime, the temperature T_{STn} drops with time, and the mode changes from A to C.

Here, water is extracted from the 130th control volume from the top of the storage tank, i.e. $n = 130$. The thresholds $\overline{T}$ and $\underline{T}$ for switching the modes for hot water supply are set at the temperature of hot water demand plus and minus 5°C, respectively.

First, the temperature distribution in the storage tank obtained for the revised system is compared with that for the conventional system in the base case. Figure 9.14a, and b show the changes in the temperature distribution in the storage tank in the charging and tapping modes, respectively, obtained on the 8th day under the condition that the inlet water temperature for heat pump shutdown is 50°C, as an example, for the revised system. As compared with Figure 9.7 in the conventional system, in the revised system, the temperature gradient changes with the vertical position, and its change is marked in the charging mode, which is different from that in the conventional system. This is because the area of tepid water becomes small. The temperature distribution changes complexly with time because of water extraction in the tapping mode. Before 20:00, the temperature distribution at the position higher than that for water extraction hardly changes with time, while the temperature distribution at the position lower than that for water extraction changes significantly. After 20:00, however, the temperature distribution at the position higher than that for water extraction also changes with time since the volume of stored hot water decreases, and the temperature gradient increases because of water extraction. As a result, the area of tepid water is small, and expands up to only about one fourth of the height of the storage tank. In addition, the temperature drop at the highest part of the storage tank is small.

Next, the performance characteristics of the revised system are compared with those of the conventional system in the base case. Figure 9.15a, and b compare the temperature distributions of the storage tank at 07:00 after heat pump operation and at 24:00 after hot water supply, and the hourly change in heat pump COP, respectively, obtained on the 8th day under the condition that the inlet water temperature for heat pump shutdown is 50°C, as an example, for the revised and conventional systems. According to Figure (a), in the conventional system, since the inlet water temperature rises up to 50°C, the temperature gradient is large in lower temperature ranges and is small in higher temperature ranges. As a result,

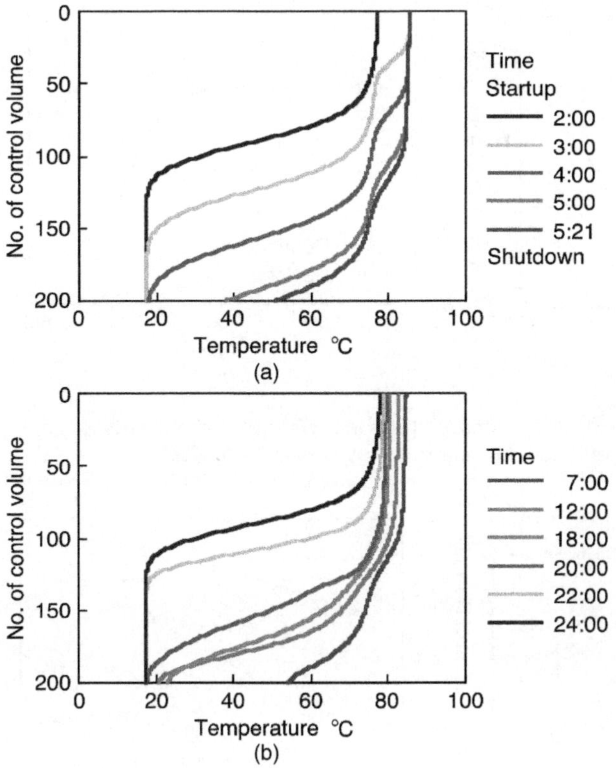

Figure 9.14 Change in temperature distribution in storage tank for revised system: (a) charging mode; (b) tapping mode.

the difference in the temperature distributions for stored and unused hot water between the conventional and revised systems is not so large. Thus, the effect of water extraction on the increases in the volumes of stored and unused hot water under the condition that the inlet water temperature for heat pump shutdown is 50°C is not so large. According to Figure (b), in the conventional system, since the area of tepid water is large, the time when the inlet water temperature rises and the heat pump COP decreases correspondingly is long. In the revised system, on the other hand, since the area of tepid water is small, the time when the inlet water temperature rises and the heat pump COP decreases correspondingly is short. In addition, the time when the heat output of the CO_2 heat pump decreases is short, and resultantly the time when the CO_2 heat pump is operated is short.

Finally, the influence of the inlet water temperature for heat pump shutdown on the system performance values are investigated. Figure 9.16a, and b compare the system performance values obtained on the 8th day for the conventional and revised systems in relation to the inlet water temperature for heat pump shutdown. Figure (a) shows the heat pump COP, and the storage and system efficiencies, while Figure (b) shows the volumes of stored and unused hot water. All the values are relative to those for the conventional system under the condition that the inlet water temperature for heat pump shutdown is 50°C. According to Figure (a), with an increase in the inlet water temperature for heat pump shutdown,

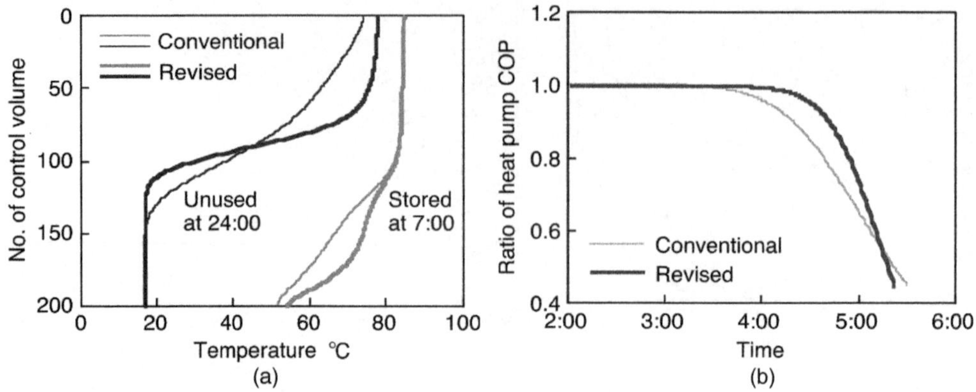

Figure 9.15 Comparison of revised and conventional systems in performance characteristics: (a) hot water stored at 07:00 and unused at 24:00; (b) hourly change in heat pump COP.

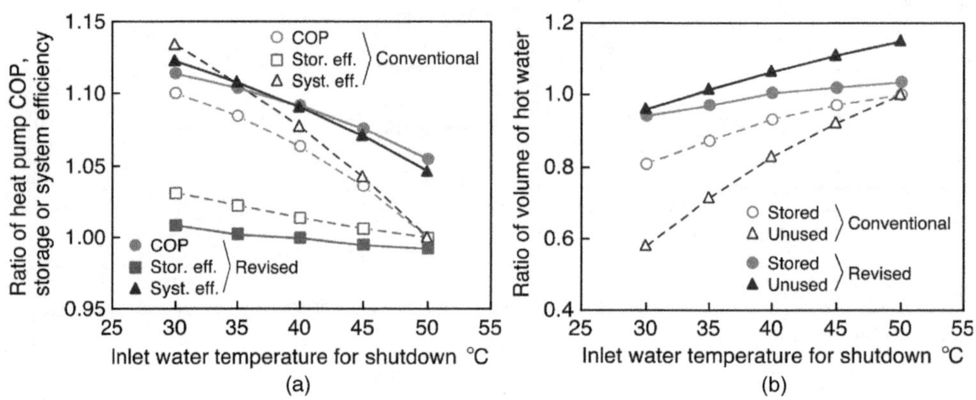

Figure 9.16 Influence of inlet water temperature for heat pump shutdown on system performance values in revised and conventional systems: (a) heat pump COP, and storage and system efficiencies; (b) volumes of stored and unused hot water.

the heat pump COP decreases in both the systems. In addition, the storage efficiency also decreases in both systems, because the average temperature in the storage tank rises. As a result, the system efficiency also decreases in both systems. On the other hand, the difference in the heat pump COP between both systems becomes large. In addition, the difference in the storage efficiency between both systems becomes small. As the result of these differences, the difference in the system efficiency between both systems becomes large, and the effect of water extraction on the increase in the system efficiency also becomes large. It should be noted that the difference in the system efficiency becomes negative under the condition that the inlet water temperature for heat pump shutdown is 30°C. According to Figure (b), with an increase in the inlet water temperature for heat pump shutdown, the volumes of stored and unused hot water in both systems increase. However, the differences in the volumes of stored and unused hot water between both systems become small, and the effect of water extraction on the increase in the volume of unused hot water also becomes small.

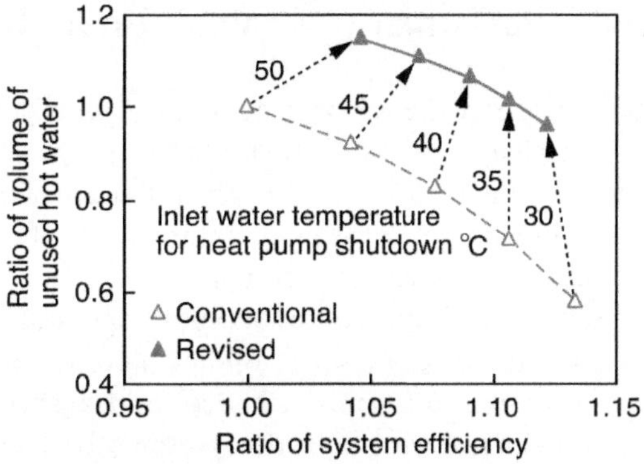

Figure 9.17 Trade-off relationships between system efficiency and volume of used hot water in revised and conventional systems.

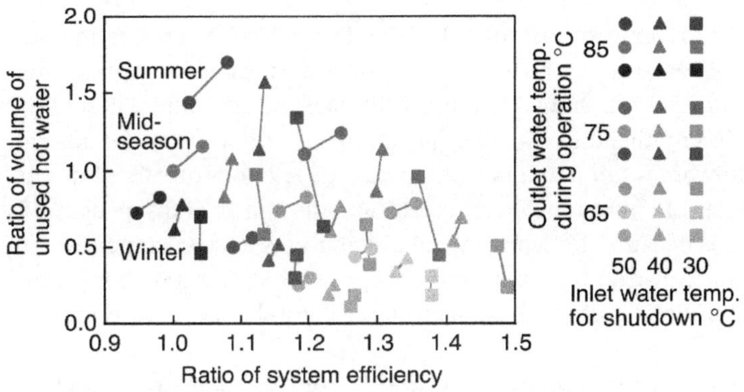

Figure 9.18 Relationships between system efficiency and volume of unused hot water in revised and conventional systems.

Figure 9.17 shows the relationships between the system efficiency and the volume of unused hot water in both systems with the inlet water temperature for heat pump shutdown as a parameter. There exist trade-off relationships between the criteria in both systems. The water extraction moves the trade-off relationship in the upper and right direction, which means performance enhancement in terms of these criteria. For example, under the condition that the inlet water temperature for heat pump shutdown is 50°C, water extraction increases system efficiency by more than 10% with the volume of unused hot water unchanged. In addition, under the condition that the inlet water temperature for shutdown is 40°C, water extraction increases the volume of unused hot water by more than 20% with system efficiency unchanged. Figure 9.18 shows the relationships between system efficiency and the volume of unused hot water in both systems for all combinations of the values for the outlet water temperature during heat pump operation, and inlet water temperature for shutdown in all seasons. The line connecting two points shows the improvement of performance by water extraction, and the upper and lower points correspond to the revised and conventional systems, respectively.

9.8 Performance Analysis Under Unsteady State [11]

In this section, the performance of a CO_2 heat pump water heating system is analyzed under a daily change in a simulated monthly hot water demand shown in Figure 9.4 in the base case by numerical simulation. The influence of the daily change in the hot water demand on the daily changes in the temperature distributions in the storage tank as well as the system performance values are investigated.

Here, as the initial condition, the temperature distribution of water in the storage tank at 00:00 on the 1st day is set as follows: Since the 1st day corresponds to the 4th representative day as shown in Figure 9.4, the numerical simulation is conducted for the repeated daily hot water demand on the 4th representative day, and the temperature distribution of water in the storage tank at 00:00 obtained by the periodic steady state is adopted. This is because the monthly performance is evaluated by reducing the influence of the initial condition on it as much as possible.

First, the daily change in the temperature distributions in the storage tank is investigated. As examples, Figure 9.19a–d show the temperature distributions in the storage tank at 00:00, 06:00, and 24:00 on the 7th, 20th, 21st, and 28th days, respectively. These days correspond to the 6th representative day with the maximum daily hot water demand of 650 L d^{-1}. Although the daily hot water demand is the same, the temperature distributions are different, and the volumes of hot water stored at 06:00 and unused at 24:00 are also different, as shown later. This is because the temperature distributions at 00:00 are different, and they affect the temperature distributions within the range of 50–85 at 06:00. This means that the temperature distributions significantly depend on the daily change in the hot water demand.

Figure 9.20a, and b show the daily changes in the volumes of hot water stored at 06:00 and unused at 24:00, respectively. The figures include the values on the consecutive 30 days (case A) and those on the six representative days obtained independently under the periodically steady state (case B). As shown in Figure (b), the tendency of the daily change in the volume of unused hot water in case A coincides with that in case B. This is because the volume of unused hot water significantly depends on the daily hot water demand on the corresponding day. However, the difference in the volume of unused hot water between cases A and B changes daily. This is because the volume of unused hot water depends not only on the daily hot water demand but also on the volume of stored hot water, and the difference in the latter between cases A and B changes daily, as shown in Figure (a). The daily change in the volume of unused hot water in case A is larger than that in case B. This feature is not suitable, because it increases the possibility of shortage in hot water supply. On the other hand, as shown in Figure (a), the daily change in the volume of stored hot water in case A is delayed for a day in comparison with that in case B. This is because the volume of stored hot water depends on the daily hot water demand on the previous day. For example, when the daily hot water demand on the previous day is small, the gradient of the temperature distribution in the storage tank at 00:00 is small, and the volume of stored hot water is also small at 06:00. The daily change in the volume of stored hot water in case A is slightly smaller than that in case B.

Figure 9.21a–c shows the daily changes in the heat pump COP, and storage and system efficiencies, respectively. Here, the heat pump COP is defined as the ratio of the total heat

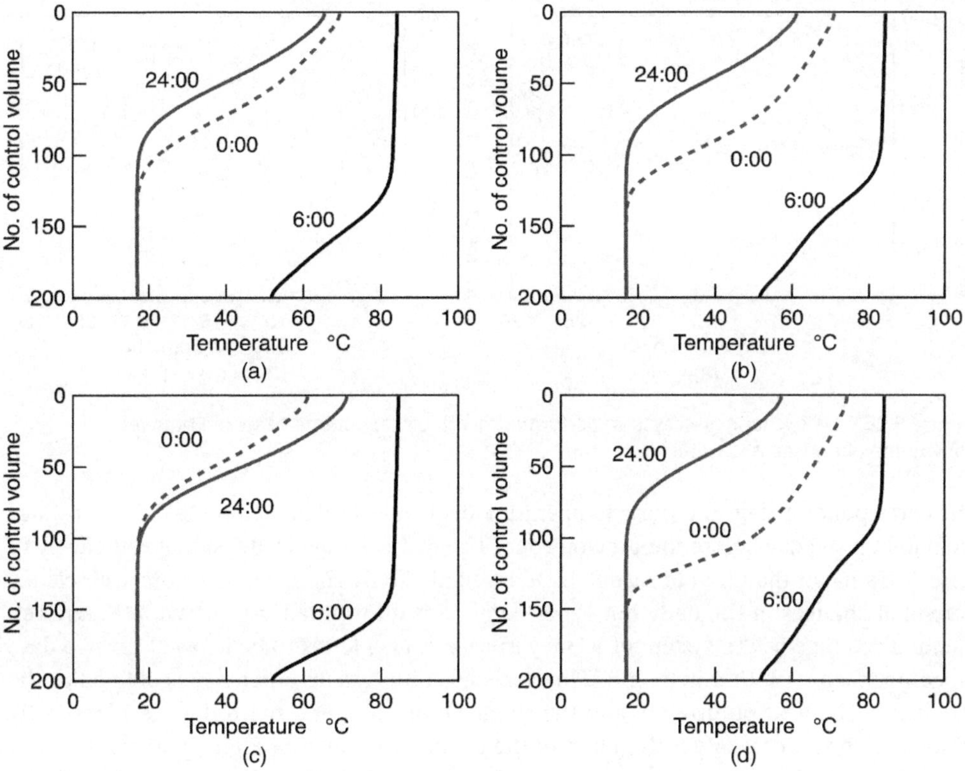

Figure 9.19 Daily changes in temperature distributions in storage tank: (a) 7th representative day; (b) 20th representative day; (c) 21st representative day; (d) 28th representative day.

output to the total power consumption on the corresponding day, the storage efficiency is defined as the ratio of the total hot water demand on the previous day to the total heat output on the corresponding day, and the system efficiency is defined as the ratio of the total hot water demand on the previous day to the total power consumption on the corresponding day, or the product of the heat pump COP and the storage efficiency. In addition, the values are relative to those obtained on the 1st day, or the 4th representative day in case B. The figures also include the values in cases A and B. As shown in Figure (a), the daily change in the heat pump COP in case A is delayed for a day in comparison with that in case B. This is because the heat pump COP significantly depends on the temperature distribution in the storage tank at 00:00, and this temperature distribution depends on the daily hot water demand on the previous day. For example, when the daily hot water demand on the previous day is small, the gradient of the temperature distribution in the storage tank at 00:00 is small, and the ratio of the operation time of the heat pump with high inlet water temperatures to the total operation time is large. The daily change in the heat pump COP in case A is slightly smaller than that in case B. As shown in Figure (b), the daily change in the storage efficiency in case A tends to be delayed for two days in comparison with that in case B. As defined previously, the storage efficiency depends on the temperature distributions in the storage tank during the daytime on the previous day and during the night time on

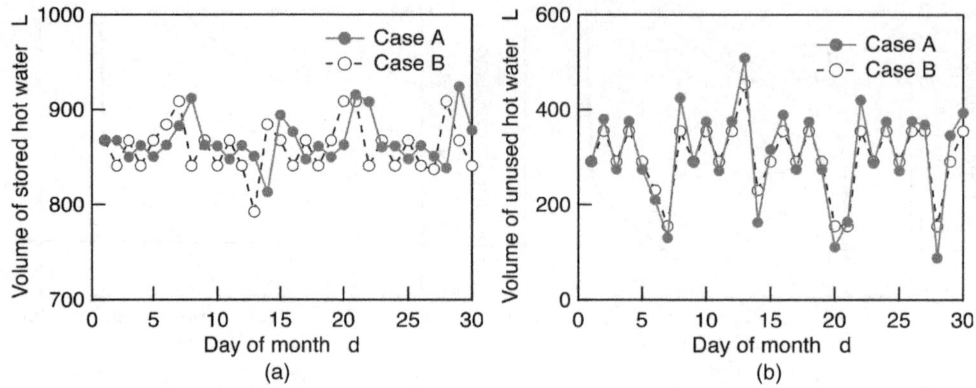

Figure 9.20 Daily changes in system performance values: (a) volume of stored hot water; (b) volume of unused hot water.

the corresponding day, and these temperature distributions depend on the daily hot water demand on the day before the previous one. The daily change in the storage efficiency in case A is smaller than that in case B. In addition, the daily change in the storage efficiency for small changes in the daily hot water demand is quite small. As shown in Figure (c), the daily change in the system efficiency in case A also tends to be delayed for two days in comparison with that in case B. This is because the system efficiency is defined as the product of the heat pump COP and the storage efficiency, and the influence of the daily change in the latter is larger than that in the former, as shown in Figures (a) and (b). The daily change in the system efficiency in case A is also smaller than that in case B. In addition, the daily change in the system efficiency for small changes in the daily hot water demand is also quite small.

Finally, monthly system performance values are evaluated. Table 9.3 shows the ratios of the monthly values of the heat pump COP, and storage and system efficiencies in case A to those in case B. The values in case A are evaluated based on the monthly total hot water demand, heat output, and power consumption, while those in case B are evaluated by averaging the daily hot water demand, heat output, and power consumption. The differences in all the values between cases A and B are within 1.0%, and are quite small. The difference in the heat pump COP is only 0.5%. The difference in the storage efficiency is only 0.9%. From the aforementioned discussions, the difference in the system efficiency between cases A and B is only 0.4%. These results show that the monthly values of the heat pump COP, and storage and system efficiencies in case A are evaluated approximately by averaging their daily values in case B.

9.9 Performance Estimation Under Unsteady State [15]

In the previous sections, several types of performance analyses are conducted by numerical simulation. However, it takes a long computing time to estimate daily changes in system performance values by numerical simulation with complex computation, and thus it is difficult to determine operating conditions optimally by numerical simulation, because

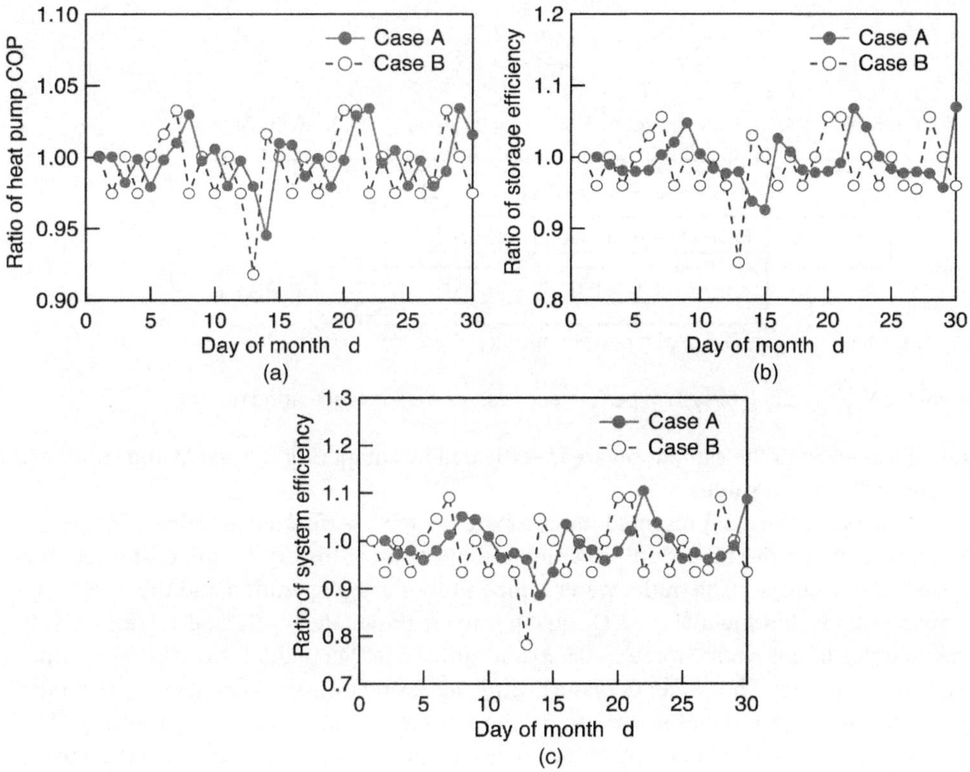

Figure 9.21 Daily changes in system performance values: (a) heat pump COP; (b) storage efficiency; (c) system efficiency.

Table 9.3 Ratio of system performance values in case A to those in case B.

Item	**Value**
Heat pump COP	1.005
Storage efficiency	0.991
System efficiency	0.996

the optimization needs to estimate daily changes in system performance values repeatedly under various operating conditions. Therefore, it is necessary to establish easier methods of estimating daily changes in system performance values accurately, and determining operating conditions optimally. In this section, a method of estimating daily changes in system performance values by neural network models is shown for a CO_2 heat pump water heating system. In addition, daily changes in system performance values are estimated under a daily change in a simulated monthly hot water demand shown in Figure 9.4, and the validity

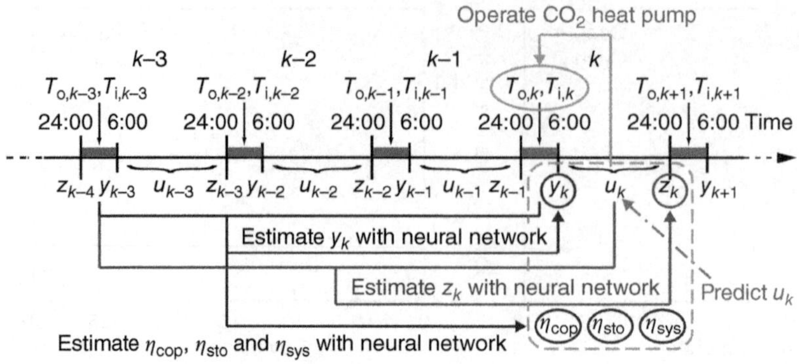

Figure 9.22 Estimation of system performance values based on operating history.

and effectiveness of the estimation are investigated by comparing estimated and simulated system performance values.

First, a procedure is presented to estimate system performance values accurately. Figure 9.22 shows the procedure in which the operational history on the past three days is used as an example. The outlet water temperature during operation and the inlet water temperature for shutdown of the CO_2 heat pump are designated by T_o and T_i, respectively. The volumes of hot water stored at 06:00 and unused at 24:00 are designated by y and z, respectively. The total hot water demand during the period from 06:00 to 24:00 is designated by u. The subscript k denotes a value on the kth day. In addition, the heat pump COP, storage efficiency, and system efficiency are designated by η_{cop}, η_{sto}, and η_{sys}, respectively. First, at 00:00 on the kth day, the volume of hot water stored at 06:00 on the kth day y_k is estimated using the outlet water temperature during heat pump operation, the inlet water temperature for heat pump shutdown, the volumes of stored and unused hot water, and the total hot water demand on the (k3)th to (k−1)th days as well as the candidates for the outlet water temperature during heat pump operation and the inlet water temperature for heat pump shutdown on the kth day. Next, the volume of hot water unused at 24:00 on the kth day z_k is also estimated using the estimated value for the volume of stored hot water y_k and the predicted value for the total hot water demand u_k on the kth day in addition to the aforementioned values. Finally, the heat pump COP η_{copk}, storage efficiency η_{stok}, and system efficiency η_{sysk} on the kth day are also estimated similarly as the volume of stored hot water y_k. This is based on the following reasons: The heat pump COP depends on the inlet water temperature, and the inlet water temperature depends significantly on the temperature distribution in the storage tank at 24:00; The storage efficiency depends on the temperature distribution in the storage tank throughout the day, and is roughly expressed by the temperature distributions in the storage tank at 06:00 and 24:00; The system efficiency is equal to the product of the heat pump COP and storage efficiency, and is also roughly expressed by the temperature distributions in the storage tank at 06:00 and 24:00.

Three-layered neural network models are used to estimate the system performance values. As aforementioned, each system performance value is estimated independently by the corresponding model. For long-term operation of existing systems, it is necessary to measure necessary data continuously and identify model parameter values repeatedly, and estimate system performance values correspondingly. Here, the estimation only for short-term operation is considered. In the input layer, the operating conditions, the volumes of stored and unused hot water, and the total hot water demand on the past days as well as the operating conditions on the current day are adopted commonly as the inputs to the model to estimate all the system performance values. The estimated volume of stored hot water and the predicted total hot water demand on the current day are adopted additionally to estimate the volume of unused hot water. In the other layers, each neuron has multiple inputs and single output, and converts the weighted sum of the inputs minus the threshold to the output by a response function. The hyperbolic tangent function is used as the response function to obtain positive and negative values from the output. Here, the value from the output ranges only from −1.0 to 1.0 by normalizing the values to the inputs and from the output in advance. The numbers of neurons for the neural network models used for the performance estimation are set as follows: The data on the past two days are used; The numbers of neurons in the input and output layers are 12 and 1, respectively, for the models to estimate the heat pump COP, storage and system efficiencies, and volume of stored hot water; The numbers of neurons in the input and output layers are 14 and 1, respectively, for the model to estimate the volume of unused hot water; The number of neurons in the hidden layer is 3 commonly for all models.

To estimate the system performance values by the neural network models, it is necessary to identify the values of model parameters, or weights and thresholds. The squared error between the estimated value and the corresponding measured value is evaluated for each pattern, and its summation for all the patterns is minimized as the objective function to identify the values of model parameters. Here, to secure the local optimality of solutions and make the convergence faster, the total error function for all patterns is minimized simultaneously. To identify the values of model parameters, the modal trimming method proposed for nonlinear programming problems is adopted as a global optimization one [19]. This method is composed of the following two procedures: A local optimal solution is searched to obtain a tentative global quasi-optimal one; A feasible solution with the value of the objective function equal to or smaller than that for the tentative global quasi-optimal one is searched to obtain an initial point for finding a better local optimal one. These procedures are repeated until a feasible solution with the value of the objective function equal to or smaller than that for the tentative global quasi-optimal one cannot be found, and the tentative global quasi-optimal one is adopted as the global quasi-optimal one. A local optimal solution is searched by a conventional gradient method. On the other hand, a feasible solution is searched by an extended Newton-Raphson method based on the Moore–Penrose generalized inverse of the Jacobi matrix of the objective function. The method can have a high possibility of deriving global optimal solutions, if it has the capability of global search for feasible ones.

It is necessary to use some system performance values to identify the values of model parameters. In applying the method of performance estimation to existing systems, measured data on system performance values must be used. Here, values obtained by numerical simulation are used in place of measured values. The heat pump is started up at 00:00 and 01:00, when the total hot water demand on the previous day is larger than or equal to and smaller than 500 $\mathrm{L\,d^{-1}}$, respectively. The outlet water temperature during heat pump operation is selected among 65, 75, and 85°C, and the inlet water temperature for heat pump shutdown is selected among 30, 40, and 50°C. The daily operating conditions are set by

Table 9.4 Operating conditions for identification and verification.

Case	Purpose	Hot water demand	Outlet water temperature during heat pump operation	Inlet water temperature for heat pump shutdown
1–54	Identification	E	CO	CI
55–63		P	CO	CI
64–66		P	VO	VI
67–69		P	CO	CI
70, 71		P	VO	VI
72	Verification	P	VO	VI

E: Each of six representative days; P: Pattern of consecutive 30 days; CO: Constant (each of 65, 75, and 85°C); CV: Variable (combination of 65, 75, and 85°C); CI: Constant (each of 30, 40, and 50°C); CI: Variable (combination of 30, 40, and 50°C).

combining these values. 72 cases are investigated by the numerical simulation. Table 9.4 shows the conditions on the outlet water temperature during heat pump operation and the inlet water temperature for heat pump shutdown in cases 1–72. Cases 1–71 are used to identify model parameter values, while case 72 is used to verify the validity of model parameter values. In cases 1–54, the numerical simulation is conducted for the periodically steady state on each of the six representative days shown in Figure 9.4a under each combination of the constant outlet and inlet water temperatures. In cases 55–63, the numerical simulation is conducted on the consecutive 30 days shown in Figure 9.4b under each combination of the constant outlet and inlet water temperatures. In cases 64–66, the numerical simulation is conducted on the consecutive days under variable outlet water temperature and each constant inlet water temperature. In cases 67–69, the numerical simulation is conducted on the consecutive days under each constant outlet water temperature and variable inlet water temperature. In cases 70–72, the numerical simulation is conducted on the consecutive days under variable outlet and inlet water temperatures.

Figure 9.23 shows the daily changes in the operating conditions and system performance values in case 70. Figure (a) shows the operating temperatures given in advance, and Figures (b) and (c) show the system efficiency, and the volumes of stored and unused hot water, respectively, estimated by the neural network models under the given operating temperatures. These figures also show the corresponding values obtained by numerical simulation. The system efficiency is shown as the ratio of the system efficiency to its value on the 1st day. The estimated system performance values coincide well with the simulated ones. This result shows that the values of model parameters are identified properly by the global optimization method, and that the system performance values are estimated with high accuracy.

Figure 9.24 shows the daily changes in the operating conditions and system performance values in case 72. Figures (a)–(c) show the same items as aforementioned. Although these simulated system performance values are not used to identify the values of model parameters, the estimated system performance values coincide well with the simulated ones. This result shows that the system performance values are estimated with high

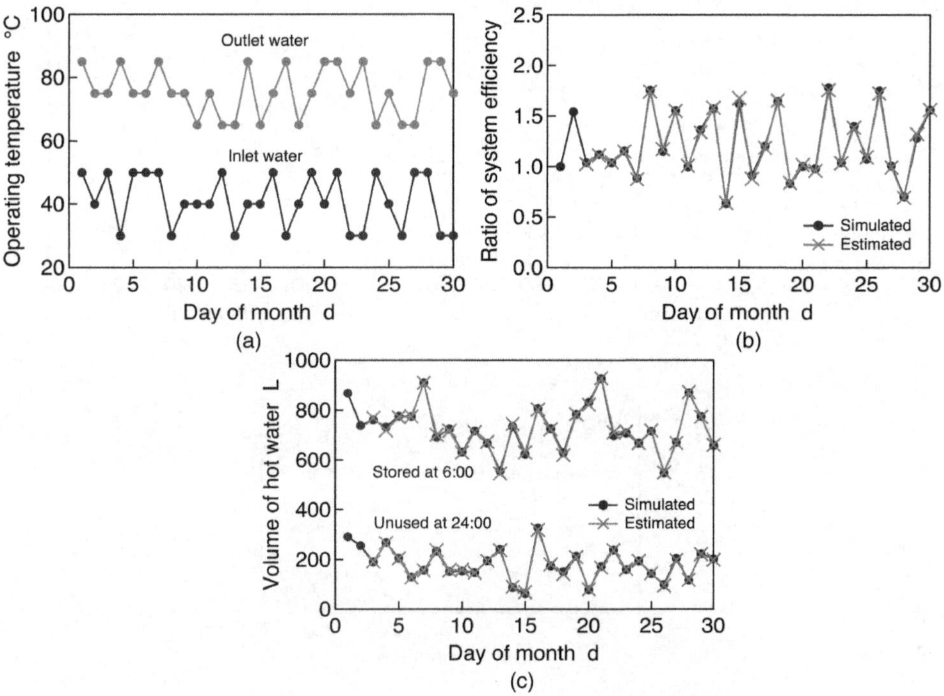

Figure 9.23 Daily changes in operation conditions and system performance values in case 70: (a) operating temperatures; (b) system efficiency; (c) volume of unused hot water.

accuracy by the same neural network models even under different daily changes in the operating conditions.

9.10 Performance Optimization Under Unsteady State [15]

In this section, a method of determining operating conditions optimally based on the system performance values obtained by the estimation is analyzed. In addition, the operating conditions are determined optimally under a daily change in a simulated monthly hot water demand shown in Figure 9.4, and the validity and effectiveness of the optimization are investigated by comparing the system performance values obtained by optimal and non-optimal operating conditions.

It is important to enhance the system efficiency and prevent the shortage in hot water supply. Thus, the system efficiency is maximized subject to a lower limit for the volume of hot water unused at 24:00. The outlet water temperature during heat pump operation and the inlet water temperature for heat pump shutdown are adopted as the variables, and their values are determined so as to attain the objective and satisfy the constraint. Here, the lower and upper limits for the outlet water temperature during heat pump operation are set at 65.0 and 85.0°C, respectively, and those for the inlet water temperature for heat pump shutdown are set at 30.0 and 50.0°C, respectively. Figure 9.25 shows a flow chart for the concrete procedure of determining the optimal operating conditions based on the estimated

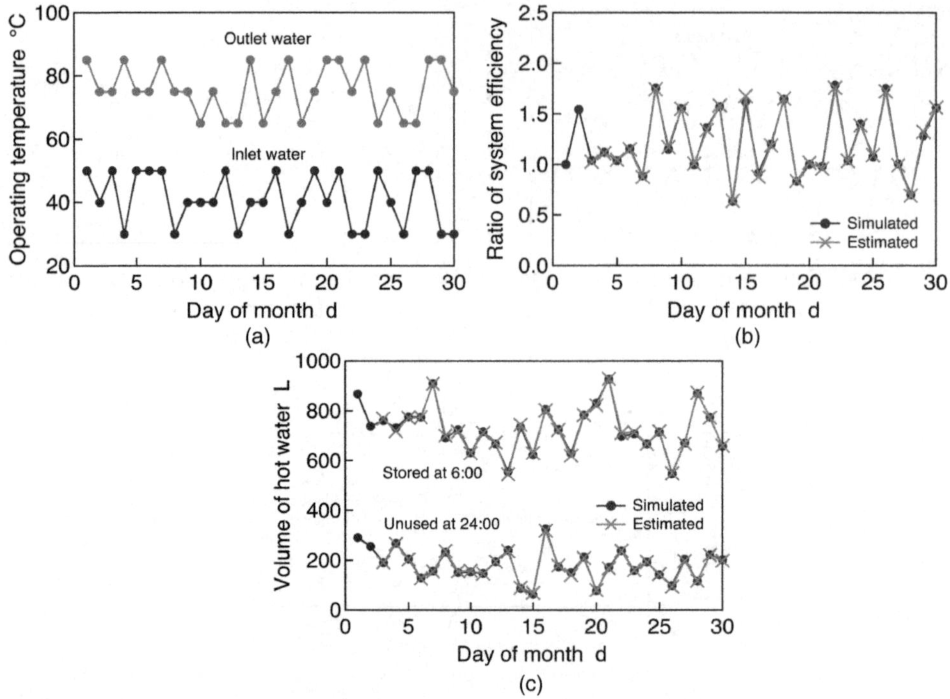

Figure 9.24 Daily changes in operation conditions and system performance values in case 72: (a) operating temperatures; (b) system efficiency; (c) volume of unused hot water.

system performance values. On each day, each system performance value is estimated for all the combinations for the outlet and inlet water temperatures. For simplicity, the outlet water temperature is selected among its discrete values set by 1°C from 65.0 to 85.0°C, and the inlet water temperature is selected among its discrete values set by 1°C from 30.0 to 50.0°C. Here, the outlet water temperature is constrained so that the stratification in the storage tank is kept. Based on this estimation, the combination of the outlet and inlet water temperatures is selected so that the estimated system efficiency has its maximum and the estimated volume of unused hot water is equal to or larger than its lower limit. In case there is no combination by which the estimated volume of unused hot water is equal to or larger than its lower limit, the combination by which the estimated volume of unused hot water is the closest to its lower limit is selected.

Before optimization results are shown, the procedure of determining the optimal operating conditions is shown using an example. Figure 9.26 shows the system efficiency and volume of unused hot water as the objective function and constraint, respectively, estimated on the 3rd day in relation to the operating conditions. This figure shows that the system efficiency decreases and the volume of unused hot water increases with increases in the operating temperatures. Based on these relationships, the operating temperatures are selected to maximize the system efficiency subject to the lower limit for the volume of unused hot water.

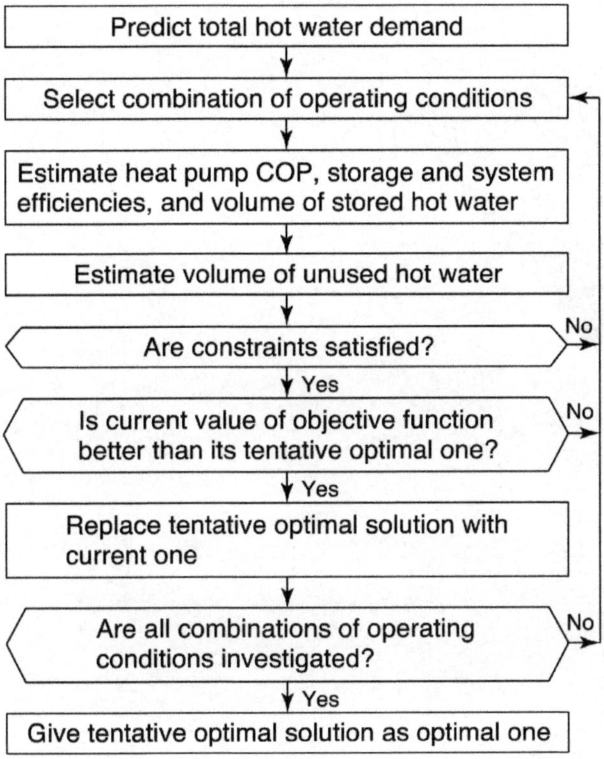

Figure 9.25 Flow chart for determining optimal operating conditions.

In the case study, the lower limit for the volume of unused hot water is changed by 50 L from 50 to 250 L in cases 73–77, respectively, and its influence on the system performance is investigated. Figures 9.27–9.29 show the daily changes in the operating conditions and system performance values in cases 73, 75, and 77, respectively. Figure (a) shows the operating temperatures determined optimally, and Figures (b) and (c) show the system efficiency, and the volumes of stored and unused hot water, respectively, estimated by the neural network models under the optimal operating temperatures. These figures also show the corresponding values obtained by numerical simulation. Although these operating conditions and the corresponding system performance values are not used to identify the values of model parameters, the estimated system performance values coincide well with the simulated ones. This result shows that the system performance values are estimated with high accuracy by the same neural network models even under daily changes in the optimal operating conditions. In case 75, as shown in Figure 9.28, although the volume of unused hot water changes around 150 L, it becomes larger than 150 L on a few days. This is because both operating temperatures attain their lower limits, or the outlet water temperature attains the temperature at the top of the storage tank on those days. As a result, the daily change in the volume of unused hot water is small. In case 73, as shown in Figure 9.27, the volume of unused hot water changes above 50 L on many days. This is also because both operating

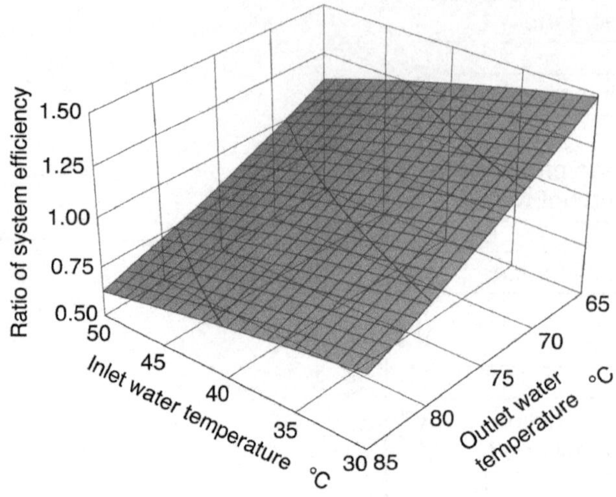

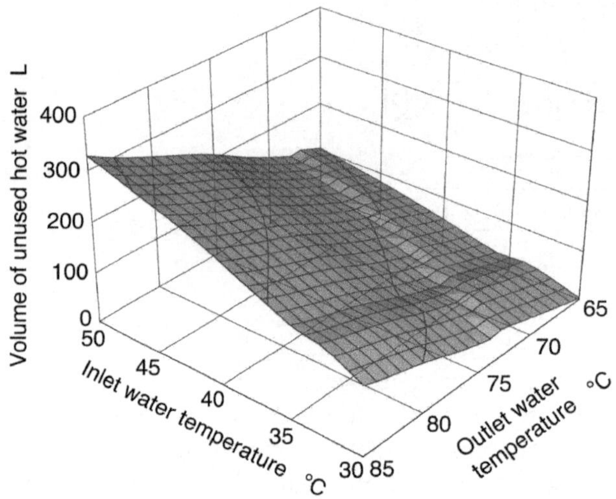

Figure 9.26 Dependence of system performance values on operating conditions: (a) system efficiency; (b) volume of unused hot water.

temperatures attain their lower limits, or the outlet water temperature attains the temperature at the top of the storage tank on those days. As a result, the daily change in the volume of unused hot water is larger. On the other hand, in case 77, as shown in Figure 9.29, the volume of unused hot water changes below 250 L on several days. This is because the operating conditions attain their upper limits on those days. As a result, the daily change in the volume of unused hot water is slightly larger. The system efficiency changes in accordance with the changes in the operating temperatures in these cases.

Figure 9.30 shows the relationship between the lower limit for the volume of unused hot water and the monthly values of the ratio of system efficiency and the volume of unused hot water. The average value is adopted for the ratio of system efficiency, and the average,

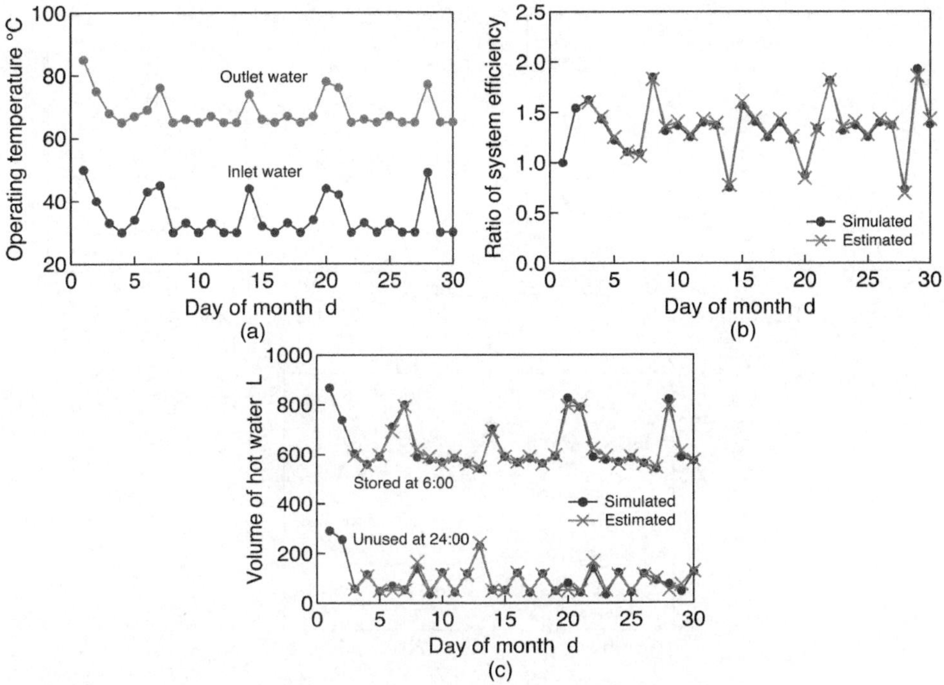

Figure 9.27 Daily changes in operation conditions and system performance values in case 73: (a) operating temperatures; (b) system efficiency; (c) volume of unused hot water.

maximum, and minimum values are adopted for the volume of unused hot water. The average values of the ratio of system efficiency and the volume of unused hot water have a trade-off relationship. However, the average value of the ratio of system efficiency and the maximum or minimum value of the volume of unused hot water do not have a trade-off relationship. This is because in case the lower limit for the volume of unused hot water is small or large, the daily change in the volume of unused hot water becomes large, and the difference between the maximum and minimum values of the volume of unused hot water also becomes large.

Figure 9.31 shows the comparison of the monthly values of the ratio of system efficiency and the volume of unused hot water in cases 70–72 in Table 9.4 and cases 73–77. The average value is adopted for the ratio of system efficiency, and the average and minimum values are adopted for the volume of unused hot water. As aforementioned, the average values of the ratio of system efficiency and the volume of unused hot water under the optimal operating conditions in cases 73–77 have a trade-off relationship. In addition, those under the non-optimal operating conditions in cases 70–72 are very close to the trade-off relationship. Thus, the optimal operation is not effective from the viewpoint of the average system performance values. On the other hand, the average value of the ratio of system efficiency and the minimum value of the volume of unused hot water under the optimal operating conditions in cases 73–77 have a trade-off relationship partly in cases 73–75. In addition, those under the non-optimal operating conditions in cases 70–72 are far from the trade-off relationship. As for the volume of unused hot water, the minimum value is more important

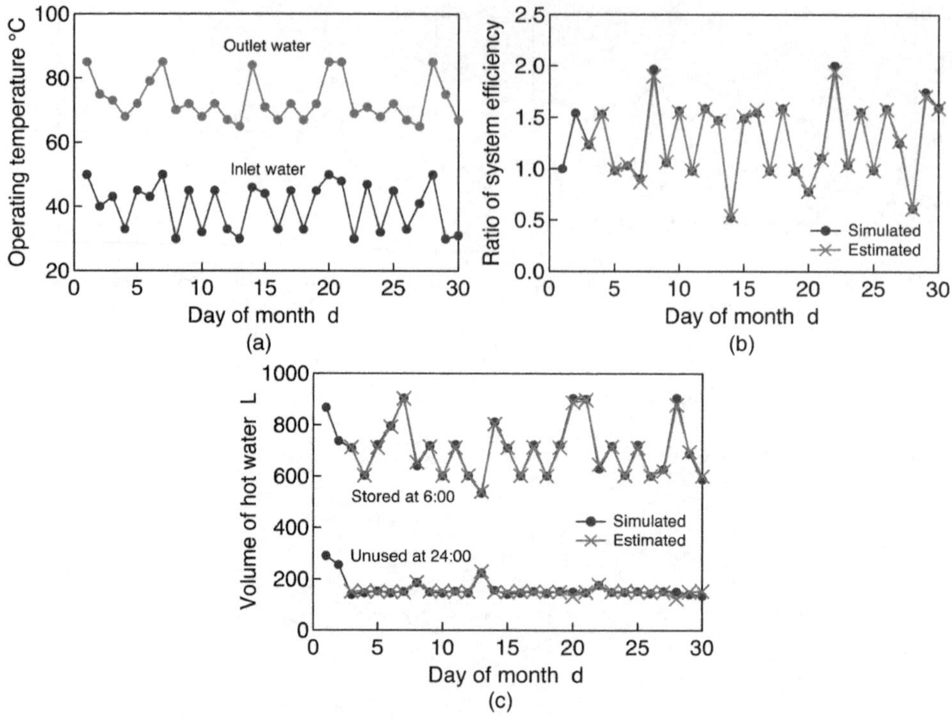

Figure 9.28 Daily changes in operation conditions and system performance values in case 75: (a) operating temperatures; (b) system efficiency; (c) volume of unused hot water.

than the average one to prevent the shortage in hot water supply. Thus, as shown by arrows, it is possible to enhance the average value of the system efficiency with the minimum value of the volume of unused hot water kept constant. The increases in the average value of the system efficiency are expected to be about 9.0%, 9.9%, and 8.2% in cases 70–72, respectively.

9.11 Other Issues on Performance Analysis and Optimization

In the previous sections, a unifunctional CO_2 heat pump water heating system with the function only of hot water supply is investigated. On the other hand, multi-functional CO_2 heat pump water heating systems with the functions of both hot water supply and bath heating have also been developed. It is necessary to distinguish these two functions for the performance analysis of a multi-functional system. This is because hot water retrieved from the top of the storage tank is returned to its bottom or side after heat exchange for bath heating, which destroys a stratified temperature distribution in the storage tank and causes three-dimensional convectional water flow, and the temperature distribution in the storage tank is essentially different from that of a unifunctional system. As a result, system performance values of the multifunctional system differ from those of the unifunctional system. We have conducted the performance analysis of the multifunctional system, and have

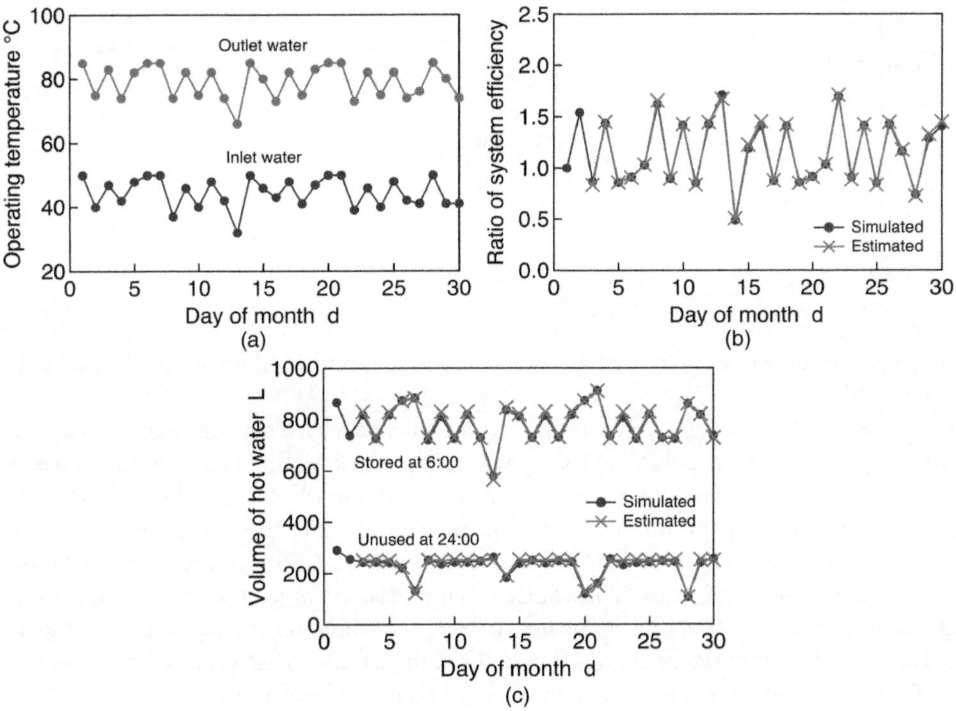

Figure 9.29 Daily changes in operation conditions and system performance values in case 77: (a) operating temperatures; (b) system efficiency; (c) volume of unused hot water.

Figure 9.30 Relationship between monthly system performance values.

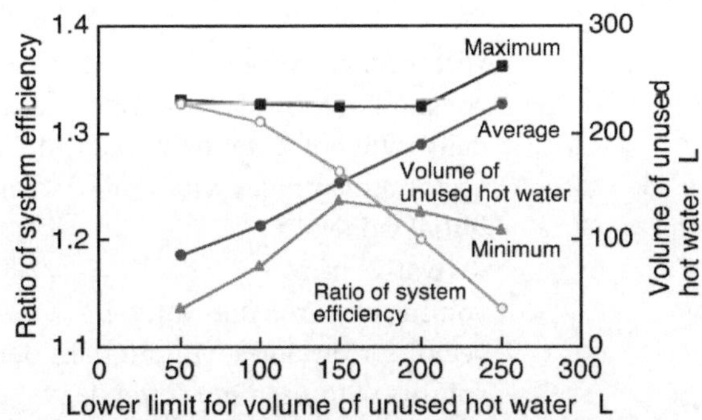

clarified the difference in the performance between unifunctional and multi-functional systems and the influence of the position for hot water return on the performance [14].

In Sections 9.9 and 9.10, methods of estimating daily changes in system performance values accurately, and determining operating conditions optimally are presented. Here, it is assumed that the total hot water demand on the current day can be predicted exactly, which is difficult actually. Thus, it is important to estimate daily changes in system performance values accurately and determine operating conditions optimally under uncertain total hot water demand. We have proposed a robust optimization method based on the minimax regret criterion to determine the operating temperatures so that the maximum regret in the estimated system efficiency is minimized while the minimum in the estimated volume of

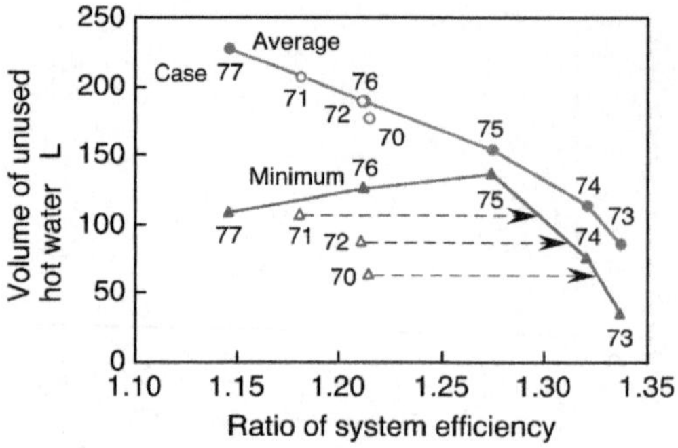

Figure 9.31 Comparison between monthly system performance values under optimal and non-optimal operating conditions.

unused hot water satisfy its lower limit under the total hot water demand predicted by its interval [20].

Solar-assisted CO_2 heat pump water heating systems have also been developed to utilize solar energy. Each system combines a CO_2 heat pump water heating system with a conventional solar heater. In this system, the temperature of an antifreeze solution for the solar heater affects the temperature distribution in the storage tank through an internal heat exchanger, and consequently the COP of the heat pump. In addition, since solar insolation depends on weather conditions significantly, it is important to predict solar insolation accurately and optimize the operation of the heat pump in consideration of the influence of solar insolation on the temperature distribution in the storage tank. These subjects are similar to the aforementioned ones, and should be investigated in the near future.

Nomenclature

$\boldsymbol{f}$	vector for equations
k	constant of formula by Runge-Kutta method
u	daily total hot water demand, L/d
$\boldsymbol{x}$	vector for variables with their derivatives
$\boldsymbol{x}_0$	initial value of $\boldsymbol{x}$
$\dot{\boldsymbol{x}}$	derivative of $\boldsymbol{x}$
y	volume of stored hot water, L
$\boldsymbol{y}$	vector for variables without their derivatives
z	volume of unused hot water, L
T_i	inlet water temperature for shutdown, °C
T_o	outlet water temperature during operation, °C
$T_{\mathrm{ST}n}$	temperature of water in nth control volume of storage tank, °C
$\overline{T}$	upper threshold of temperature, °C
$\underline{T}$	lower threshold of temperature, °C
t	time, s
t_0	initial time, s
Δt	sampling time interval, s
η_cop	heat pump COP
η_sto	storage efficiency
η_sys	system efficiency

Subscripts

HP	CO_2 heat pump
k	index for days
MV	mixing valve
$[r]$	number of applications of formula by Runge-Kutta method
ST	storage tank
(s)	number of repeats for convergence calculation by Newton-Raphson method

Abbreviations

COP	coefficient of performance
CV	control volume

References

1 Hashimoto, K. (2006). Technology and market development of CO_2 heat pump water heaters (ECO CUTE) in Japan. *IEA Heat Pump Centre Newsletter* 24 (3): 12–16.

2 Cecchinato, L., Corradi, M., Fornasieri, E., and Zamboni, L. (2005). Carbon dioxide as refrigerant for tap water heat pumps: a comparison with the traditional solution. *International Journal of Refrigeration* 28 (8): 1250–1258.

3 Stene, J. (2005). Residential CO_2 heat pump system for combined space heating and hot water heating. *International Journal of Refrigeration* 28 (8): 1259–1265.

4 Fernandez, N., Hwang, Y., and Radermacher, R. (2010). Comparison of CO_2 heat pump water heater performance with baseline cycle and two high COP cycles. *International Journal of Refrigeration* 33 (3): 635–644.

5 Minetto, S. (2011). Theoretical and experimental analysis of a CO_2 heat pump for domestic hot water. *International Journal of Refrigeration* 34 (3): 742–751.

6 Yokoyama, R., Okagaki, S., Ito, K., and Takemura, K., (2006). Performance Analysis of a CO_2 Heat pump water heating system by numerical simulation with a simplified model. Proceedings of the 19th International Conference on Efficiency, Cost, Optimization, Simulation and Environmental Impact of Energy Systems, pp. 1353–1360.

7 Yokoyama, R., Shimizu, T., Ito, K., and Takemura, K. (2007). Influence of ambient temperatures on performance of a CO_2 heat pump water heating system. *Energy* 32 (4): 388–398.

8 Yokoyama, R., Okagaki, S., Wakui, T., and Takemura, K., (2007). Effect of storage tank configurations on performance enhancement of a CO_2 heat pump water heating system. *Proceedings of the 20th International Conference on Efficiency, Cost, Optimization, Simulation and Environmental Impact of Energy Systems*, vol. I, pp. 595–602.

9 Yokoyama, R., Okagaki, S., Wakui, T., and Takemura, K. (2008). Influence of operation temperatures on performance of a CO_2 heat pump water heating system. *Journal of Environmental Engineering* 3 (1): 61–73.

10 Yokoyama, R., Wakui, T., Kamakari, J., and Takemura, K. (2010). Performance analysis of a CO_2 heat pump water heating system under a daily change in a standardized demand. *Energy* 35 (2): 718–728.

11 Yokoyama, R., Kohno, Y., Wakui, T., and Takemura, K. (2010). Performance analysis of a CO_2 heat pump water heating system under a daily change in a simulated demand. *Transactions of the JSRAE* 27 (4): 355–364.

12 Yokoyama, R. (2011). Exergy analysis of CO_2 heat pump water heating systems. *Proceedings of the 7th International Symposium on Environmentally Conscious Design and Inverse Manufacturing*, pp. 120–125.

13 Ohkura, M., Yokoyama, R., Nakamata, T., and Wakui, T. (2015). Numerical analysis on performance enhancement of a CO_2 heat pump water heating system by extracting tepid water. *Energy* 87: 435–447.

14 Yokoyama, R., Ohkura, M., Nakamata, T., and Wakui, T. (2018). Numerical analysis for performance analysis of a multi-functional CO_2 heat pump water heating system. *Applied Sciences* 8 (10): 1829, pp. 1–22.

15 Yokoyama, R., Kato, R., Wakui, T., and Takemura, K. (2013). Performance estimation and optimal operation of a CO_2 heat pump water heating system. *International Journal of Thermodynamics* 16 (2): 62–72.

16 Yokoyama, R., Takeuchi, S., and Ito, K., (2005). Thermoeconomic analysis and optimization of a gas turbine cogeneration unit by a systems approach. *Proceedings of the ASME Turbo Expo 2005*. Paper GT2005-68392, pp. 1–7.

17 Heat pump water heaters using carbon dioxide refrigerant, standard no. JRA 4050: 2011(2011). Japan Refrigeration and Air Conditioning Industry Association (in Japanese).

18 Ukaji, M., Sawachi, T., Akimoto, T. et al. (2004) Study on low energy and resource saving technologies for autonomous housing (part 6, basic schedule for the verification of energy consumption in daily human activities). *Proceedings of the SHASE Annual Conference*, pp. 209–212 (in Japanese).

19 Yokoyama, R. and Ito, K. (2005). Capability of global search and improvement in modal trimming method for global optimization. *JSME International Journal, Series C* 48 (4): 730–737.

20 Kato, R., Yokoyama, R., Wakui, T., and Takemura, K. (2013). Optimization of operating conditions of a CO_2 heat pump water heating system (robust optimization based on minimax regret criterion). *Proceedings of the 29th Conference on Energy, Economy, and Environment*, pp. 489–442 (in Japanese).

10

Transcritical CO_2 Heat Pump Space Heating

Feng Cao and Yulong Song

School of Energy and Power Engineering, Xi'an Jiaotong University, Xi'an 710049, China

10.1 Attempts Toward Space Heating Used a Transcritical CO_2 Heat Pump

In order to overcome performance deterioration with the water feed temperature increasing, tests on modified transcritical CO_2 systems were carried out regarding the performance improvement achieved by the introduction of the internal heat exchanger (IHX), multi-compression with flash tank or economizer, ejector, parallel compression, cascade and subcooler-based cycles, etc.

Referring to the modifications for direct-heat-type systems, the idea of parallel compression becomes one of the most obvious ways to improve the performance of recirculating-heat-systems due to the slight benefit from an IHX as well as the high gas-cooler outlet temperature, which is unfavorable for an ejector's properties and entrainment ratio. Three typical modes of parallel compression cycles were shown in Figure 10.1, in which the main parameters, such as the system coefficient of performance (COP), discharge pressure and medium pressure, were found to be very similar among the three cycles [1], except for a slight advantage that could be observed in the single flash tank-based cycle over the other two cycles. Apart from the performance improvement, the optimal discharge pressure was found to decline significantly and a corresponding correlation used for predicting the optimal discharge pressure was proposed based on the simulation results. Additionally, it is worth noting that the various ranges of gas-cooler outlet temperatures and evaporating temperatures are quite wide (30 to 60°C and −45°C to 5 °C, respectively), which makes this investigation suitable for space heating applications. Based on the results of this study, the parallel compression system can still perform well even when the water return temperature is higher than 50°C and ambient temperature is lower than −20°C.

Actually, as is known, the deterioration caused by the increase of water inlet temperature and thereafter the gas-cooler outlet temperature should be offset by the augment of discharge pressure. Although reduction of the optimal discharge pressure can be achieved by using the parallel compression cycle, the optimal discharge pressure is still more than 12 MPa [1, 2], which would be considered as an unsafe value in industrial and domestic

Transcritical CO_2 Heat Pump: Fundamentals and Applications,
First Edition. Xin-Rong Zhang and Hiroshi Yamaguchi.

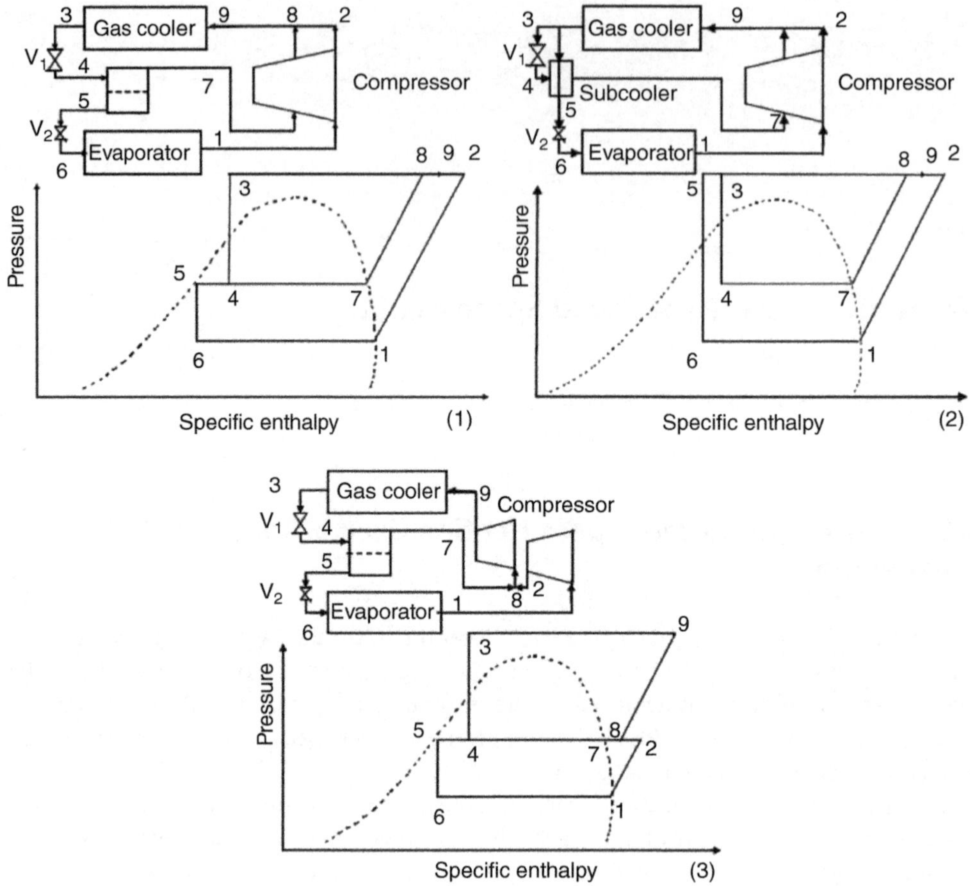

Figure 10.1 The sketch maps and P-h diagrams of three kinds of parallel compression cycles.

applications, when the gas cooler outlet temperature is higher than 40°C. Scholars are still working to find other solutions.

Facing the problem mentioned above, the cascade system is worth focusing on since the temperature difference between heat source and a heat sink is divided into two parts in the cascade system, which decreases nonlinear losses of the cycles and enhances the cycle performances in each temperature section. Based on this unique advantage, it seems that the cascade system should be more suitable for running conditions with extremely low ambient temperature (evaporating temperature) and very high water supply temperature.

It would be possible to transform the R134a cycle from the subcooler to the high-temperature stage and the CO_2 cycle from a direct heating capacity provider to the low-temperature stage with a subcritical running mode, as shown in Figure 10.2 [3]. Thanks to the prominent flow characteristics of CO_2 in very low temperatures, this cascade rig can operate stably when the CO_2 evaporating temperature is down to −40°C. There is only one optimizable value (optimal medium temperature) in the cascade model, since the transcritical CO_2 subsystem is transformed into the subcritical running type, which causes the optimal discharge pressure to no longer exist. Plentiful system parameters are

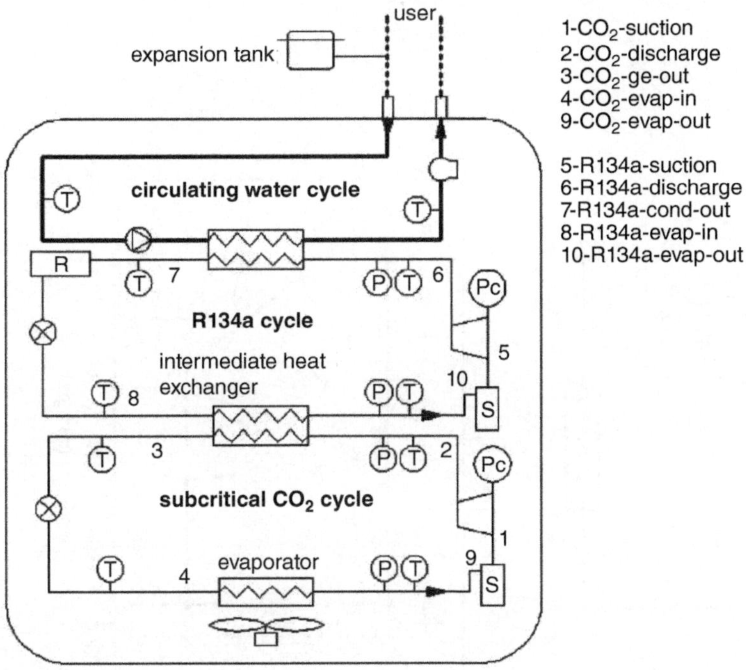

Figure 10.2 The layout of an R134a/CO_2 cascade system [3].

analyzed in detail under the variable CO_2 evaporating temperature from −40 to −30°C and R134a condensing temperature from 30 °C to 50°C. Additionally, focusing on the only optimizable parameter, a mass of test data and corresponding correlations are shown regarding the optimal medium temperature and the heat transfer temperature difference in the cascade heat exchanger.

However, because the heating capacity transport is finally achieved by refrigerant R134a, the cascade system is more suitable for running conditions with a lower ambient temperature and lower water supply temperature since the augment in the expected water supply temperature must correspond to the increase of the R134a discharge pressure. In terms of space heating fields with very high water supply temperatures, a more prominent solution is still waiting to be found.

Instead of auto-cooling structures like the economizer cycle, scholars are looking for other devices that can be installed as the pre-cooler for return water with high temperatures. Above all, a thermoelectric module was suggested as the pre-cooler of a transcritical CO_2 system [4], as shown in Figure 10.3 (1). Obviously, if introducing the thermoelectric module to subcool the refrigerant CO_2 after the gas-cooler, the expansion-valve inlet temperature can be significantly reduced. Thereafter, system performance will be enhanced. Based on energetic and exegetic analyses, and optimization, the modification results in 25.6% and 15.4% increases in the system COP improvement and discharge pressure reduction, respectively. Because the introduction of a subcooler is equivalent to the reduction of gas-cooler outlet temperature, the optimal discharge pressure can be significantly reduced as well as the discharge temperature, which further enhance the system reliability.

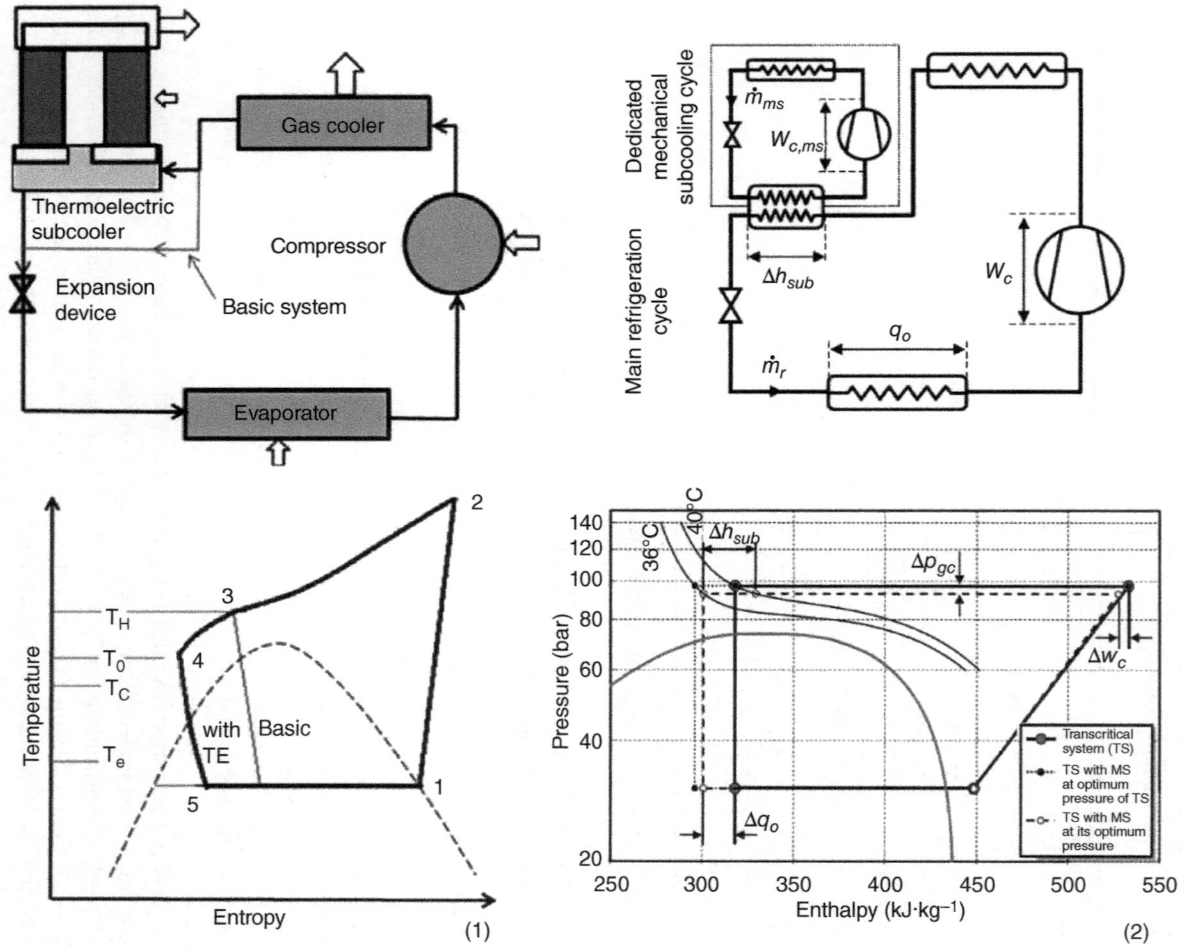

Figure 10.3 Layout and P-h diagram of the combined transcritical CO_2 system with a subcooler.

However, the thermoelectric module is a possibility for introducing a more practical subcooler into a recirculating heat-type transcritical CO_2 heat pump. Other ideas include the use of a CO_2 subcooler or, in another form, water precooler. A refrigeration cycle (instead of the thermoelectric module) was employed at the same position of the transcritical CO_2 heat pump [5, 6], as shown in Figure 10.3 (2). The system performance, especially COP, showed remarkable improvement over the baseline cycle no matter what the evaporating temperature was, and the heating capacity of the subcooler cycle can be utilized too. Besides, the higher the gas-cooler outlet temperature is, the higher the COP improvement potential, which makes the CO_2 subcooler (or water precooler in some other studies) based cycle appropriate to the application field of space heating. Additionally, the whole system's performance must be improved if the subcooler's COP under medium heat transfer temperature as evaporating temperature and water temperature as condensing temperature is higher than the CO_2 system's COP under ambient temperature as evaporating temperature.

Although being an excellent solution for the running conditions of space heating that are too severe for the standard transcritical CO_2 heat pump system, the R134a subcooler-based transcritical CO_2 heat pump system will not always be the best solution with the highest COP among all the running conditions around the world's space heating fields. Since the transcritical CO_2 subsystem is still directly bearing the heat transport from ambient with low temperature to the recirculating water with high temperature, the deterioration will be the biggest problem when the temperature difference between heat source and heat sink increases sharply.

10.2 Thermodynamic Analysis of the Subcooler-Based CO_2 Heat Pump

The typical configuration of a subcooler-based transcritical CO_2 heat pump is shown in Figure 10.4, which is comprised of a subcooler cycle, a transcritical CO_2 cycle, and a recirculating water loop including a three-way valve, a mixing tank, and a water pump. The warm feed water (return water from space heating user) at a temperature of 40–50°C flows through the three-way valve and is then split into two streams. The first stream of feed water flows into the condenser of the subcooler cycle where it is heated up by the high-temperature refrigerant. Thereafter this stream of hot water is channeled into the mixing tank. Another stream of feed water flows into the evaporator of the subcooler cycle where it is cooled down by the low-temperature refrigerant. Cold water from the subcooler cycle evaporator is then channeled into the gas-cooler of the transcritical CO_2 cycle, where the cold water subcools the CO_2 and hence is heated up by the high-temperature CO_2. This stream of hot water is pumped into the mixing tank where it is mixed with the first stream of hot water from the subcooler cycle. In the system, water is used as an intermediate fluid to link the transcritical CO_2 cycle and subcooler cycle.

Figure 10.5 shows a temperature-entropy (T-S) diagram of the two cycles in the combined system. It shows that the evaporation temperature in the subcooler cycle increases from ambient temperature up to around 15°C which is much higher than the ambient temperature in winter. This guarantees the performance of the subcooler cycle. In terms of the transcritical CO_2 cycle, CO_2 at the gas cooler outlet is subcooled directly or indirectly from

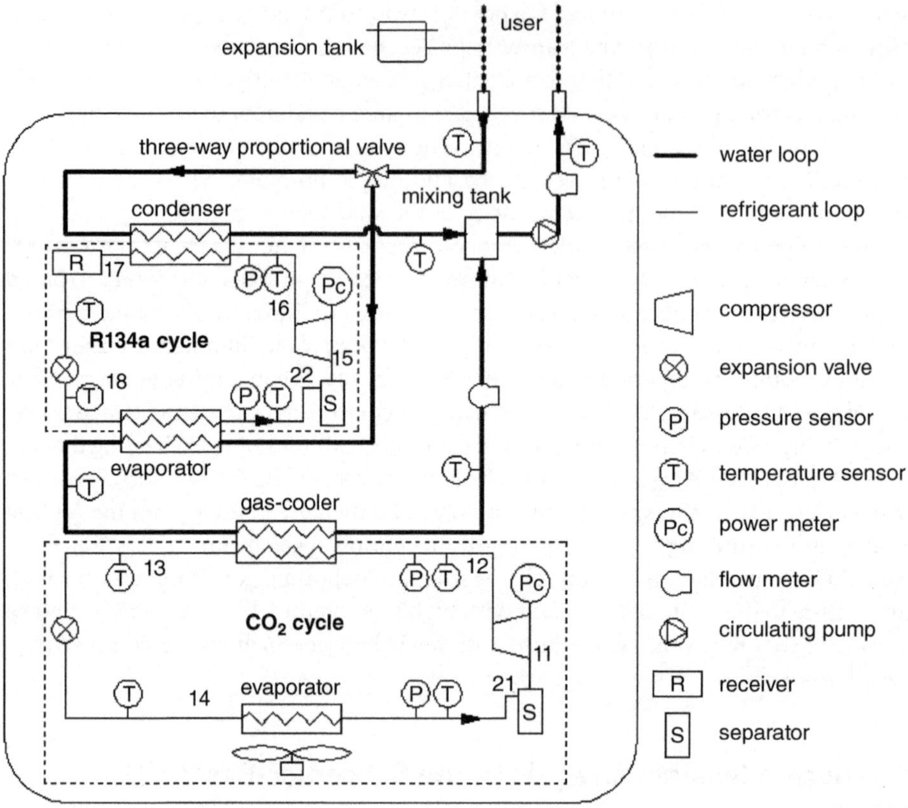

Figure 10.4 Schematic drawing of the combined R134a and transcritical CO_2 system [3].

point 19 to point 13 by cold water from the evaporator of the subcooler cycle. It is well known that subcooling CO_2 in the gas cooler improves the system COP of the transcritical CO_2 cycle. Hence the transcritical CO_2 cycle has the potential to offer a better system COP via subcooling.

In the combined system, the cooling capacity of the transcritical CO_2 cycle is expressed by:

$$\dot{q}_{c-com} = \dot{m}_{CO_2-com}(h_{21} - h_{14}) \tag{10.1}$$

The total heating capacity is determined by thermal energy rejected in the condenser of the R134a cycle plus part of the thermal energy rejected in the gas cooler. It can be calculated by

$$\dot{q}_{h-cas} = \dot{m}_{R134-com}(h_{16} - h_{17}) + \dot{m}_{CO_2-com}(h_{12} - h_{19}) \tag{10.2}$$

The relationship of the R134a flow rate in the R134a cycle and CO_2 flow rate in transcritical CO_2 cycle can be determined by energy balance between the R134a evaporator and the gas cooler in the transcritical CO_2 cycle as expressed by:

$$\dot{m}_{R13a-com}(h_{22} - h_{18}) = \dot{m}_{CO_2-com}(h_{19} - h_{13}) \tag{10.3}$$

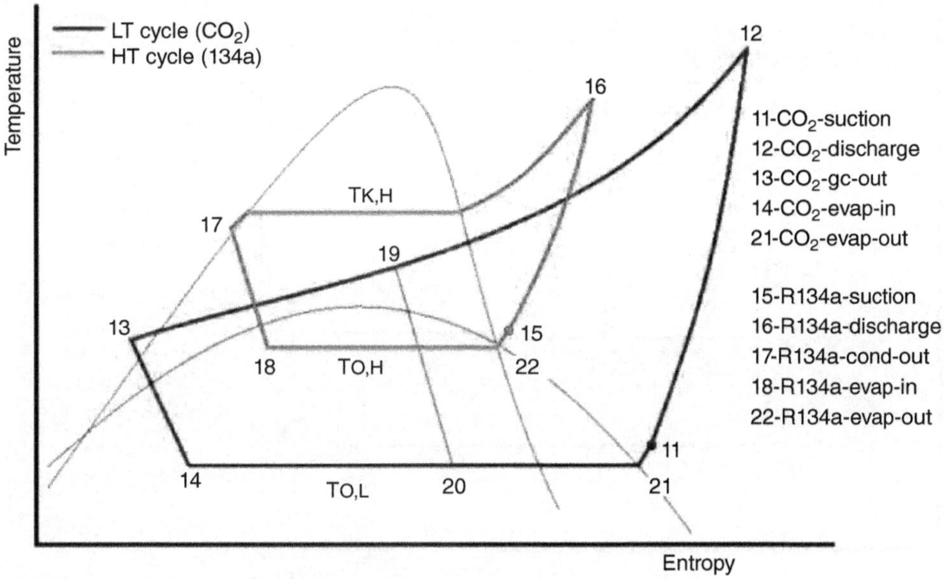

Figure 10.5 T-s diagram of the R134a/CO_2 combined system [7].

The total power consumption is equal to the sum of R134a compressor power consumption, CO_2 compressor power consumption, and electrical fan power consumption. It can be determined by:

$$\dot{W}_{total-com} = \dot{m}_{R134a-com}(h_{16} - h_{15}) + \dot{m}_{CO_2-com}(h_{12} - h_{11}) + \dot{W}_{fan-com} \tag{10.4}$$

The energy balance in the R134a and transcritical cycles can be expressed by:

$$(h_{16} - h_{17}) = (h_{16} - h_{22}) + (h_{22} - h_{18}) \tag{10.5}$$

$$(h_{12} - h_{13}) = (h_{12} - h_{21}) + (h_{21} - h_{14}) \tag{10.6}$$

The cascade system COP can be calculated by:

$$COP = \frac{\dot{q}_{h-com}}{\dot{W}_{total-com}} \tag{10.7}$$

10.3 Comparison Between the Subcooler-Based CO_2 System and the Cascade Cycle

Figure 10.2 shows the schematic drawing of an R134a/CO_2 cascade heat pump system which consists of an R134a refrigeration cycle, a subcritical CO_2 refrigeration cycle and water circulating cycle. The two cycles are coupled thermally with a heat exchanger which acts as condenser of the transcritical CO_2 refrigeration cycle and evaporator of the R134a refrigeration cycle. This heat exchanger is called an evaporative condenser. The CO_2 cycle works as the low-temperature cycle (LT) and the R134a works as the high-temperature cycle (HT). CO_2 absorbs thermal energy from the ambient air in the evaporator of the LT cycle.

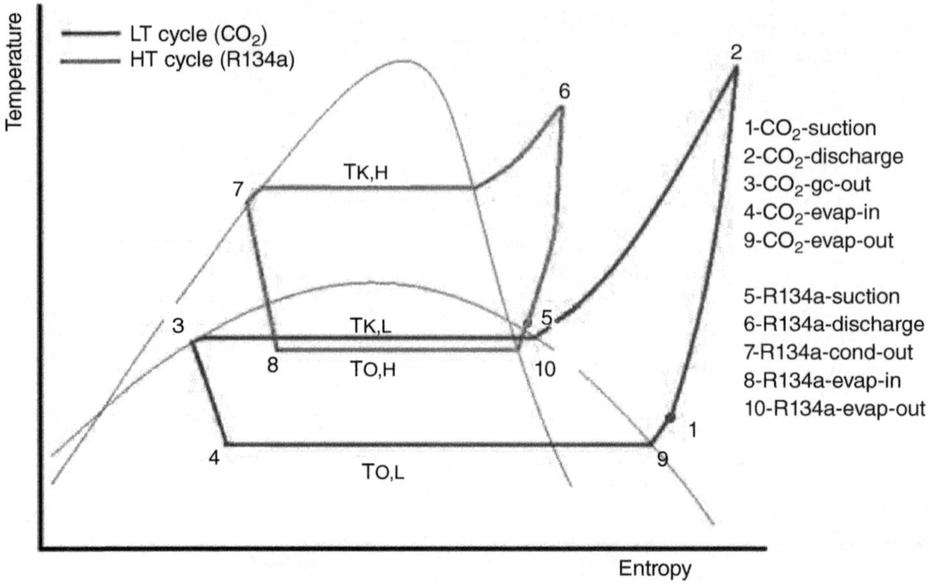

Figure 10.6 T-s diagram of the R134a/CO_2 cascade system [7].

The thermal energy in CO_2 in the condensing side is transferred to R134a in the evaporating side of the evaporative condenser. The warm feed water (return water from space heating) flows though the R134a cycle condenser where it is heated up by R134a before being supplied to users. It is obvious that the system outlet water temperature is determined by the condensing temperature of the HT cycle in the cascade system. This system outlet temperature is defined as the supply water temperature in the cascade system.

Figure 10.6 shows the temperature-entropy (T-s) diagrams of the cascade systems. In both systems, refrigerant R134a and CO_2 at the evaporator outlet are saturated vapor at each cycle. The refrigerant is heated up in the suction line, compressor motor, etc. and then becomes superheated vapor at the compressor inlet. In the cascade system, the cooling capacity is determined by the CO_2 cycle as expressed by:

$$\dot{q}_{c-cas} = \dot{m}_{CO_2-cas}(h_9 - h_4) \tag{10.8}$$

The heating capacity is determined by the R134a cycle and is calculated by:

$$\dot{q}_{h-cas} = \dot{m}_{R134-cas}(h_6 - h_7) \tag{10.9}$$

The relationship of the R134a flow rate in the HT cycle and CO_2 flow rate in LT cycle can be determined by energy balance in the evaporative condenser as expressed by:

$$\dot{m}_{R13a-cas}(h_{10} - h_8) = \dot{m}_{CO_2-cas}(h_2 - h_3) \tag{10.10}$$

The total power consumption is equal to the sum of R134a compressor, CO_2 compressor and electrical fan power consumption. It can be determined by:

$$\dot{W}_{total-cas} = \dot{m}_{R134a-cas}(h_6 - h_5) + \dot{m}_{CO_2-cas}(h_2 - h_1) + \dot{W}_{fan-cas} \tag{10.11}$$

The energy balance in the HT and LT cycle can be expressed by:

$$(h_6 - h_7) = (h_6 - h_{10}) + (h_{10} - h_8) \tag{10.12}$$

$$(h_2 - h_3) = (h_2 - h_9) + (h_9 - h_4) \tag{10.13}$$

The cascade system COP can be calculated by:

$$COP = \frac{\dot{q}_{h-cas}}{\dot{W}_{total-cas}} \tag{10.14}$$

Figure 10.7 shows the performance of the two systems under different ambient temperatures. The trends are very similar in the two systems. As the ambient temperature decreased from 0°C to −20°C, the COP of the cascade and combined systems dropped by up to 22% and 27%, respectively, under the studied feed water and supply water temperatures. It was found that the performance of the combined system was lower than that of the cascade system at a low ambient temperature and was higher than that of the cascade system at a high ambient temperature. This could be explained by the working mechanism of the two systems. In the combined system, as the ambient temperature decreases, the pressure ratio of the transcritical CO_2 sub-system increased sharply since the CO_2 discharge pressure varies very little in the transcritical region. Therefore, the combined system performance drops very quickly. However, in the cascade system, as the ambient temperature decreases, the CO_2 compressor pressure ratio also increases. However, the CO_2 condensing temperature can be adjusted by varying the evaporating temperature in the R134a cycle, and hence the CO_2 discharge compressor pressure does not increase sharply as happened in the combined system. Besides, the declines in R134a evaporating temperature in the cascade system and combined system thereafter the augment of R134a pressure ratio in both systems were close. Therefore, the COP of the cascade system does not drop as fast as that of the combined system. It is clear that the combined system works better at high ambient temperatures while the cascade system works better at low ambient temperatures. In addition, the COP of the combined system is higher than that of the cascade system for large differences between the feed water and supply water temperatures. This is mainly because the performance of the cascade system dropped much faster than that of the combined system as the supply water temperature increases.

As the feed water temperature increased from 40 to 50°C, the COP of the cascade system did not change significantly; however, the COP of the combined system dropped by up to 8%. The supply water and ambient temperatures also showed different effects on both systems. As the supply water temperature increased from 55 to 75°C, the COP of the cascaded and combined system dropped by up to 21% and 11%, respectively. It was also found that the combined system performed better at high ambient temperatures and high temperature differences between feed water and supply water. But the cascade system performed better at low ambient temperatures and small temperature differences between feed water and supply water. An empirical correlation was proposed to identify the best performing region for the two systems using the operating condition coefficient. If the coefficient value was larger than 0.263, the combined system performed better. Otherwise, the cascade system performed better. This provides engineers and researchers with a guideline to select the most suitable system for any specific operating conditions.

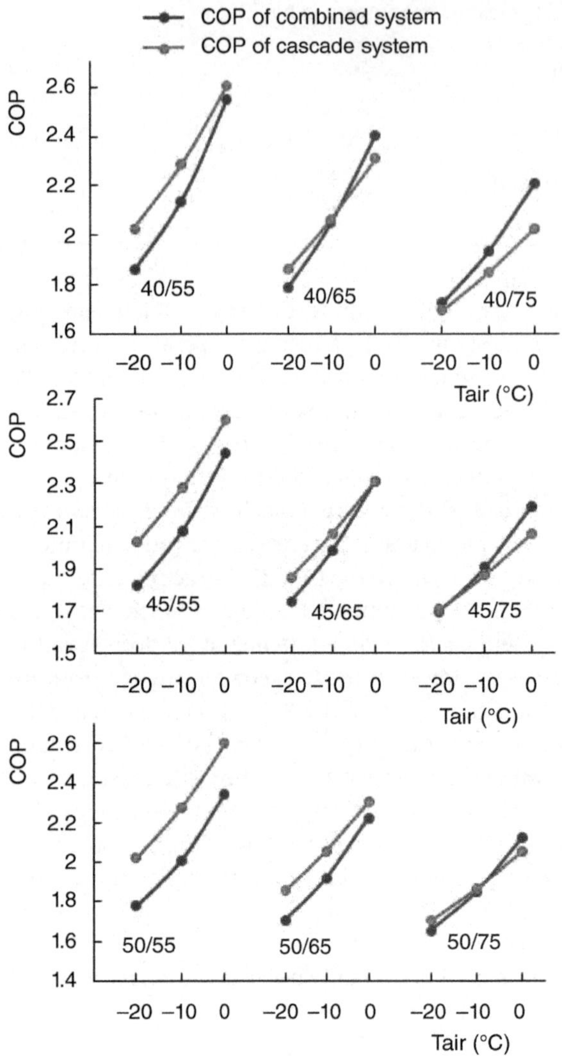

Figure 10.7 COP of the combined and cascade systems under different ambient temperatures [3].

10.4 Optimal Discharge Pressure

The pressure-enthalpy (P-h) diagrams of the two subunits in the water-precooler-based system are shown in Figure 10.8. As displayed in Figure 10.8a, being precooled by the introduction of the R134a subunit, the water temperature at the gas cooler inlet is lower. Thereafter, the CO_2 refrigerant temperature at the gas-cooler outlet can be cooled by the cold water further from state point 5 to state point 3. It is well known that lower CO_2 gas cooler outlet temperature corresponds to a higher COP of the transcritical CO_2 cycle in the case of other conditions keeping unchanged, that is, the transcritical CO_2 subsystem has very much potential to offer a better system COP via this kind of water-precooler under the condition of high return water temperature.

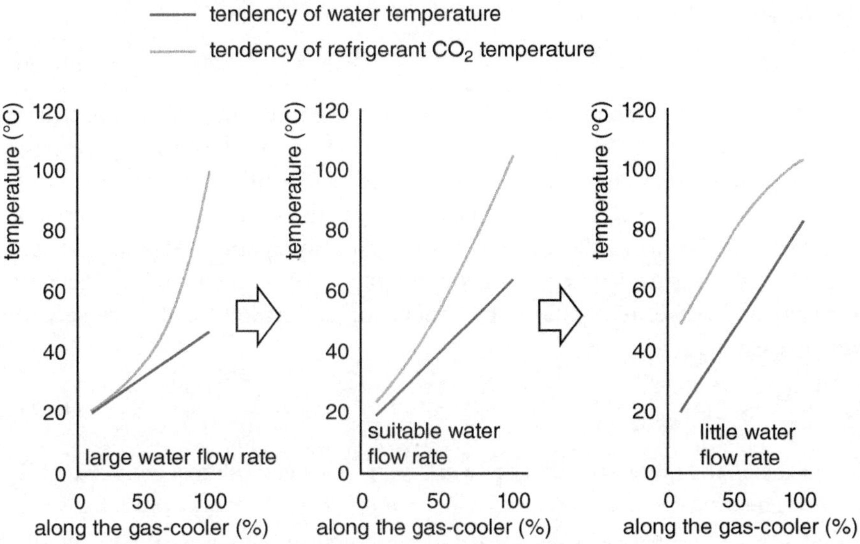

Figure 10.8 Sketch map of CO_2 and water temperature distribution [8].

Compared with the single running transcritical CO_2 system under return water with very high temperature (that is, state point 5 as the CO_2 gas-cooler outlet), the introduction of the R134a subunit subcools indirectly the refrigerant CO_2 in the gas cooler, which brings out additional heating capacity of CO_2 subunit. However, this portion of additional heating capacity to water is equal to the cooling capacity to water in the evaporator of the R134a subunit. This portion of capacity is under the title of Qh2 in both the R134a and CO_2 subsystems. According to the first law of thermodynamics, the relationship of heating capacity, cooling capacity and power dissipation can be written as (see Figure 10.8):

$$Q_{c1} + W_{co_2} = Q_{h1} \tag{10.15}$$

$$Q_{h1} + W_{134a} = Q_{h3} \tag{10.16}$$

The total heating capacity of the water-precooler-based transcritical CO_2 system is equal to all the heating capacity to water minus all the cooling capacity to water as follows:

$$Q_{h-tot} = Q_{h1} + Q_{h2} + Q_{h3} - Q_{h2} = Q_{h1} + Q_{h3} = Q_{h1} + Q_{h2} + W_{134a} \tag{10.17}$$

And COP of the water-precooler-based transcritical CO_2 system can be written as:

$$COP_{\text{sub}} = \frac{Q_{h-tot}}{W_{tot}} = \frac{Q_{h1} + Q_{h2} + W_{134a}}{W_{CO_2} + W_{134a}} \tag{10.18}$$

While COP of the single running transcritical CO_2 system in low water return temperature (state point 3 as the CO_2 gas-cooler outlet) can be written as:

$$COP_{sta} = \frac{Q_h}{W_{CO_2}} = \frac{Q_{h1} + Q_{h2}}{W_{CO_2}} \tag{10.19}$$

It is obvious that the introduction of the R134a subunit amount to an electrical heating element whose power equals the R134a power dissipation in order to decrease the CO_2

water inlet temperature. Thus, we have:

$$COP_{sub} < COP_{sta} \tag{10.20}$$

It is well known that there is an optimal discharge pressure in the standard transcritical CO_2 system at which maximal COP can be reached. The existence of the optimal discharge pressure can be explained from Figure 10.3. Due to the particular physical property of CO_2 refrigerant in the transcritical region, the heating capacity of the standard system increases sharply first and then gently, while the increase rate of power dissipation keeps almost constant as the discharge pressure increases. As the heating COP is calculated by the ratio of heating capacity and power dissipation, the conclusion is that COP will increase with discharge pressure when:

$$\frac{\partial Q_h}{\partial P_d} / \frac{\partial W_{CO_2}}{\partial P_d} > COP_{sta}|_{P_d} \tag{10.21}$$

As mentioned above, the increased rate of power dissipation keeps almost constant while the increase rate of heating capacity decreases with an increase in discharge pressure. Thus, we can see that $\frac{\partial Q_h}{\partial P_d} / \frac{\partial W_{CO_2}}{\partial P_d}$ decreases as the discharge pressure increases. When the discharge pressure rises to some value at which

$$\frac{\partial Q_h}{\partial P_d} / \frac{\partial W_{CO_2}}{\partial P_d} = COP_{sta}|_{P_d} \tag{10.22}$$

can be obtained, this discharge pressure must be the optimal value. Besides, if the discharge pressure keeps on increasing, the result that

$$\frac{\partial Q_h}{\partial P_d} / \frac{\partial W_{CO_2}}{\partial P_d} < COP_{sta}|_{P_d} \tag{10.23}$$

must be observed subsequently. In the water-precooler-based system, however, the introduction of the R134a subunit brings about a decrease of total COP, as mentioned above. Therefore the discharge pressure must rise to higher value than the optimal value of the standard system in order to achieve the result that:

$$\frac{\partial Q_h}{\partial P_d} / \frac{\partial W_{CO_2}}{\partial P_d} = COP_{sub}|_{P_d} \tag{10.24}$$

And this higher value of the discharge pressure is the new optimal value of the transcritical CO_2 subunit in the water-precooler-based system. That is, the optimal discharge pressure of the water-precooler-based system is higher than that of the standard system when the same temperature of water at the CO_2 gas cooler inlet is insured in both systems, which does not contradict the conclusion of Llopis, R. *et al.* [5]. The result can be written as:

$$P_{d,opt,sub} > P_{d,opt,sta} \tag{10.25}$$

That is, the optimal discharge pressure of the water-precooler-based system is always higher than that of the standard system in order to reach the maximum COP in each system.

In conclusion, similar to the standard system, the optimal discharge pressure is also the most remarkable parameter for the subcooler-based system. It can be concluded that the optimal discharge pressure of the subcooler-based system is much higher than that of the standard system in all operating conditions. Moreover, an empirical correlation for the subcooler-based transcritical CO_2 system is proposed to evaluate the optimal discharge pressure by the ambient temperature, water return temperature and water supply temperature as the independent variables, as shown below [9].

$$P_d = 34.5 + 1.135 * T_{w,f} + 1.1 * (T_{w,s} - T_{w,f}) + 0.7 * T_{air} \tag{10.26}$$

10.5 Optimal Medium Temperature

Similar to the medium temperature in the cascade refrigeration systems, the name "medium temperature" was employed to describe the water temperature between the R134a evaporator and the CO_2 gas cooler. Moreover, the section of return water entered into the R134a condenser was called the R134a water flow while the other flow was called the CO_2 water flow. As can be seen in Figure 10.5, the CO_2 temperature in the gas-cooler exit will be quite high, and thereafter the heating capacity (Q_{hco2}) will be low due to the high water return temperature if the transcritical CO_2 system was employed alone in space heating field. However, the actual medium temperature is quite low due to the pre-cooling by the R134a cycle, which enhanced the heating performance of the CO_2 subunit in the specific working condition.

Actually, similar to the intermediate temperature in the cascade system, the medium temperature must be a remarkable parameter of the subcooler-based transcritical CO_2 system. As a combined system, the evaporating temperature after the performance of the R134a subunit increases, while the CO_2 gas-cooler outlet temperature increases after the heating performance deteriorates with the increase in medium temperature. Since COPs in both subunits are meaningful properties to assess if the subcooler-based transcritical CO_2 system is effective and the medium temperature is also a significant property to the system optimization, it is essential to discuss if an optimal medium temperature exists in the subcooler-based transcritical CO_2 system.

The medium temperature can be adjusted by regulating the CO_2 water flow rate under certain conditions. Subsequently, the evaporating temperature, suction density, and thereafter the R134a mass flow rate, would decrease with the medium temperature. Though the inlet water temperature decreased, the outlet water temperature of the CO_2 subunit increased faster since the water flow rate reduced. Thus, the outlet water temperature of the R134a subunit thereafter the R134a condensing temperature declined slightly to keep the water delivery temperature constant by augmenting slightly the water flow rate in the R134a condenser. Consequently, the power dissipation of the R134a subunit declined due to the decreased refrigerant mass flow rate and condensing pressure. Besides, the CO_2 gas cooler outlet temperature declined, and the heating capacity of the CO_2 subunit increased with the decrease of the medium temperature when other parameters (such as CO_2 evaporating temperature, discharge pressure, power dissipation of CO_2 subunit, etc.) remained constant. Based on the mathematical theory that the value of a fraction increases when its denominator and numerator reduce by the same quantity, moreover, with the increasing Q_{hco2}, it can be observed that the COP_{sys} increased definitely when the medium temperature decreased.

However, the declining CO_2 water flow rate caused not only the decrease of medium temperature, but also the deterioration of heat transfer performances. Additionally, the water temperature increased rapidly in the gas cooler when the water flow rate was low, which caused the increase of average water temperature, the decline of the log mean temperature difference and thereafter the decline of the heating capacity in the gas cooler. The variation of the temperature of water and CO_2 with the decreased water flow rate were sketched in Figure 10.8 [8]. It can be seen that the water temperature increased faster with less water flow rate, and the CO_2 gas cooler outlet temperature increased. The heating capacity declined as a result of the low medium temperature caused by the low water flow rate.

As detailed above, it can be noted that there must be an optimal medium temperature which corresponds to the maximum system COP in the subcooler-based transcritical CO_2 system.

The optimal value of the medium temperatures under different operating conditions was shown in Figure 10.9. It can be seen that the optimal medium temperature increased obviously with the water return temperature. A relative explanation can be concluded that the water outlet temperature of the R134a evaporator (medium temperature) increased naturally with the augment in water inlet temperature of the R134a evaporator (water return temperature). Additionally, the cooling capacity of the R134a subunit increased with the water return temperature due to the increasing R134a evaporating temperature and the increasing mass flow rate thereafter the performances of the R134a subunit, thus the increased amplitude of medium temperature was lower than that of water return temperature. As shown in Figure 10.9, the medium temperature increased by about 6°C when the water return temperature increased from 40°C to 50°C. Besides, the medium temperature decreased with the increase of the water delivery temperature. Because the expected increase in CO_2 water outlet temperature can be acquired only by the decline in water flow rate, which caused the decline in water outlet temperature in the R134a evaporator. While the medium temperature declined, the ambient temperature decreased. Since the CO_2 mass flow rate and heating capacity in the CO_2 subunit decreased sharply with the ambient temperature decline, the CO_2 water flow rate had to be decreased in order to reach the accepted water outlet temperature, which caused the decline in medium temperature.

An equation was established to determine the optimal medium temperature from the ambient temperature, water return temperature and water delivery temperature [8]:

$$T_{om} = -0.02 \cdot T_{w,s} \cdot (1 - T_{air})^{\frac{1}{16}} + 0.15 \cdot T_{air} + 0.239 \cdot T_{w,f}{}^{1.2} \quad (10.27)$$

$$T_{om} = -0.02 \cdot T_{w,s} \cdot (1 - T_{air})^{\frac{1}{4}} + 0.15 \cdot T_{air} + 0.232 \cdot T_{w,f}{}^{1.2} \quad (10.28)$$

The Eq. (10.27) was employed at the range of $-10 \leq T_{air} < 0$°C and the Eq. (10.28) was employed at the range of $-20 \leq T_{air} < -10$°C. The error of this equation was less than 3% within the whole test range in this study, which was believed to be accurate in industry applications.

10.6 Conclusion and Prospects

Based on the brief perspective on the theoretical gist and experimental validation of the subcooler-based system, the subcooler-based transcritical CO_2 heat pump system,

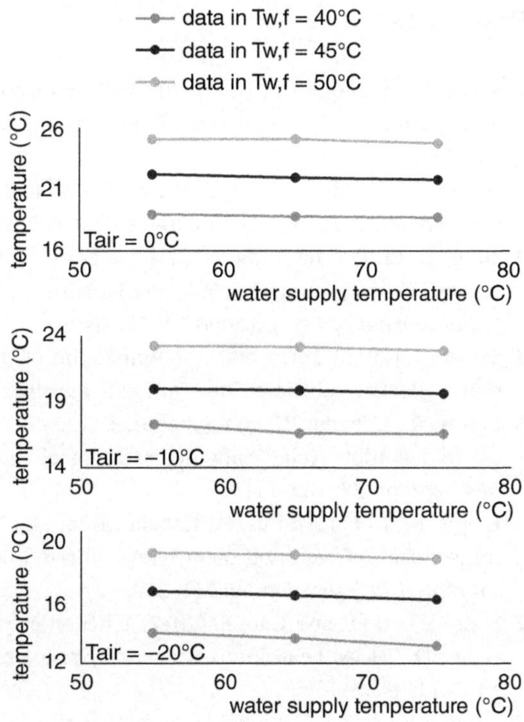

Figure 10.9 The optimal value of the medium temperatures [8].

which enhanced remarkably the overall performance by using the subcooler's cooling capacity to decline the CO_2 temperature before throttling point and its heating capacity to assist the heating supply, was finally proposed as an almost ideal solution under the space heating conditions. According to plentiful experimental and theoretical studies, the subcooler-based system was found more suitable to be adopted in running conditions with higher ambient temperature, lower water inlet temperature and higher temperature difference between water inlet and outlet compared with a same-scale cascade system. Besides, unlike the cascade system that has only one quantity that can be optimized, there are two optimal terms named the optimal discharge pressure and optimal medium temperature in the subcooler-based system.

However, what we should recognize is that the research mentioned above is still in its elementary stage, which shows that more effort should be devoted to the investigation on the fields of performance improvement and control strategies development, etc. For instance, the cooling performance of the transcritical CO_2 system could be remarkably improved by employing an adaptable two-phase ejector to save a considerable part of power consumption. Thus, the proposed transcritical CO_2 system could be designed under the additional consideration of the cooling effort in the summer. As a frontier technology with great practicality nowadays and excellent prospect in the future, the subcooler-based transcritical CO_2 system is believed to have remained as the key topic in the next decade.

References

1 Sarkar, J. and Agrawal, N. (2010). Performance optimization of transcritical CO_2 cycle with parallel compression economization. *International Journal of Thermal Sciences* 49: 838–843.

2 Andrea, C., Fabio, E., and Giovanni, F. (2014). Experimental analysis of R744 parallel compression cycle. *Applied Energy* 135: 274–285.

3 Song, Y., Li, D., and Yang, D. (2017). Performance comparison between the combined R134a/CO_2 heat pump and cascade R134a/CO_2 heat pump for space heating. *International Journal of Refrigeration* 74: 592–605.

4 Sarkar, J. (2013). Performance optimization of transcritical CO_2 refrigeration cycle with thermoelectric subcooler. *International Journal of Energy Research* 37: 121–128.

5 Llopis, R., Cabello, R., and Sanchez, D. (2015). Energy improvements of CO_2 transcritical refrigeration cycles using dedicated mechanical subcooling. *International Journal of Refrigeration* 55: 129–141.

6 Llopis, R., Nebot-Andres, L., Cabello, R. et al. (2016). Experimental evaluation of a CO_2 transcritical refrigeration plant with dedicated mechanical subcooling. *International Journal of Refrigeration* 69: 361–368.

7 Song, Y., Li, D., and Cao, F. (2017). Theoretical investigation on the combined and cascade CO_2/R134a heat pump systems for space heating. *Applied Thermal Engineering* 124: 1457–1470.

8 Song, Y. and Cao, F. (2018). The evaluation of the optimal medium temperature in a space heating used transcritical air-source CO_2 heat pump with an R134a subcooling device. *Energy Conversion and Management* 166: 409–423.

9 Song, Y. and Cao, F. (2018). The evaluation of optimal discharge pressure in a water-precooler-based transcritical CO_2 heat pump system. *Applied Thermal Engineering* 131: 8–18.

Index

Transcritical CO_2 Heat Pump: Fundamentals and Applications,
First Edition. Xin-Rong Zhang and Hiroshi Yamaguchi.

h

i